AF411772

Reproductive Biology of the Great Apes
Comparative and Biomedical Perspectives

CONTRIBUTORS

Delwood C. Collins
Charles Faiman
Dian Fossey
Gene B. Fuller
Biruté M. F. Galdikas
Kenneth G. Gould
Charles E. Graham
Alexander H. Harcourt
William C. Hobson
O. Ray Kling

David E. Martin
Patrick R. McGinnis
Paul I. Musey
Ronald D. Nadler
John R. K. Preedy
Francisco I. Reyes
R. V. Short
Kelly J. Stewart
Caroline E. G. Tutin
Jeremy S. D. Winter

Kristina Wright

Reproductive Biology of the Great Apes

Comparative and Biomedical Perspectives

Edited by

CHARLES E. GRAHAM
Reproductive Biology Division
Primate Research Institute
New Mexico State University
Holloman Air Force Base, New Mexico

ACADEMIC PRESS

A Subsidiary of Harcourt Brace Jovanovich, Publishers

New York   London   Toronto   Sydney   San Francisco   1981

COPYRIGHT © 1981, BY ACADEMIC PRESS, INC.
ALL RIGHTS RESERVED.
NO PART OF THIS PUBLICATION MAY BE REPRODUCED OR
TRANSMITTED IN ANY FORM OR BY ANY MEANS, ELECTRONIC
OR MECHANICAL, INCLUDING PHOTOCOPY, RECORDING, OR ANY
INFORMATION STORAGE AND RETRIEVAL SYSTEM, WITHOUT
PERMISSION IN WRITING FROM THE PUBLISHER.

ACADEMIC PRESS, INC.
111 Fifth Avenue, New York, New York 10003

United Kingdom Edition published by
ACADEMIC PRESS, INC. (LONDON) LTD.
24/28 Oval Road, London NW1   7DX

Library of Congress Cataloging in Publication Data
Main entry under title:

Reproductive biology of the great apes.

    Includes bibliographies and index.
    1. Apes--Reproduction.   2. Mammals--Reproduction.
I. Graham, Charles E.   [DNLM: 1. Pongidae--Phys-
iology. 2. Reproduction. WL 737.P96 G738r]
QL737.P96R45   599.88'40416   80-89417
ISBN   0-12-295020-8

PRINTED IN THE UNITED STATES OF AMERICA

81 82 83 84   9 8 7 6 5 4 3 2 1

CONTENTS

Chapter 1   MENSTRUAL CYCLE OF THE GREAT APES

Charles E. Graham

Chapter 2   ENDOCRINOLOGY OF PREGNANCY IN APES

Charles Faiman, Francisco I. Reyes, Jeremy S. D. Winter, and William C. Hobson

Chapter 6 THE MALE APE GENITAL TRACT
AND ITS SECRETIONS

David E. Martin and Kenneth G. Gould

Chapter 7 COMPARATIVE ASPECTS OF APE STEROID
HORMONE METABOLISM

*Kristina Wright, Delwood C. Collins, Paul I. Musey,
and John R. K. Preedy*

Chapter 8 LABORATORY RESEARCH ON SEXUAL
BEHAVIOR OF THE GREAT APES

Ronald D. Nadler

CONTRIBUTORS

Numbers in parentheses indicate the pages on which the authors' contributions begin.

DELWOOD C. COLLINS (69,163),   Departments of Medicine and Biochemistry, Emory University School of Medicine, Atlanta, Georgia 30322

CHARLES FAIMAN (45,83),   Department of Endocrinology and Metabolism, University of Manitoba, Winnipeg, Manitoba R3E 0Z3, Canada

DIAN FOSSEY (265),   Section of Neurobiology and Behavior, Langmuir Laboratory, Cornell University, Ithaca, New York 14850

GENE B. FULLER (83),   Reproductive Biology Division, Primate Research Institute, New Mexico State University, Holloman Air Force Base, New Mexico 88330

BIRUTÉ M. F. GALDIKAS* (281),   Orangutan Research and Conservation Project, Pangkalan Bun, Kalimantan Tengah, Indonesia

KENNETH G. GOULD (105,127,375),   Yerkes Regional Primate Research Center, Emory University, Atlanta, Georgia 30322

CHARLES E. GRAHAM (1,69,407),   Reproductive Biology Division, Primate Research Institute, New Mexico State University, Holloman Air Force Base, New Mexico 88330

ALEXANDER H. HARCOURT (265,301),   Department of Applied Biology, University of Cambridge, Cambridge CB2 3DX, England

WILLIAM C. HOBSON (45,83),   Reproductive Biology Division, Primate Research Institute, New Mexico State University, Holloman Air Force Base, New Mexico 88330

O. RAY KLING (69),   Department of Obstetrics, University of Oklahoma, College of Medicine, Health Science Center, Oklahoma City, Oklahoma 75190

*Present address: Department of Archaeology, Simon Fraser University, Burnaby, British Columbia, Canada.

DAVID E. MARTIN (105,127,343),   College of Allied Health Sciences, Georgia State University, Atlanta, Georgia 30303, and Yerkes Regional Primate Research Center, Emory University, Atlanta, Georgia 30322

PATRICK R. McGINNIS (239),   102 Putnam Street, Watertown, Massachusetts 02172

PAUL I. MUSEY (163),   Department of Medicine, Emory University School of Medicine, Atlanta, Georgia 30322

RONALD D. NADLER (69,191),   Yerkes Regional Primate Research Center, Emory University, Atlanta, Georgia 30322

JOHN R. K. PREEDY (163),   Department of Medicine, Emory University School of Medicine, Atlanta, Georgia 30322

FRANCISCO I. REYES* (45,83),   Departments of Obstetrics and Gynaecology and Physiology, University of Manitoba, Winnipeg, Manitoba, R3E 0Z3, Canada

R. V. SHORT (319),   Medical Research Council Unit of Reproductive Biology, Centre for Reproductive Biology, Edinburgh EH3 9EW, Scotland

KELLY J. STEWART (265),   Sub Department of Animal Behavior, University of Cambridge, Madingley, Cambridge CB3 8AA, England

CAROLINE E. G. TUTIN (239),   Department of Psychology, University of Stirling, Stirling FK9 4LA, Scotland

JEREMY S. D. WINTER (45,83),   Department of Paediatrics, University of Manitoba, Winnipeg, Manitoba, R3E 0Z3, Canada

KRISTINA WRIGHT (163),   Department of Medicine, Emory University School of Medicine, Atlanta, Georgia 30322

Present address: Department of Obstetrics and Gynecology, State University of New York, Downstate Medical Center, Brooklyn, New York 11203.

FOREWORD

This valuable compilation of information concerning the reproductive biology and sexual behavior of mankind's closest living relatives, the great apes, is remindful of the heroic effort that was made to salvage some of the archaeological treasures of Egypt from the rising waters of the Aswan dam. Study of these rare and disappearing species represents exploration of a "last frontier" in reproductive biology that is providing an exciting glimpse of the evolution of human reproduction as well as insight into the specialized physiological mechanisms peculiar to apes and humans.

From the study of numerous laboratory mammals much has been learned over the past half century concerning both the basic principles of reproductive physiology that are common to a wide variety of mammals and the special mechanisms that apply to individual species. The stimulatory action of estrogen on the female reproductive tract, the follicle-stimulating action of FSH, the ovulation-inducing action of LH, the masculinizing action of testosterone, and the obligatory role of progesterone in the establishment of pregnancy are examples of basic principles that could have been revealed by the study of any or all of the common laboratory animals. On the other hand, we have learned to our dismay that during evolution each species has developed features that are peculiar to its own reproductive process, and that these specializations limit our ability to extrapolate findings from one species to another.

Thus, all species exhibit minor to major differences in the anatomy of their reproductive systems, in timing and control of hormone-induced events, in alterations of peptide or protein hormone structure that bestow immunologic specificity, and in the quantity and metabolism of hormones secreted. It is the details of this process that have been most susceptible to evolutionary modification. As our knowledge deepens, so has the need for detailed information concerning those species most closely related to *Homo sapiens*. The elucidation of reproductive functions has now progressed to the point that even the monkeys have limitations as a model for humans. Unquestionably the Day of the Ape has arrived! The effort and cost necessary to study these difficult species are justified.

When I first learned that this volume on the reproductive biology of the great apes was in preparation, it brought to mind an ancient parable originally applied in a rather different context, "The last shall be first and the first last." The apes, though last to be utilized as laboratory subjects for reproductive studies, stand first as the animal of choice when the ultimate objective is transference of the information gained to humans. Although the old adage that "better late than never" also holds true in this instance, there are other compensating aspects to this long neglect of the apes. Most important is the fact that research on these special species is now greatly facilitated by the revolutionary improvements that have been made in the methods and equipment available for basic biological research. Moreover, these modern laboratory studies on captive apes benefit from the many preceding years of close observation of the social organization, living habits, and sex behavior of the great apes in their natural habitats and from the pioneering experimental studies of Robert M. Yerkes and his colleagues.

The reasons for this long neglect of the apes as laboratory animals for research on reproduction are not hard to find. If the principal criteria for suitable laboratory animals—ease of handling, ready availablility in large numbers, and low cost—are kept in mind, then the adult apes rate rather low on all counts. Anyone who has ever observed these powerful and forbidding adults in a zoo will appreciate that the necessary serial sampling of blood and urine is not something for the faint of heart. Although modern techniques have overcome some of these difficulties, only a limited number of laboratories are properly equipped for the conduct of basic biological research on reproduction in apes. Likewise, the opportunities for testing in apes the unending stream of ideas that flow from comparative studies in other species cannot be fully accommodated. This book gives consideration to such limitations in servicing the types of problems for which apes are suited.

The growing body of recently derived information about the reproductive biology of the chimpanzee, gorilla, and orangutan has finally reached a critical mass sufficient to generate a full review. The present volume is a ground-breaking endeavor that will broaden and enhance reproductive science. It will also contribute to the obliteration of a long-standing gap in our knowledge of the reproductive process in the nonhuman primates and will provide a preliminary measure of the apes as model animals for their less brachiating and more depiliated human counterparts who abound in the jungles of the world's cities. It is certain that specialists concerned with the control of human fertility will find this material on the great apes of keen interest. A much larger body of reproductive biologists and anthropologists will be eager to learn in what ways ape reproductive phenomena agree with or differ from those of other laboratory animals, or shed light on human evolution.

Roy O. Greep

PREFACE

This volume is an up-to-date and thorough review of all aspects of great ape reproduction, in a form that we expect will be of use to human physiologists, comparative anatomists and zoologists, behaviorists, taxonomists, anthropologists, primatologists, and breeders of apes. Particular thought has been given to the needs of biomedical scientists working in the area of control of human fertility and gynecological disease who may discover in these pages useful applications of apes as research models for the human. The phylogenetic affinity and obvious physical similarities of apes to man suggest the hypothesis that apes resemble man more closely in the anatomy, physiology, and behavior of reproduction than do monkeys and lower mammals. As the reader will discover, this assumption has generally proved to be correct.

To the extent that the apes share with man resemblances not possessed by other animals, they are of truly unique value for biomedical studies. In drawing attention to the potential of chimpanzees as models for man, Robert M. Yerkes pointed out that "The greater the degree of similarity (of structure and function), the safer the inference that what is true for the one animal will apply also to the other." In no field, with the possible exception of behavioral research, are apes of more potential value than in the field of reproductive biology. This fact is illustrated by the early appearance of papers on chimpanzee reproduction from 1930 to 1945, and a strong resurgence of activity during the last decade which provided the main incentive for preparing the present volume.

The availability of orangutans and gorillas for biomedical research is severely limited. Comparative study of the reproduction of these great apes, however, is of considerable academic interest, as well as of importance to the conservation of apes in the natural state and in captivity.

Reproduction is a dominant factor in the daily physical and behavioral activity of primates. Therefore it is necessary for every student of ape biology to be cognizant of the impact of reproductive behavior or physiology on his or her area of study. Reproduction also plays an important role in species survival and evolution; investigation of the evolution of reproductive patterns is therefore a proper anthropological endeavor. Since the reproductive organs are composed of soft tissues not preserved in the fossil record, comparative study of living primates, including the great apes, is the only pragmatic approach to this problem.

This book is intended not only as a reference work, but also as an opportunity to identify new areas ripe for investigation. As with other aspects of ape biology, there are many areas of reproduction that are totally or incompletely studied, particularly in orangutans and gorillas. Much new comparative data of significance to human sexual biology therefore await discovery.

Although every effort has been made to make this volume as comprehensive as possible, I am aware that reproduction of the pigmy chimpanzee has been almost ignored, except in the context of breeding. This is not an accidental omission, but reflects the lack of opportunity for study and acquisition of objective data on this species. This obscurity of *Pan paniscus* has led most of us to refer to *Pan troglodytes* as "the chimpanzee."

I was fortunate to have the opportunity to study reproductive biology of the great apes during 13 years spent at Yerkes Regional Primate Research Center under the sympathetic directorship of Geoffrey Bourne. I also enjoyed the collaboration of a number of colleagues who contributed their expertise to make those studies possible; many of them are contributors to this volume. To all these colleagues I express my special thanks.

I am no less grateful to those other contributors with whom I have never previously worked, but who in every case have prepared excellent and comprehensive reviews of their topics. All have been commendably patient with me during the long process of editing. I especially want to thank A. H. Harcourt, who helped me organize and edit the contributions on field studies, in order to achieve a uniformity of presentation that would facilitate interspecies comparison.

Support from PHS Grant RR00165 to the Yerkes Regional Primate Center is gratefully acknowledged. Also, the Ford Foundation generously and foresightedly underwrote much of the cost of my earlier studies, in the days when chimpanzees were generally regarded as exotic and impossible subjects. This book should help to change that view, by showing that much data of theoretical and practical value are obtainable from study of the great apes.

Chapter 1

MENSTRUAL CYCLE OF THE GREAT APES

Charles E. Graham

I. EXTERNALLY VISIBLE PARAMETERS OF THE MENSTRUAL CYCLE

The temporal aspects of the great ape menstrual cycle may be analyzed on the basis of two types of externally visible parameter, menses, and in the case of gorilla and chimpanzee, the labial or sexual swelling. Like the human female, the orangutan lacks any externally quantifiable cyclic swelling.

1

Copyright © 1981 by Academic Press, Inc.
All rights of reproduction in any form reserved.
ISBN 0-12-295020-8

A. Menses and Intermenstrual Interval

In most instances menses is overt, i.e., readily visible, in the caged chimpanzee. This is not usually the case in the orangutan or gorilla; in these species, however, occult menstrual blood may be detected in voided urine by using a chemical test strip (Hemastix, Ames Co., Elkhart, Iowa).

1. *Chimpanzee*

The mean length of the menstrual cycle for 653 cycles of 22 chimpanzees of all ages was given by Young and Yerkes (1943) as 37.3 ± 0.14 days with a median of 35 days, a mode of 33 days, and a total range of 22 to 187 days. The authors made the important point that a mean cycle length for a heterogeneous group of animals must be interpreted with caution since they found that various factors had a significant effect upon cycle length. In particular, adolescent cycles tended to be longer than cycles in mature individuals. Implantation bleeding, which mimics menses, has occasionally been noted (Nissen and Yerkes 1943; Graham, unpublished observations).

2. *Gorilla*

The first record for the gorilla was made by Noback (1939), based on a gorilla estimated to be nine years old when the first menses (menarche) was observed. Eight "cycles" ranging from 36–72 days (mean: 49 days) were observed. These data may be misleading because it is possible that two long intermenstrual intervals each represent two cycles. Since prolonged adolescent cycles in the chimpanzee and man frequently occur, a similar explanation may apply to this adolescent gorilla. When the two prolonged intervals are excluded to take into account the possibility of missed menses, the mean length was 43 days, although Noback thought even this figure might be an overestimate. Menses usually lasted for three days. The duration of the menstrual cycle derived from this study is significantly longer than the more reliable data based on labial swelling (see Section I,B,2).

3. *Orangutan*

The intermenstrual interval has been used by the author to determine the length of several menstrual cycles in captive orangutans at Yerkes Primate Center. The interval between first day of menses of three successive menstrual cycles in one subject of estimated age 13 years was

24, 25, and 26 days. In a 12-year-old subject, cycle lengths of 29, 25, and 32 were recorded.

Nadler (1977) has recorded eight additional cycles in three females aged 16–19 years, ranging from 26–32 days in duration: the median cycle length was 30.5 days. Number of days of menses ranged from one to four with a median of 2.8 days.

The only other reliable record of intermenstrual interval in the orangutan is that of Aulmann (1932), who observed cycles of 32, 32, 32, and 31 days in a 9½-year-old subject.

B.   Genital Swelling

1.   Chimpanzee

Changes in size and turgidity of the conspicuous sexual skin of the female chimpanzee (Fig. 1) provide a valuable indication of the progress of the reproductive cycle (Clark and Birch, 1948; Yerkes and Elder, 1936a, b; Graham, 1970). The maximal size of the sexual swelling varies from individual to individual, and in the same individual under different circumstances, as we shall see later. In some cases, it may be exceedingly large, completely surrounding the vagina and the anus and having a volume of up to 1.4 liters (Elder and Yerkes, 1936). The function of the swelling has not been experimentally determined, although it may be suspected of serving as a visual cue to the male of the sexual state of the female.

At the Yerkes Center the size of the swelling is scored daily on a scale of 0 to 4. Because of the variation from animal to animal in the size of maximal swelling, the successful determination that maximal swelling and turgidity has developed depends on the observers' familiarity with each animal; for an accurate determination, it is necessary to train the animal to present the swelling for palpation. Erikson (1963) attempted to increase the accuracy of such estimates of the sexual swelling by making clay and plaster of paris models of the sexual swelling of the anesthetized animal. Earlier, Tinklepaugh (1930) had devised an instrument to measure the contours of the sexual swelling by mechanical means, but the techniques of both investigators suffered from the serious limitation of being inapplicable to routine use on the unrestrained animal. This difficulty has not been surmounted.

The typical chimpanzee sexual cycle (Fig. 2) may be divided into four major phases (Young and Yerkes, 1943), the first of which is characterized by a period of quiescence of the sexual skin beginning the day after cessation of menses and usually lasting about seven days (pre-

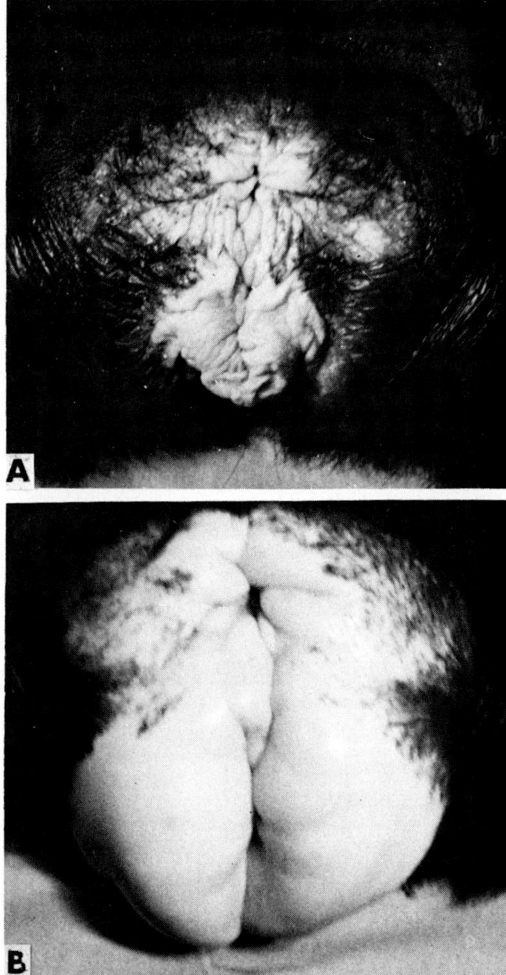

Fig. 1. Sexual swelling of chimpanzee. (A) completely detumesced; (B) during maximal swelling. (Graham *et al.*, 1972; reproduced with permission of the J. B. Lippincott Co.)

swelling phase). The sexual skin then begins to swell, reaches maximum, and remains there for several days; this period (the swelling phase) lasts for about 17 days. Detumescence then begins, is rapid, and is followed by a period of quiescence of the sexual skin; this period (about 10 days) is termed the postswelling phase. The last period is menses and is typically of three days' duration. The swelling and postswelling phases may be subdivided, as illustrated in Fig. 2. Six definable points occur during the sexual cycle: beginning and end of tumescence,

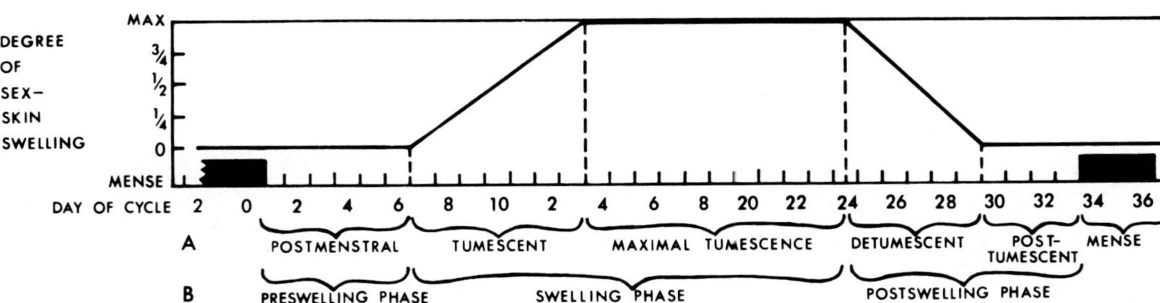

Fig. 2. The phases of the chimpanzee reproductive cycle, showing alternative terminologies. (A) after Yerkes and Elder (1936); (B) after Young and Yerkes (1943). (Graham, 1970; reproduced with permission of S. Karger AG.)

beginning and end of detumescence, and beginning and end of menstruation. The existence of these easily recognizable reference points taken in conjunction with close physiological similarities with man make the chimpanzee of unique value in reproduction research.

Young and Yerkes emphasized the variability both of total cycle length and of the duration of the constitutent phases; not all cycles fitted the typical pattern illustrated in Fig. 2. They calculated correlation coefficients for each of the six possible combinations of the basic four phases of the cycle, and showed that within the cycle the length of each phase varied independently of the others. There was only a slight correlation between frequency of variability in phase duration and infertility.

A number of factors can affect the pattern of sexual swelling of chimpanzees, including their age. Adolescent animals have longer than average cycles, with prolonged preswelling and swelling phases (Fig. 2), and a shortened postswelling phase, suggesting a short and inadequate luteal phase (Young and Yerkes, 1943). For example, in the first six months after the menarche, the mean duration of the preswelling, swelling, and postswelling phases of 16 chimpanzees were 20, 26, and 6 days, respectively, compared with 3, 18, and 12 days, eight years after menarche. The first cycle after pregnancy tended to revert to the adolescent condition; the length of the same three phases of the first cycle after pregnancy were 11, 23, and 9. By the second cycle after pregnancy, the typical adult temporal relationships were restored.

A seasonal factor affecting the length of the cycle was also noted by Young and Yerkes (1943). This was reflected in increased length of the preswelling phase most obvious in the months of November through February. In Atlanta, Georgia, the intermenstrual interval is sometimes so long that the term "winter amenorrhea" seems appropriate to describe what may be regarded as an interruption of cyclicity. This suggests the concept that a prolonged preswelling phase should be interpreted as a delay in the initiation of a new cycle, rather than prolongation of the cycle.

Perhaps at this point a distinction should be made between the usual convention of equating the intermenstrual interval with menstrual cycle duration, in contrast to a more physiological view that the intermenstrual interval could include more or less than a single, complete ovulatory cycle. By thinking in terms of the ovulatory cycle, we can analyze variations in the length of the intermenstrual interval in terms of variations in length of the ovulatory cycle (sum of follicular and luteal phases), and presence and duration of other components. The latter might include a period of ovarian quiescence, or the development and atresia of a crop of follicles preceding the initiation of an ovarian cycle sufficiently normal to be characterized by the development of sexual swelling and menses. At the present time, the precise physiological events associated with longer-than-normal cycles are unknown; this is a field for future investigation.

Savage–Rumbaugh and Wilkerson (1978) have described the morphology of the sexual swelling and its cyclic changes in the pygmy chimpanzee *Pan paniscus.*

In an adolescent pygmy chimpanzee, five irregular cycles were observed during a period of six months' observation; finally, a full swelling developed. An adult was studied through six swelling cycles. Although cycles lengths were not specified, graphic data indicated a range of 28 to 37 days.

2. Gorilla

The lowland gorilla (*Gorilla gorilla gorilla*) possesses a labial swelling, first reported by Noback (1939). As the labial swelling develops, so the lips of the labia part to expose the urogenital cleft. Noback measured the urogenital cleft, which varied in a consistent way from 1–4 cm; the interval from the beginning of swelling detumescence to menses varied from four to 10 days. This is consistent with other data indicating variability in length of the luteal phase in adolescent chimpanzees and girls. Variation in the size of the urogenital cleft was noted for two months preceding the menarche.

Nadler (1975a) subsequently conducted a more detailed study of nine subjects, of which eight were adults. Perineal changes were assessed by two methods; both applied to gorillas trained to mount the cage wire for examination. The first method utilized an estimate (on a three-point scale) of the amount of exposure of the pink labial mucosa; in some females, however, the labial mucosa was not exposed during labial swelling, and in these the thickness and smoothness of the labia was the subject of evaluation, since these features became more pronounced as labial swelling developed. Fifty-six cycles were evaluated in seven adult subjects by this method, giving a median cycle length (range) of 30.8 (13–55) and a mean length ($\pm$ S.D.) of 31.1 ($\pm$ 7.5) days (Fig. 3).

The second method utilized an objective measurement of the length of the urogenital cleft after the animals had been trained to mount the front of their cage. This method reduced variability noted in the first approach. Twenty-five cycles in six adults and one subadult yielded a median cycle length (range) of 31.8 (24–42) and a mean ($\pm$ S.D.) of

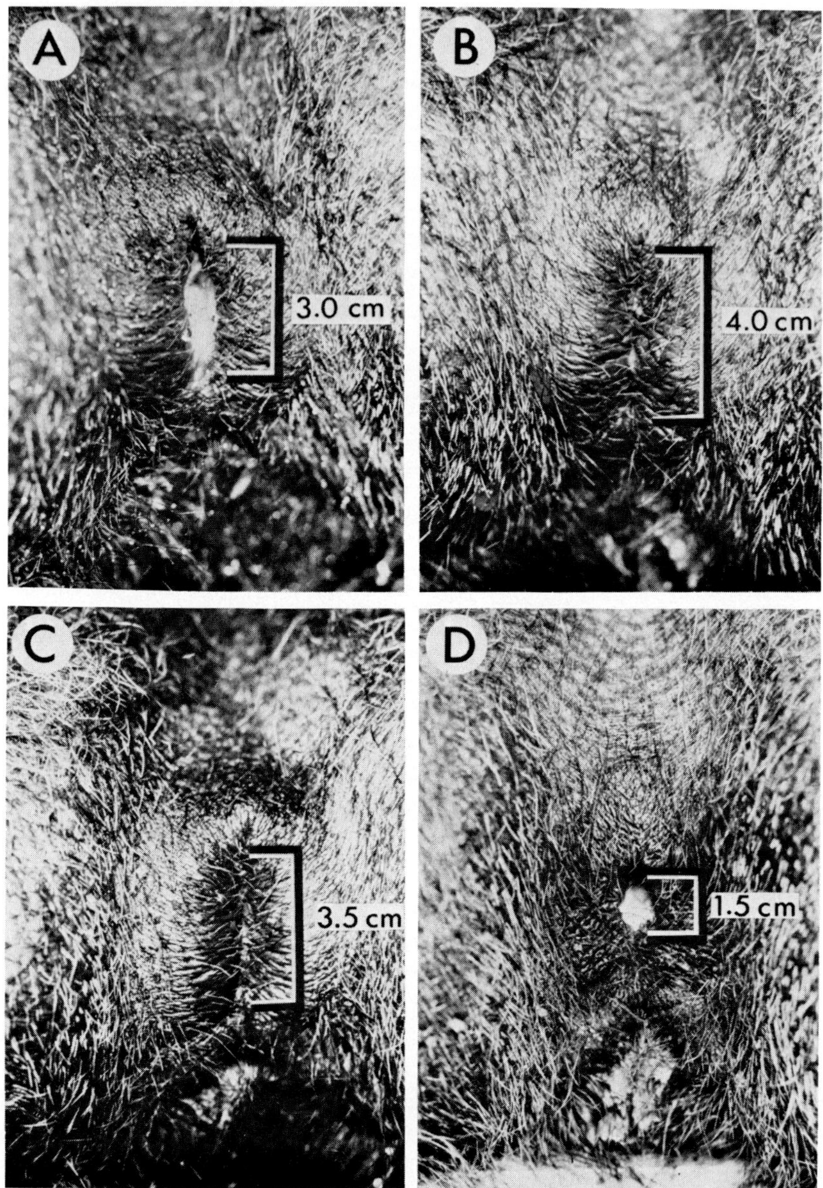

Fig. 3. Labial tumescence during different phases of the menstrual cycle in a female lowland gorilla. (A) Moderate tumescence, with wrinkling of the labia, associated with the late follicular phase; (B) Maximal tumescence, with relatively smooth labia, associated with the peri-ovulatory phase; (C) Submaximal tumescence on the first day of detumescence; (D) Complete detumescence, with considerable wrinkling and regression of the labia, associated with the late luteal phase. (Nadler *et al.*, 1979; reproduced with permission of the Williams and Wilkins Co.)

32.5 ($\pm$ 4.6). The subadult had longer cycles than the other subjects
(40.7 $\pm$ 3.5); recalculation of the data excluding this subject yielded
$\bar{\chi}$ $\pm$ S.D. = 32.4 $\pm$ 2.3. Maximal labial swelling persisted for an
average of 2.2 days in the gorilla compared with 10 days in the chimpanzee. Menses were observed in six cycles during this study, and the
interval between labial detumescence and menses was eight to 11 days.

3. Orangutan

The orangutan lacks a detectable genital swelling during the menstrual cycle.

C. Sexual Behavior

This subject is discussed throughly in Chapters 8–12, but evidence
concerning the period of sexual receptivity in apes will be summarized
here, because it is likely to be a cyclic phenomenon related to the time
of ovulation.

1. Chimpanzee

Sexual activity tends to increase during the swelling phase, and only
during the period of maximum swelling is the female *Pan troglodytes*
highly receptive (Yerkes and Elder, 1936b). Since the duration of maximum swelling is relatively long, and copulations occur at other times
in the cycle also, copulatory activity is of restricted value in character-
izing the menstrual cycle of chimanzees, especially in view of the wealth
of other data available.

Captive pygmy chimpanzees show some copulation throughout the
cycle, but 68% of copulations occurred during the period of maximal
swelling (Savage–Raumbaugh and Wilkerson, 1978). A detailed account
of copulatory behavior in captive pygmy chimpanzees has been given
by these authors, but the number of subjects studied is limited.

2. Gorilla

Nadler (1975b) has shown that captive lowland gorillas have a restricted period of receptivity associated with the brief period of labial
tumescence; during this restricted period females presented to males
and copulated with them more frequently than at other times. Tijskens
(1971) studied the incidence of mating in two captive female mountain
gorillas (*G. g. beringei*), approximately 10 years of age. The incidence
of copulation was monitored in these females, but the number of hours

monitored daily was not recorded. Adequate data for analysis were obtained from only one female. Mating was noted on one or two consecutive days at approximately monthly intervals; occasionally a bimonthly interval was noted, suggesting that in some cycles copulation did not occur or was not observed. Nine cycles comprising 25–42 days between successive copulations were recorded. This gives a mean and standard deviation for cycle length of 32.8 ± 5.7. This is exceedingly close to the figure obtained by Nadler based on interval between labial swelling (32.4 ± 2.3), suggesting that menstrual cycle length in the two subspecies is very similar. A number of studies are cited by Tijskens that provide confirmatory data on duration of the reproductive cycle of the gorilla. The longer cycle length based upon intermenstrual interval in a single adolescent animal cited earlier is likely to be a function of immaturity.

3. Orangutan

Studies by Nadler (1977 and Chapter 8, this volume) show that orangutans, if paired for only a 30-minute period daily, usually copulated daily throughout the cycle. When paired for five hours, copulation frequency was higher during 10 days at midcycle interval than during the rest of the cycle. However, the midcycle increase in copulatory frequency was not sufficiently defined to be of value as an accurate cycle parameter. Females rarely showed proceptive sexual behavior under the prevailing experimental conditions.

II. ENDOCRINOLOGY

A. Sex Steroid and Gonadotropin Patterns

1. Chimpanzee

The earliest attempt to measure estrogens in the urine of chimpanzees utilized a mouse bioassay method and succeeded in showing a greater excretion of estrogenic activity at midcycle than during menses (Allen *et al.*, 1935). Using an improved method of extraction, followed by bioassay, Fish *et al.* (1941) showed clearly that there is a midcycle and a midluteal peak of estrogen excretion, as in human females. Androgens were also assayed and showed constant levels throughout the cycle; no modern study of androgen levels during the chimpanzee menstrual cycle has been published.

Figure 4 shows the urinary excretion pattern of estrogen and pregnanediol during 11 menstrual cycles in relation to sexual-swelling regression. This data incorporates several cycles previously published (Graham *et al.* 1972, 1977). Most notable is the luteal rise of estrone that in individual animals reached 50 to 100% of the late follicular peak. These composite data confirm the pattern of ovarian steroid excretion and its correlation with sexual swelling established in the original study by Graham *et al.* (1972). In this study, carefully validated physico-chemical assays and rigorous identification of the steroids were employed to describe in detail the excretion of the three classic estrogens and pregnanediol throughout the menstrual cycle of three chimpanzees. The following compounds were identified: estrone, estriol, estrone sulfate, estrone glucosiduronate, estradiol-17β glucosiduronate, pregnanediol, and pregnanediol glucosiduronate.

The daily outputs of estrone, estradiol-17β, estriol, and pregnanediol in 24-hour urine samples were correlated with the sexual swelling and occurrence of menses. The values for estrone and estradiol-17β were similar to those of the human menstrual cycle, but the values for estriol

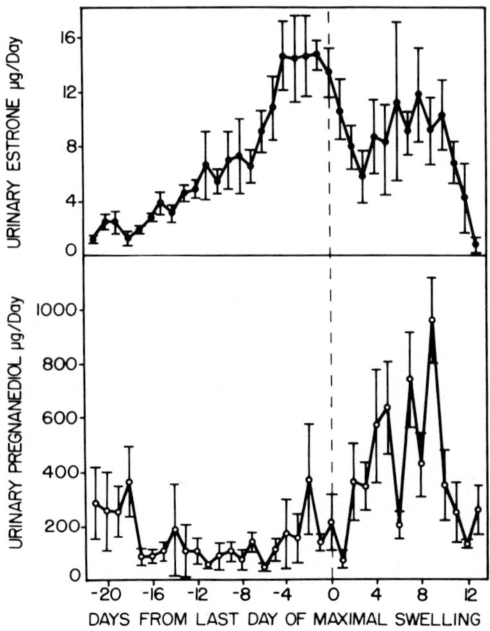

Fig. 4. Daily urinary estrone and pregnanediol excretion throughout 11 chimpanzee cycles related to regression of the sexual swelling. Day 0 is the last day of maximal sexual swelling.

were considerably lower. This pattern of changes was shown most clearly in estrone excretion, and was reflected to a varying extent by the less abundant metabolites, estradiol-17β and estriol.

Rising estrogen levels were correlated approximately with increasing size of sexual swelling, and estrogen levels fell at the time of sexual-swelling detumescence. This correlation is discussed more fully in Section IV,B.

Based upon other studies (Wright *et al.*, Chapter 7, this volume), pregnanediol was known to be the chief metabolite of progesterone. Urinary pregnanediol levels were low until sexual-swelling detumescence, after which a rise occurred with a peak midway between the time of swelling detumescence and onset of menses, coinciding with the second estrone peak. The urinary pregnanediol levels were lower than in the human, both in absolute terms, and relative to plasma progesterone levels measured in the chimpanzee as described below.

Urinary ovarian steriod excretion has been studied in other primates. In the rhesus monkey and baboon, the principle urinary estrogen is estrone, as in the chimpanzee; unlike the chimpanzee and human, however, estriol is not excreted in detectable quantities during the menstrual cycle. Both rhesus monkey and baboon lack a regular, definite luteal estrone elevation in the urine in contrast to chimpanzees and human females. Some individual rhesus monkeys and baboons showed an apparent small luteal elevation in estrone (Hopper and Tullner, 1970; Stevens *et al.*, 1970); however, the apparent elevation in most instances really represented a recovery from a temporary depression to very low levels after the ovulatory peak, and luteal values usually did not rise much above early follicle levels. A similar interpretation applies to plasma estradiol levels in the rhesus monkey.

As discussed in greater detail by Wright *et al.* (Chapter 7, this volume) monkeys excrete androsterone rather than pregnandiol as the major metabolite of progesterone during the menstrual cycle, although pregnanediol has been recovered from rhesus monkeys and baboons after administration of 4-[^{14}C]progesterone. Thus, chimpanzees resemble women more closely than monkeys in some aspects of their ovarian steroid metabolism.

Circulating hormone levels also have been studied in the chimpanzee. Plasma progesterone levels, measured by a competitive protein-binding technique, were <1.0 mg/ml until shortly before swelling detumescence; after detumescence, values rose to approximately 15 mg/ml plasma (Graham *et al.*, 1972). Subsequent studies by the same group using radioimmunoassay described plasma progesterone levels throughout the menstrual cycle in more detail, with confirmatory results (Graham *et al.*, 1974; Graham, 1976).

Reyes *et al.* (1975) determined levels of estradiol-17β, estrone, estriol, progesterone, LH, FSH, and prolactin by radioimmunoassay in serum samples obtained without anesthesia from six cycles without conception in four chimpanzees, and a fertile cycle in three of the same animals. The same group had previously measured serum gonadotropins from four additional cycles, one of which resulted in conception (Howland *et al.*, 1971). In both studies, the hormone patterns of the menstrual cycles were similar to those of women, with high levels of FSH in the early follicular phase, followed by rising estradiol-17β concentration to a peak at or just before a midcycle LH and FSH peak (Fig. 5). In most cycles there was a luteal estradiol-17β rise in association with the luteal progesterone peak. Prolactin levels fluctuated but showed no clear pat-

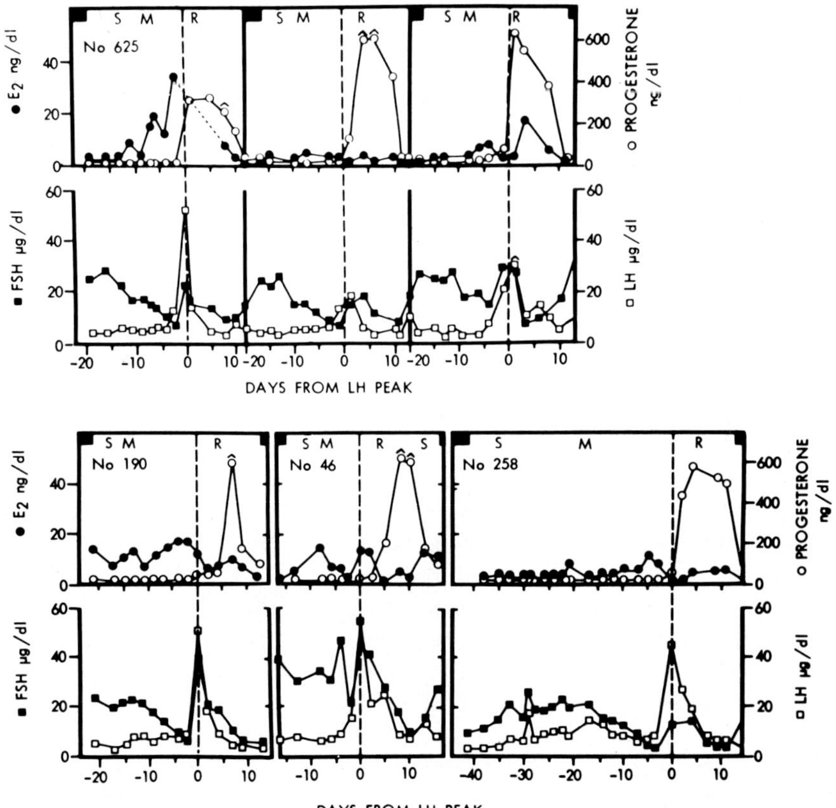

Fig. 5. Plasma gonadotropins and ovarian steroids during six menstrual cycles of four chimpanzees. Day 0 = actual or presumed LH peak; Bar = menses; S = start, M = maximum, R = receding sex-skin swelling. (Reyes *et al.*, 1975; reproduced with permission of the J. B. Lippincott Co.)

tern during the menstrual cycle. The sampling protocol did not permit complete characterization of the late follicular estradiol peak.

Sexual swelling began to detumesce one to seven days after the midcycle gonadotropin peak in the two studies summarized here. This variability may be a function of the animals selected, or related to the criterion used to define swelling detumescence. The latter explanation is likely, since there was considerable variability in the interval between swelling regression and menses in one of these studies (Howland *et al.*, 1971), in contrast to the very regular interval of 10–11 days reported by Elder and Yerkes (1936).

The circulating hormone patterns in these chimpanzees were remarkably similar to the human, but different in several details from rhesus monkeys and baboons. For example, LH was frequently elevated for more than two days at midcycle in the chimpanzee, whereas an elevation of one to two days is characteristic of rhesus monkeys and baboons. The patterns and levels of estradiol-17β and progesterone also closely resembled human menstrual cycle values, whereas monkeys and baboons lack regular midluteal elevation of estradiol that exceeds early follicular values, and in the rhesus monkey, estradiol and progesterone levels are somewhat lower (Wright *et al.*, Chapter 7, this volume).

2. Gorilla

The urinary excretion of sex hormones has been studied in the menstrual cycle of four gorillas. As in the chimpanzee, estrone was the most abundant urinary estrogen identified; it showed well-defined follicular and midluteal elevations. Estradiol and estriol showed a comparable pattern, although the quantities excreted were considerably smaller. The follicular estrogen peaks were approximately correlated in time with the plasma LH peak, and preceded labial regression by one to four days. A well-developed luteal elevation of pregnanediol developed, reaching 700 μg/24 hr, and was detectable as the glucuronide at maximum concentration of 1800–2000 μg/24 hr.

Nadler *et al.* (1979) have described circulating hormone levels in four menstrual cycles of three gorillas (Fig. 6). LH and FSH were measured in a human radioimmunoassay system, and estradiol-17β, progesterone, and testosterone were also measured by radioimmunoassay. The changes in hormone levels were correlated with changes in labial swelling. All cycles exhibited a midcycle LH peak, and FSH increased at midcycle, although not as conspicuously as LH. Prolactin showed no rhythmicity. The 17β-estradiol showed a midcycle peak (200–500 pg/

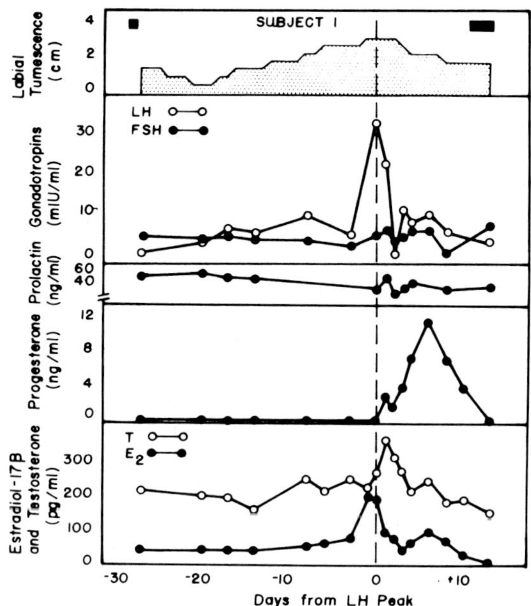

Fig. 6. Plasma LH, FSH, prolactin, progesterone, estradiol, and testosterone and labial tumescence during the menstrual cycle of a female lowland gorilla. Data are normalized to the day of the LH peak. Black bars indicate days of menstruation; T = testosterone; E_2 = estradiol-17β. (Nadler *et al.*, 1979; reproduced with permission of the William and Wilkins Co.)

ml) that preceded the LH peak and a luteal phase elevation that approximated the time course of progesterone. Progesterone was barely detectable during the follicular phase (<0.5 ng/ml), rose on the day of the LH peak or a day later, peaked six to seven days after the LH peak (10.8–12.2 ng/ml), and decreased to low levels at menses. Testosterone exhibited a midcycle peak of 350–500 pg/ml compared to follicular and luteal phase values of 150–300 pg/ml. Labial swelling was minimal at menses, rose after six to 16 days, and reached maximum nine to 35 days after menses, in every case on the day before the LH peak, in association with the midcycle elevation in 17β-estradiol. Detumescence occurred zero to three days after the LH peak, when 17β-estradiol was decreasing and progesterone was increasing.

Although lack of a gorilla gonadotropin standard prohibits comparison of absolute values, the pattern of gonadotropin secretion and pattern of absolute levels of estradiol-17β and progesterone are remarkably similar to the chimpanzee and human. A number of human studies have reported a midcycle elevation of testosterone excretion (e.g., Guerrero *et al.*, 1976), and Hess and Resko (1973) reported a similar rise in testosterone in rhesus monkeys, although it occurred on the day of the midcycle estradiol peak, rather than one or two days later as in gorillas and women. Nadler *et al.* (1979) have postulated that the midcycle testosterone peak, by virtue of its association with the presumed time of ovulation and the restricted period of female sexual proceptivity, may play an important role in the control of gorilla sexual behavior.

In common with the woman, the gorilla exhibits a number of differences from the rhesus monkey, including a midcycle estradiol peak that occurs one to several days previous to the LH peak, a pronounced midluteal elevation in estradiol, relatively higher concentrations of progesterone during the luteal phase, generally lower concentrations of testosterone during the cycle, and a midcycle testosterone peak that occurs one to several days later, in relation to the LH peak. On the other hand, the length of the menstrual cycle and of the luteal phase are more similar in rhesus monkeys and women.

Perhaps due to its shorter duration, maximal labial swelling in the gorillas was more closely related in time to the midcycle estradiol peak than was maximal genital swelling in chimpanzees. Moreover, in all gorilla subjects, the LH peak occurred on the day after the first day of maximal labial swelling. If these observations in a few cycles are confirmed on a larger scale, it will show that progression of the menstrual cycle is more predictable on the basis of genital swelling in the gorilla than in the chimpanzee.

3. *Orangutan*

Menstrual cycle hormone levels have been measured in a single study, performed by the Yerkes group (Collins *et al.*, 1975). Urinary estrone, estradiol-17β, estriol, pregnanediol, and androsterone were identified and measured during three menstrual cycles of two orangutans (Fig. 7). The pattern of excreted hormones was comparable to the human and chimpanzee, with the exception of androsterone. Of particular note was the conspicuous luteal estrone rise. The values for estrone and estradiol-17β were similar to the human and chimpanzee, but levels of estriol were higher than chimpanzee but lower than women. Pregnanediol and androsterone were significantly lower than in the human female.

Androsterone in the orangutan is of particular interest because of a marked rise in excretion from 120–800 μg/day in the follicular phase to 95–1140 μg/day in the luteal phase. Women have considerably higher androsterone levels, but show no luteal rise (Kirschner and Lipsett,

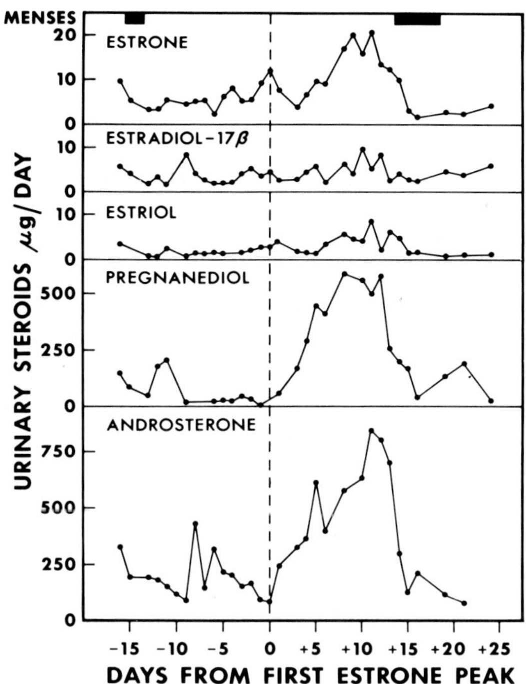

Fig. 7. Urinary steroid excretion in an orangutan throughout the menstrual cycle (Collins *et al.*, 1975; reproduced with permission of the J. B. Lippincott Co.)

1964). Rhesus monkeys show a relatively larger output of androsterone, with a luteal peak in some cycles, explicable on the basis that androsterone is the major metabolite of progesterone in this species (Plant *et al.*, 1969; Plant,1974). However in the orangutan, administration of [14C] progesterone does not permit recovery of [14C]androsterone, indicating that the luteal peak of androsterone is not derived from progesterone metabolism (YoungLai, Graham, and Collins, unpublished). Therefore, it is possible that some other steroid secreted by the corpus luteum is the precursor of the luteal androsterone increment in this species.

B.   Ovarian–Pituitary–Hypothalamic Axis

This subject has been studied only in the chimpanzee among the great apes. Injection of synthetic luteinizing hormone releasing hormone (LHRH), and measurement of subsequent release of gonadotropins into peripheral blood provides a useful means of testing the capacity of the adenohypophysis to secrete gonadotropins under various conditions. Hobson and Fuller (1977) found that LH peaks occurred after intravenous injection of 125 mg LHRH in all the chimpanzees studied, in contrast to rhesus monkeys, which frequently failed to respond. Peri-ovulatory chimpanzees gave the largest responses (LH 560, FSH 210% control) compared with midfollicular animals (LH 290, FSH 190% control) or luteal animals (LH 370, FSH 140% control). The mean peak LH response of male chimpanzees was 285% of control. The chimpanzee responses were in the lower part of the human range. Rhesus monkeys do sometimes respond to acute LHRH administration, but their LH responses are usually less than 200% of control values; Hobson and Fuller pointed out that the inconsistency of their responses restricts the utility of rhesus monkeys as models for the study of the control of gonadotropin release. Chimpanzee and human responses to acute administration of LHRH are shown in Table 1. Peak and mean LH and FSH elevations in the chimpanzee are within or slightly below the human range in the midfollicular phase. Although peak levels of LH tend to be less than human values, mean levels are relatively higher, due to a more gradual return of peak values to baseline in the chimpanzee. Early follicular chimpanzees showed no return to gonadotropin levels to baseline within two hours, as reflected in Table 1 by a value of 0 for BR (relative return to baseline, or Baseline Return).

Graham *et al.* (1979) verified the midfollicular responses reported by Hobson and Fuller (1977), and compared responses of early follicular and postpartum chimpanzees (low estrogen levels: $E_2 < 45$ pg/ml) with midfollicular chimpanzees or menstruating chimpanzees administered estrogen (elevated estrogen levels: $E_2 = 73.6$ pg/ml and 283 pg/ml, respectively). Whereas under conditions where estrogen levels were low, the release of gonadotropin was monophasic, elevated estrogen levels were associated with a biphasic increment in gonadotropin levels, with an initial increase beginning within five minutes of administration of LHRH followed by a plateau or slight decline, and a second increase beginning 60 minutes after LHRH administration (Fig. 8). This second phase of release was most clearly exhibited by LH, but when 10 mg of a potent analog of LHRH, des-Gly[10]-(D-Trp[6], Pro[9]-NEt) LHRH was administered, the second phose of release was greatly exaggerated for both gonadotropins. It was evident that this agonistic analog did not alter the response during the first hour. This observation suggested that either elevated estrogen levels or increased potency of the releasing stimulus could induce a second phase of gonadotropin release. Males showed a monophasic response to LHRH, and when doses of LHRH ranging from 10–500 mg were tested, no dose response or second phase of LH release was observed. A biphasic response also has been observed

TABLE 1

Comparison of chimpanzee and human gonadotropin responses to LHRH.

| Species | Luteinizing hormone (LH)[a] | | | | | | Follicle stimulating hormone (FSH)[a] | | | | | | n | LHRH Dose |
| | Early follicular | | | Mid-follicular | | | Early follicular | | | Mid-follicular | | | | |
	Peak Δ	$\overline{X}\Delta$	% BR	Peak Δ	$\overline{X}\Delta$	% BR	Peak Δ	$\overline{X}\Delta$	% BR	Peak Δ	$\overline{X}\Delta$	% BR		
Chimpanzee														
Graham *et al.*, 1979	236	177	30	104	77	0	62	56	0	34	24	31	7	100 μg
Hobson *et al.*, 1977[b]	—	—	—	150	132	13	—	—	—	72	60	27	15	125 μg
Woman														
Thomas *et al.*, 1973[b]	176	88	77	180	138	42	—	—	—	—	—	—	4	50 μg
Nillius and Wide, 1972[b]	106	67	72	187	118	72	12	6	77	40	20	73	7	100 μg
Yen *et al.*, 1972[b]	650	374	43	374	288	14	87	59	29	86	46	0	6	150 μg

[a] Values are peak increment (peak Δ) and mean increment ($\overline{X}\Delta$) expressed as % of mean pre-LHRH levels, and relative return to baseline; e.g.,

$$\frac{\text{peak } \Delta - \text{2h post-LHRH value}}{\text{peak } \Delta} \times 100 = \text{Baseline Return (BR)}.$$

[b] Values interpolated from published data.

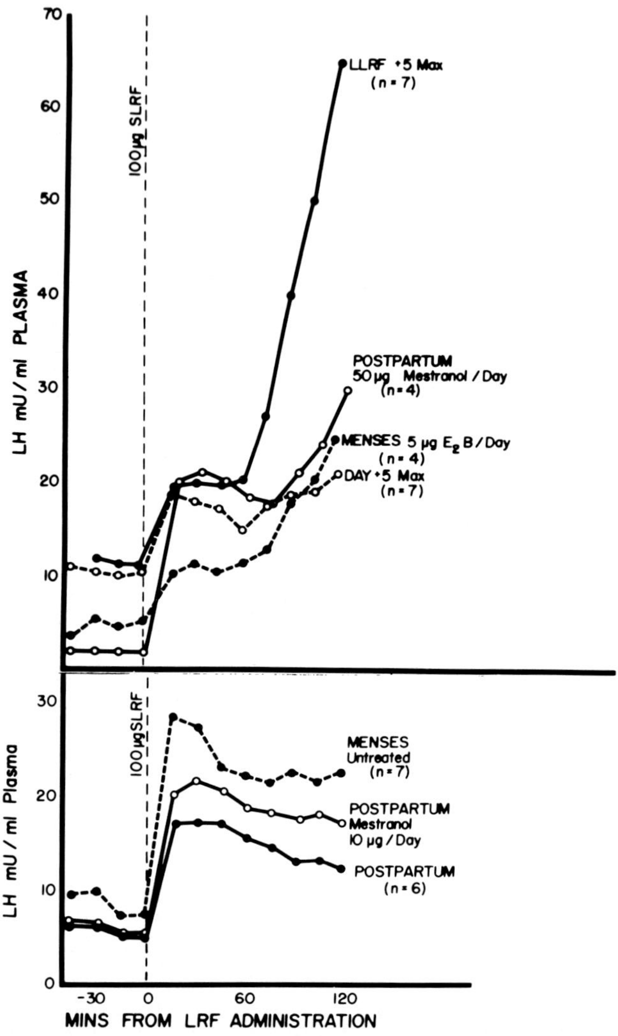

Fig. 8. LHRH responses in chimpanzees. Lower block shows monophasic responses in animals not having elevated estrogen levels. In the upper block, animals with elevated estrogen levels exhibit a biphasic elevation of LH. Note the particularly large response of animals that received a potent LHRH analog (LLRF). Estradiol benzoate was administered for two days prior to LHRH administration, and mestranol for seven days prior in those groups indicated as receiving these hormones.

in postpartum chimpanzees given 500 mg LHRH, although not at lower doses (Graham, unpublished data). In men and women, a single injection of LHRH normally induces a dose-dependent monophasic response. Infusion of LHRH can elicit both the initial gonadotropin rise, and also a second phase of release commencing about one hour after the beginning of infusion. The chimpanzee data shows that the two phases of release can be differentially controlled and provides an excellent model in which to dissect the control systems regulating them. A working hypothesis consistent with current studies is that the second phase of gonadotropin release requires that the LHRH stimulus must exceed a certain threshold, and that circulating estrogen levels can lower this threshold.

The physiological significance of these observations are as follows: The preovulatory rise in circulating estradiol levels is now recognized to be causally related to the subsequent ovulatory gonadotropin surge (such gonadotropin surges can be elicited by exogenous estrogen in various primates including the chimpanzee.) There is evidence from studies on women that the preovulatory gonadotropin surge partly depends on mobilization of a so-called "second pool" of gonadotropin under the influence of elevated estrogen levels (Hoff *et al.*, 1977). The facilitation of the second phase of gonadotropin release by estrogen in chimpanzees is consistent with this interpretation, although the true nature of the so-called "second pool" has yet to be resolved.

The larger and more replicable responses of chimpanzees make the species preferable for studies designed to explore the mechanisms controlling gonadotropin release in the human. Hobson's group has gone on to use the chimpanzee to study LHRH analogs that are or appear to be competitive inhibitors of LHRH, (Hobson *et al.*, 1978; Gosselin *et al.*, 1979a). The LHRH antagonist (D-Phe^2, D-Trp^3, D-Phe^6)-LHRH (35 mg) administered in the early follicular phase did not affect basal gonadotropin levels, but considerably reduced the LH response to two 10-μg doses of LHRH spaced 90 minutes apart.

Further studies by the same group (Gosselin *et al.*, 1979b) investigated the anti-ovulatory potential of a second LHRH analog, (D-PGlu^1, D-Phe^2, D-Trp^{3,6})-LHRH. This analog significantly inhibited the ability of exogenous LHRH to induce LH elevations in adult female chimpanzees. This drug was active at a dose level of 1.5–7.0 mg.

The activity of this analog was also evaluated in a model system that utilized estradiol benzoate-induced gonadotropin surges in the chimpanzee. Two E$_2$B injections (25 μg/kg, subcutaneous) were given eight hours apart on Day 3 of menses, then either vehicle (initial menstrual cycle) or 5 mg of analog (following cycle) were given subcutaneously at six-hour intervals beginning 48 hours and ending 96 hours after the first dose of E₂B. Serum levels of LH and FSH were depressed by E₂B during the first 48 hours. Subsequently, two peaks developed at 72 hours and 96 hours followed by a return to baseline levels at 168 hours. The first gonadotropin surge was significantly ($p < 0.05$) delayed and/ or diminished by analog. A partial inhibition of the second surge was also observed in some animals.

These studies establish the utility of the chimpanzee for investigating the biological properties of LHRH analogs that are inhibitors of LHRH and that may have clinical value for controlling gonadotropin secretion, and show that LHRH antagonists may have value for the suppression of ovulation in higher primates.

LHRH agonists and antagonists are also likely to prove valuable as experimental tools to dissect the two phases of gonadotropin release.

III.   CYCLIC CHANGES IN SECONDARY SEX ORGANS

A.   Genital Swelling

1.   *Chimpanzee*

The sexual skin of the chimpanzee is a hairless, unpigmented, and circumvaginal area of epidermis and subcutaneous tissue that usually extends to surround the anus. The sexual swelling appears in these areas during the follicular phase in a regular manner that makes it valuable as a temporal parameter of the cycle, as described in Section I,B,1. It also appears intermittently during the first trimester of pregnancy in some animals. Harcourt discusses the possible function of the swelling in Chapter 12.

A number of studies showed that estrogen administration could induce swelling in adolescent adult and castrate female chimpanzees (e.g., Clark, 1947; Zuckerman and Fulton, 1934). Swelling could not be so induced in intact or castrate males, in contrast to baboon males, in which estrogen administration can induce swelling (Clark, 1945, 1949). Experiments in the ovariectomized chimpanzee indicated that progestins could at least temporarily inhibit estrogen-induced swelling (Clark and Birch, 1946, 1948).

The experiments by Clark and Birch were based on a small but unspecified number of incompletely described experiments with two animals. In order to determine more exactly the ovarian hormone control of sexual swelling, Graham *et al.* (1972) undertook a more systematic study. As mentioned in Section II,A,1, urinary estrone levels increased as the sexual swelling developed, and fell at about the time that swelling began to regress. An increase in pregnanediol levels was associated with swelling detumescence. Oral administration of 200 μg/day of the estrogen mestranol to two adult ovariectomized chimpanzees resulted in development of a normal sexual swelling; either withdrawal of mestranol, or commencement of oral progestin administration after swelling was fully developed, replically resulted in swelling detumescence (Fig. 9). Menses followed detumescence almost immediately when regression was induced by withdrawal of mestranol; however, when regression was induced by progestin administration (4 mg chlormadenone acetate/day), menses was delayed, more in conformity with the interval between swelling regression and menses in the normal cycle. On the basis of these observations, it was concluded that (1) adequate levels of circulating estrogen of exogenous or endogenous origin can induce sexual swelling, and (2) increased levels of progestin of exogenous or endogenous origin can inhibit estrogen-induced swelling. The regression of swelling in the normal cycle is likely to be due to a changing estrogen/progestin ratio brought about by the cessation of follicular estrogen secretion and commencement of luteal progesterone secretion. The absence of sexual swelling during the luteal estrogen rise can be attributed to the inhibitory effect of the elevated progesterone levels at that time. The menstrual cycle pattern of circulating estradiol-17β or progesterone described earlier is consistent with this interpretation.

It is not certain from this work that estrogen and progestin are the only classes of steroid hormone that can affect the sexual swelling. Gilbert and Gillman (1945) found that the sexual swellings of baboons can be inhibited by administration of androgens or corticosteroids. Individuals working closely with chimpanzees often comment that stress (e.g., fights, anesthesia) frequently results in temporary or premature swelling regression, suggesting an adrenal effect on the swelling; however, proper documentation for such an effect is lacking.

It should be noted that there is no evidence in the chimpanzee that estrogens or progestins act directly on the sexual swelling. Their effects could conceivably be mediated through the hormone system controlling water balance. During deturgescence of the sexual swelling of baboons and pigtail monkeys, there is a marked polyuria (Naidoo, 1971; Krohn and Zuckerman, 1937). The endocrine control of sexual swelling in the baboon has been extensively studied, and the role of ovarian hormones is comparable to the condition in the chimpanzee (Gillman and Gilbert, 1956a).

Hyaluronidase metabolism also appears to play a role in diuresis

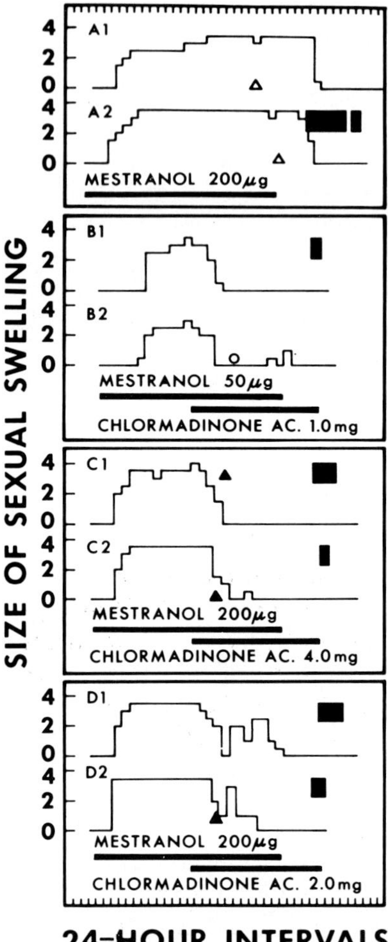

Fig. 9. Sexual swelling patterns during the administration of mestranol alone or mestranol and chlormadenone acetate to a chimpanzee. Atrophic endometrium, open circles; proliferative, open triangles; secretory, closed triangles. (Graham *et al.*, 1972; reproduced with permission of the J. B. Lippincott Co.)

associated with sexual-swelling detumescence. Vasopressin (an antidiuretic hormone) has its effect in part by increasing the permeability of the collecting ducts of the kidney to water (Ginetzinsky, 1958). Histochemical studies indicate that during diuresis, increased permeability is due to dissolution of hyaluronic acid-rich intracellular cement between the collecting duct cells. Consistent with this interpretation, during diuresis, hyaluronidase levels in the urine are very low, whereas during water retention, the opposite is true (Ginetzinsky, 1958).

Hyaluronic acid has been also implicated in fluid retention in tissues (Pearce and Watson, 1949; Ragan and Meyer, 1949; Ludwig *et al.*, 1950). Swelling of the sexual skin in intact baboons and in ovariectomized estrogen-treated animals is accompanied by an increase in the dry-weight concentration of hyaluronic acid, and reduction in urine volume and total urinary nonsulfated glycosaminoglycans (hyaluronic acid and lower molecular weight compounds). During detumescence, there is a precipitous decline in the sexual swelling content of hyaluronic acid, and an increase in urine volume and glycosaminoglycan content (Naidoo, 1971; Naidoo and Qureshi, 1974). These data are consistent with a role for hyaluronic acid in retention of water in the baboon sex swelling; at first sight, it seems possible that antidiuretic hormone could be one of the controlling factors in regulating the hyaluronic acid and water content of the swelling in a manner analogous to the mechanism of water retention by the kidney. However, neither thyroidectomy, adrenalectomy, nor hypophysectomy of baboons prevented water retention induced by estrogen or delayed excretion of water after withdrawal of estrogen (Gillman and Gilbert, 1956b), indicating that the effect of estrogen is not mediated by vasopressin or adrenal hormones. Sexual swelling is not part of a general antidiuretic effect, since it occurs partly at the expense of water content in other tissues; although the total body weight of the baboon increases due to the very large size of the sexual swelling, most of the body actually loses weight (Clarke, 1940). Moreover, large doses of antidiuretic hormone did not inhibit spontaneous swelling regression (Gillman and Gilbert, 1956c). Therefore, sexual swelling in baboons and presumably in chimpanzees is apparently due to an organ-specific, estrogen-induced effect on hyaluronic acid and water accumulation.

2. Gorilla and Orangutan

The gorilla and orangutan lack a sexual swelling of the type possessed by the chimpanzee. The gorilla possesses a labial swelling (Noback, 1939), which presumably results from localized hyperaemia or edema. The gorilla sexual swelling was discussed previously in Section, I,B,2. Maximal labial swelling persists for two to three days and is associated with increased frequency of copulation (Nadler, 1975a). The orangutan lacks a sexual swelling during the menstrual cycle, but labial swelling develops during pregnancy and is useful for pregnancy diagnosis. The endocrine control of labial swelling in these species has not been investigated, although the temporal correlations with endocrine events in the gorilla described by Nadler *et al.* (Section II, A, 2 and 1979) compare favorably with the chimpanzee, suggesting that ovarian steroids play a similar role in the two species.

B. Vagina

1. Chimpanzee

Studies of vaginal exfoliation in the chimpanzee have previously been thoroughly reviewed by this author (Graham, 1970). The pattern of exfoliation was not consistent between studies, and the patterns of cornification did not reliably reflect the physiological status of the animal, in contrast to many other species. The indistinct patterns of cornification noted were attributed to the pooling of several days' production of exfoliated cells in the lower vagina and vestibule (the vestibule is the passage through the sexual swelling, distal to the vagina), and contamination by vestibular cells that probably lack marked hormone dependent changes (Graham, 1970). In view of the difficulty of obtaining smears in adult unrestrained subjects, and the indistinct patterns of cornification observed in unrestrained animals, it was concluded that vaginal smears present no advantage over the excellent data obtainable from sexual swelling for monitoring progress of the menstrual cycle.

2. Gorilla

Vaginal smear changes have been observed in three menstrual cycles of a single lowland gorilla (Noback, 1939). A definite cyclic pattern of exfoliation was noted. An increase in the exfoliated cells occurred during the first part of the cycle, reaching a maximum at the beginning of the copulative phase, which was maintained until the following premenstrum. During the copulative phase, lightly stained superficial cells with pyknotic nuclei were abundant. This study was conducted in a menarchal gorilla that easily could be handled. It may not be generally feasible to obtain smears from adult gorillas in this way without lengthy training and some physical risk to the observer. Therefore, monitoring of labial swelling and menses is recommended as a more practical and adequately validated method of following the menstrual cycle.

C. Endometrium

1. *Chimpanzee*

Histological changes in the uterus throughout the chimpanzee menstrual cycle have been described by Graham (1973). Histological changes approximated those of the human uterus as described by Noyes *et al.* (1950), except in one important detail: like human beings, unmated chimpanzees show spontaneous decidualization of the endometrium (Fig. 10). In the other species that have been studied, including the rhesus monkey, decidualization does not develop in unmated females; in rats and mice it is known that decidualization only occurs when a blastocyst reaches the uterus. Since humans and chimpanzees show a luteal estrogen increment and spontaneous uterine decidualization, whereas no other species are known to show these features, it was hypothesized that these two phenomena might be causally related.

In order to determine whether elevated estrogen levels are a factor in decidualization, two experimental approaches were explored (Graham *et al.*, 1978). In one approach, intact chimpanzees were administered estrogen inhibitors during the luteal phase in an attempt to inhibit decidualization. However, no inhibition occurred. In a second

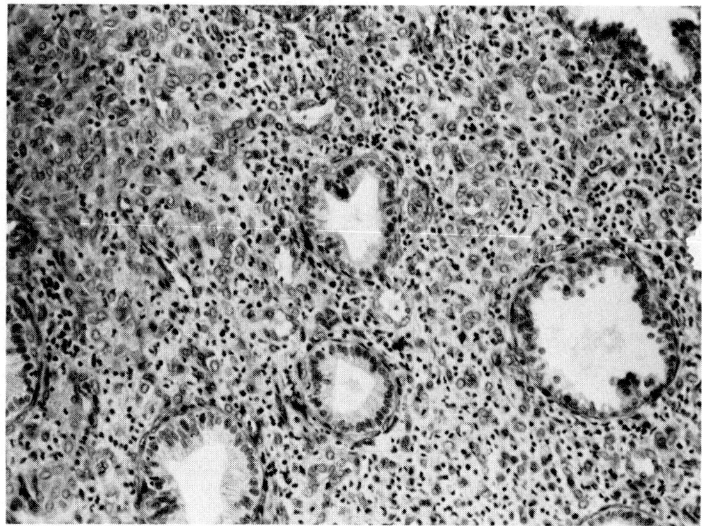

Fig. 10. Endometrium of unmated chimpanzee late in the luteal phase, showing spontaneous endometrial decidualization. Hematoxylin and eosin, ×250. (Graham, 1973; reproduced with permission of S. Karger AG.)

experiment, ovariectomized chimpanzees were administered estrogen followed by progesterone with or without additional estrogen in order to mimic the luteal estrogen peak of the intact animal and to determine whethe] a second estrogen surge was necessary for induction of decidualiz tion. Both protocols resulted in the induction of a fully decidualized endometrium. Thus, there is no evidence that luteal estrogen secretion is required for spontaneous decidualization. Therefore, another explanation has to be sought for the physiological basis of spontaneous endometrial decidualization in both man and the chimpanzee, as well as for the function of the luteal estrogen peak.

A condition termed *membranous dysmenorrhea* in chimpanzees has recently been reported from the Dutch Primate Center TNO. In women this condition is characterized by pain associated with passage of one or more large, decidualized endometrial fragments or casts (Parsons and Sommers, 1962). The etiology of the condition is unknown. Six uterine casts were recovered from four chimpanzees, three of which became pregnant shortly afterwards (Solleveld and van Zwieten, 1978). One of the animals passed endometrial casts in three consecutive menstrual cycles.

Subsequently, three similar uterine casts were obtained at the Yerkes Primate Center from two unmated chimpanzees (Graham, unpublished). Interestingly, a similar case has been documented in a wild chimpanzee observed in Tanzania by Dr. Jane van Lawick–Goodall. The chimpanzee, Gigi, had been recorded as commencing cyclicity in 1965 and experienced regular cycles until about 1975, when her cycles became more irregular. She had not been observed to become pregnant. In 1978 she was observed to be menstruating heavily, and a piece of tissue was recovered from the vicinity and was forwarded to the author. The formalin-fixed tissue was found to be an empty sac formed of heavily decidualized endometrium that macroscopically and microscopically exactly resembled uterine casts recovered in the laboratory.

2. Gorilla

In collaboration with R. D. Nadler, the endometrium of a gorilla was sampled in two cycles by the author (Graham, unpublished). The vagina of the gorilla is extremely small, preventing visualization of the cervix in some subjects even with use of a small speculum. Because of the small size of the genitalia, a Vabra-aspirator (Cooper Laboratories, Wayne, N.J.) was utilized as a curette, using gentle suction from an attached syringe. By this means a small, but adequate sample could be obtained in one subject.

On the first day of labial tumescence, the endometrium showed a mixed proliferative-secretory pattern, since mitotic figures and early secretory vacuoles were present. By the second day of labial tumescence, basal vacuoles were well-established (Fig. 11). Assuming similar peri-ovulatory temporal correlations to women, these limited observations are consistent with the interpretation that ovulation occurred during labial tumescence.

D.   Uterine Cervix

Changes in human cervical mucus that facilitates sperm penetration occur at the time of ovulation. Notably the mucus becomes thinner and less viscous, and when allowed to dry, forms elaborate fern patterns reflecting a change in the chemical composition at this time. Cervical mucus is relatively scanty in the chimpanzees, and it was found that ferning was more frequent in the early follicular phase, although it was sometimes also associated with the time of sexual-swelling detumescence (Gould *et al.*, 1976). Intravaginal artificial insemination in the chimpanzee has been successful in the absence of ferning mucus (Martin *et al.*, 1978).

Analysis of chimpanzee cervical mucus with scanning electron microscopy and energy dispersive x-ray analysis showed a bimodal elevation of sodium and potassium concentration during the follicular phase, with the second elevation approximately coincident with the time of swelling regression. The K/Na ratio increased at this time to greater than unity, possibly due to the ovulatory fall in estrogen levels or in estrogen/progestin ratio. These preliminary data suggest that changes in electrolyte concentration in chimpanzee cervical mucus could serve as peri-ovulatory parameters, although the need for anesthesia to obtain mucus may limit the utility of such information (Gould *et al.*, 1976).

IV.   INDICES OF OVULATION TIME IN APES

A.   General Considerations

Time of ovulation is one of the most physiologically significant parameters of the menstrual cycle, yet is one of the most difficult events to time precisely. In almost all species, including the chimpanzee, the evidence concerning time of ovulation is indirect.

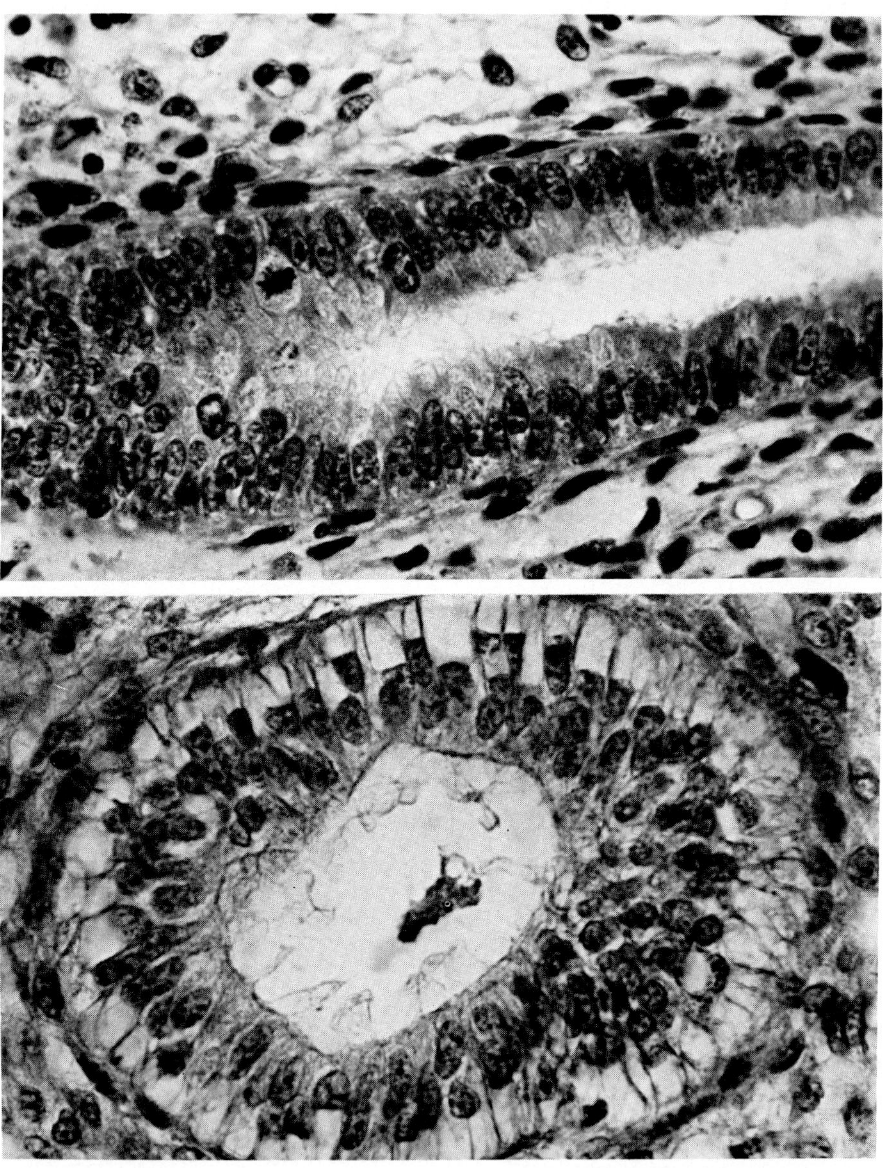

Fig. 11. Endometrium of gorilla. (A) First day of labial tumescence, showing predominantly proliferative pattern but with suggestion of early formation of secretory vacuoles; (B) Second day of labial detumescence, showing marked formation of secretory vacuoles. Hematoxylin and eosin.

The only direct method of determining the exact time of ovulation is by direct and continuous observation of the ovaries by endoscopy or laparotomy. This has been achieved in isolated cases in some primates, but not in the chimpanzee. This method is open to the criticism that the experimental procedure may affect the time of ovulation.

All other methods are indirect and imprecise, being based upon assumptions concerning the time interval between pre- or post-ovulatory events that in most cases are linked by hormonal control mechanisms to ovulation itself. Endocrine changes are the most accurate indicators of impending ovulation, especially the preovulatory estrogen peak and the ovulatory LH peak that it triggers. The luteal rise in progesterone values and the associated progesterone dependent events, such as development of secretory endometrium and basal body temperature shift, are of some retrospective value and are used widely in some species, including man. These progesterone-dependent events are not necessarily absolute proof that ovulation has occurred. Rather, they only indicate that there has been an acute surge of LH.

B. Peri-ovulatory Correlations

1. Chimpanzee

a. Sexual Swelling. Earlier in this chapter it was shown that the chimpanzee and human menstrual cycle ovarian steroid excretion patterns are similar. During the chimpanzee menstrual cycle, development of the sexual swelling is associated with increasing secretion and excretion of estrogens in the follicular phase, and swelling detumescence with decreased levels of estrogens and increased levels of progesterone in plasma and urine (Graham *et al.*, 1972).

These relationships and various studies suggest that regression of sexual swelling is associated both with collapse of the follicle and a consequent fall in the secretion of estrogen at ovulation, and with the subsequent development of the corpus luteum and its secretion of progesterone. Since sexual-swelling detumescence seems to be so closely related functionally with ovulation and initiation of luteinization, it has the potential of a very valuable, highly visible marker of the occult endocrinological events. Consequently, the main thrust of this analysis is to investigate the probable time of ovulation with respect to sexual-swelling detumescence. As early as 1930, Zuckerman postulated that ovulation occurs at about the time of swelling detumescence, based upon studies in baboons and rhesus monkeys. Yerkes and Elder postulated such a relationship for the chimpanzee in 1936. For the purpose of this discussion, the last day of maximal swelling is defined as Day) of the cycle.

b.   Timed Matings.   The earliest work on the time of ovulation was based on early studies of Robert Yerkes and his associates who reported the outcome of timed matings. Of particular interest are conceptions resulting from a single insemination on a known cycle day. The reports of these studies were re-analyzed by Dr. D. Martin and are discussed more fully in Chapter 14. The data (see Table 2) include one artificial insemination performed on Day -3. Fertile matings occurred on the last six days of maximal swelling.

Young and Yerkes (1943) examined the ovaries of five animals at laparotomy. The expected time of detumescence was estimated from the mean of the length of the swelling phase of previous cycles. One animal was examined five days before expected detumescence, and a large but unripe follicle was observed. Two other animals were examined two days and one day before swelling regression was expected; each had a follicle estimated to be due to rupture within 24 hours. An adolescent animal had a ruptured follicle during maximal swelling four days before detumescence was anticipated based on previous cycle records; however, this subject was typical of adolescent chimpanzees in having consistently experienced a short interval between detumescence and menses, suggesting inadequate luteal function with the attendant possibility of insufficient progestone secretion to inhibit sexual swelling promptly after ovulation. These data provide no real evidence on the relationship between ovulation and swelling detumescence because of the variability from cycle to cycle in the duration of sexual swelling, and the lack of objective criteria by which to judge the stage of development of follicles.

One animal, examined on the evening of the day of detumescence (Day $+1$), had a recently ovulated follicle, showing that follicular rupture can be closely associated in time with detumescence.

c.   Endoscopy.   Graham *et al.* (1973) performed a small series of endoscopies on chimpanzees before and after swelling detumescence.

TABLE 2

Time of successful insemination of chimpanzees related to last day of maximal sexual swelling (Day 0).

Cycle day	-5	-4	-3	-2	-1	0
No. of fertile inseminations: term pregnancy	1	2	1	1	2	2
Conception followed by abortion	—	—	—	1	—	1

Although the numbers are insufficient for a definitive statement, it is suggestive that only follicles were observed before detumescence, and only corpora lutea on and after the first day of swelling detumescence (Table 3).

Two endoscopies on Day −1 revealed unruptured follicles that appeared as large, pellucid, and slightly raised areas with a poorly defined outline. On Day 0, a well-defined follicle protruding from the surface of the ovary was observed (Fig. 12A). A convoluted blood vessel was noted at the base of this follicle, and a small vessel traversed the follicular dome, as described by Jewett and Dukelow (1971) in *Macaca fasicularis* within eight hours of ovulation. On Days +1, +2, and +3 early corpora lutea protruded from the surface of the ovary (Fig. 12B–D). Table 3 shows that secretory endometria and plasma progesterone values in excess of 4.0 ng/ml were associated with corpora lutea after Days +2 and +3, respectively.

A slight elevation of plasma progesterone is seen on Days +1 and +2, in association with early corpora lutea, suggesting that ovulation is closely associated with the time of sex-swelling detumescence. (In the same study, it was established that endoscopy did not alter the interval between detumescence and menses, and hence did not affect the time of ovulation.) It is notable that these endoscopy and laparotomy studies show no evidence of a corpus luteum in the normal adult earlier than Day +1.

d.   Endometrial Histology and Plasma Progesterone Levels.   A more comprehensive study of the relationship between plasma progesterone levels, endometrial histology, and sexual swelling was made by Graham (1976). Thirty-four menstrual cycle endometrial biopsies and corre-

TABLE 3

Correlations of ovarian morphology with sexual swelling, uterine histology, and plasma progesterone levels.[a]

Cycle day	−1	0	+1	+2	+3	+4	+5	+6
Uterus:	Proliferative				Secretory			
Ovary[b]:	F F	F O	CL	CL	CL	CL		CL
Plasma Progesterone (ng/ml)	0.26	0.14 0.17	1.08	0.6		4.95		4.51

[a] Graham *et al.*, 1973; with permission from Wistar Institute Press.

[b] F, unovulated follicle; CL, corpus luteum; O, neither mature follicle nor corpus luteum evident.

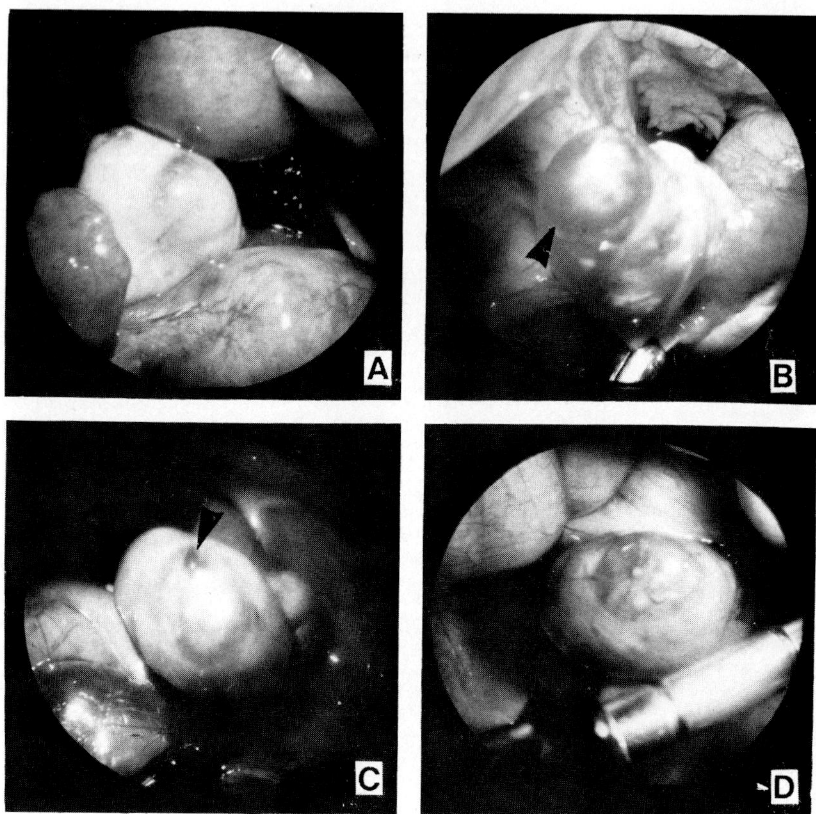

Fig. 12. Peri-ovulatory ovarian morphology as seen by endoscopy. (A) Ovary with follicle on Day 0, showing vascularity of follicular membrane; (B) Corpus luteum (arrow) on Day +1; (C) Corpus luteum with unhealed rupture point (arrow) observed on Day +4; (D) Corpus luteum on Day +6. Day 0 = last day of maximal sexual swelling. (Graham *et al.*, 1973; reproduced with permission of the Wistar Institute Press.)

sponding plasma samples for radioimmunoassay of progesterone were obtained from 19 intact adult chimpanzees (Fig. 13; Graham *et al.*, 1974).

These more comprehensive data indicate that progesterone levels can rise up to 4 ng/ml one to two days before swelling detumescence (insufficient data is available to evaluate the significance of a single elevated progesterone value on Day −3). The early elevation of progesterone is consistent with reports that progesterone can rise significantly during the LH surge in women (e.g., Johanssen and Wide, 1969), since the LH surge in chimpanzees usually occurred on the last two days of maximal swelling (see Section C). The endometrium first

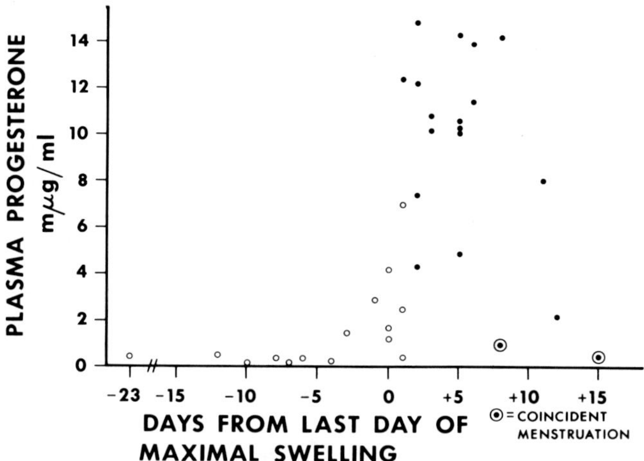

Fig. 13. Plasma progesterone, endometrial morphology, and sexual swelling during chimpanzee menstrual cycles. The last day of sexual swelling is designated Day 0. Open circles, proliferative endometrium; solid circles, secretory endometrium (Graham, 1976; reproduced with permission of Gustav Fischer Verlag.)

showed secretory changes on the first day of swelling detumescence, and by Day +2 all endometria were secretory.

Although early studies concluded that secretory changes occur 36–48 hours after ovulation in women (e.g., Noyes *et al.*, 1950), the more recent demonstration that glycogen vacuoles appear in the cytoplasm 40 hours after the LH peak, indicates that the interval between ovulation and secretory transformation is as short as 24–48 hours (Delaforge *et al.*, 1970). If similar time relationships occur in the chimpanzee, the endometrial and progesterone data support the interpretation that ovulation can occur on the last two days of maximal sexual swelling, or early on the day of detumescence.

e. Luteinizing Hormone. In a cross-sectional study of the Yerkes adult female chimpanzee colony, LH was determined by radioimmunoassay in 300 randomly selected menstrual cycle plasma in the laboratory of J. D. Neill, in collaboration with the author, and the values were correlated with the sexual swelling (Fig. 14). These previously unpublished data showed a midcycle rise of LH on Day −1 with a peak on Day 0. Since progesterone levels can rise before the LH peak is attained in women (although only after the LH rise begins, as previously mentioned), this observation is consistent with the previous study of progesterone levels and endometrial development. This study did not reveal a cyclic pattern of FSH levels.

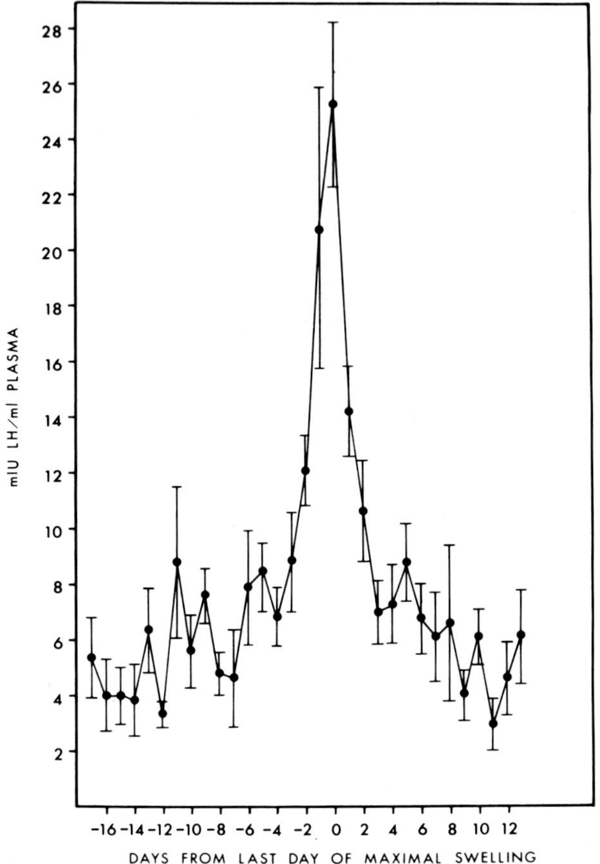

Fig. 14. 300 random LH values determined by radioimmunoassay, normalized to the last day of maximal sexual swelling (Day 0). Note that mean LH levels (± SE) are considerably elevated on the last two days of maximal sexual swelling.

In women, the rise in LH precedes the peak by approximately 12 hours, and the peak in LH precedes ovulation by 12–24 hours (Yussman and Taymor, 1970; Yussman *et al.*, 1970). Since the majority of chimpanzee LH samples were collected in the morning, the assumption of a similar temporal relationship between the two species indicates that ovulation would most often occur late on the last day of maximal swelling, or early on the morning of detumescence.

f. *Steroid Hormones.* In cycles in which chimpanzee urinary steroids were monitored, mean estrogen levels rose to a peak five days before detumescence and began to fall on the last day of maximal swelling, with a much larger fall on the first day of detumescence (Fig. 4). This fall thus closely follows the LH peak. Urinary pregnanediol rose significantly on Day +3. These relationships fall within the range of variability reported for the human (Burger *et al.*, 1968; Johanssen *et al.*, 1971; Newton *et al.*, 1971; Klopper, 1957). The correlation of plasma progesterone levels was mentioned in Section IV,B,1,d.

g. Basal Body Temperature. In women, increasing estrogen levels in the follicular phase result in a depression of basal body temperature that is reversed as progesterone levels begin to rise and estrogen levels fall at the time of the LH rise. The low point of the temperature curve, the nadir, usually occurs on the same day as the LH peak in woman (Morris *et al.*, 1976). Using a remote temperature telemetry system, Graham *et al.* (1977) detected nadirs in 9 out of 11 cycles monitored, most of them on the same days on which elevated LH levels were observed in the separate study described above, namely Days −1 and 0 (Fig. 15). In three cycles in which urinary steroid levels were assayed, the nadir occurred on the first day estrone levels began to fall from the midcycle peak (Fig. 16). This data again suggest that peri-ovulatory time correlations are similar in women and chimps.

The study indicated that a 0.3°C temperature rise after the nadir, an alternative criterion of the time of the temperature shift, was not closely correlated to swelling detumescence, and had a very variable temporal relationship to the nadir of one to five days (Fig. 15).

h. Time of Ovulation in Chimpanzee. All of the peri-ovulatory correlates we have observed in the chimpanzee are closely comparable in their temporal characteristics with the human data. Because of the consistent pattern, it is reasonable to assume that the interval between ovulation and the measurable peri-ovulatory parameter are also similar in the two species. This assumption is borne out by the limited direct

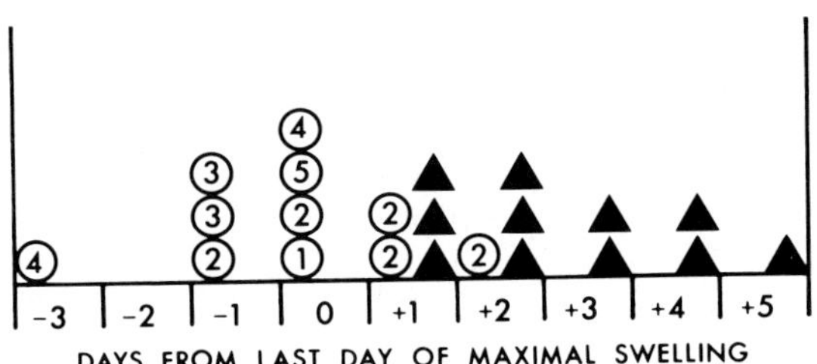

Fig. 15. Incidence of nadirs (circles), interval from nadir to 0.3°C rise (enclosed value), and incidence of 0.3°C rise (triangles) in 11 chimpanzee menstrual cycles. Nadirs clearly show a closer relationship to the last day of maximal swelling than do the 0.3°C rises.

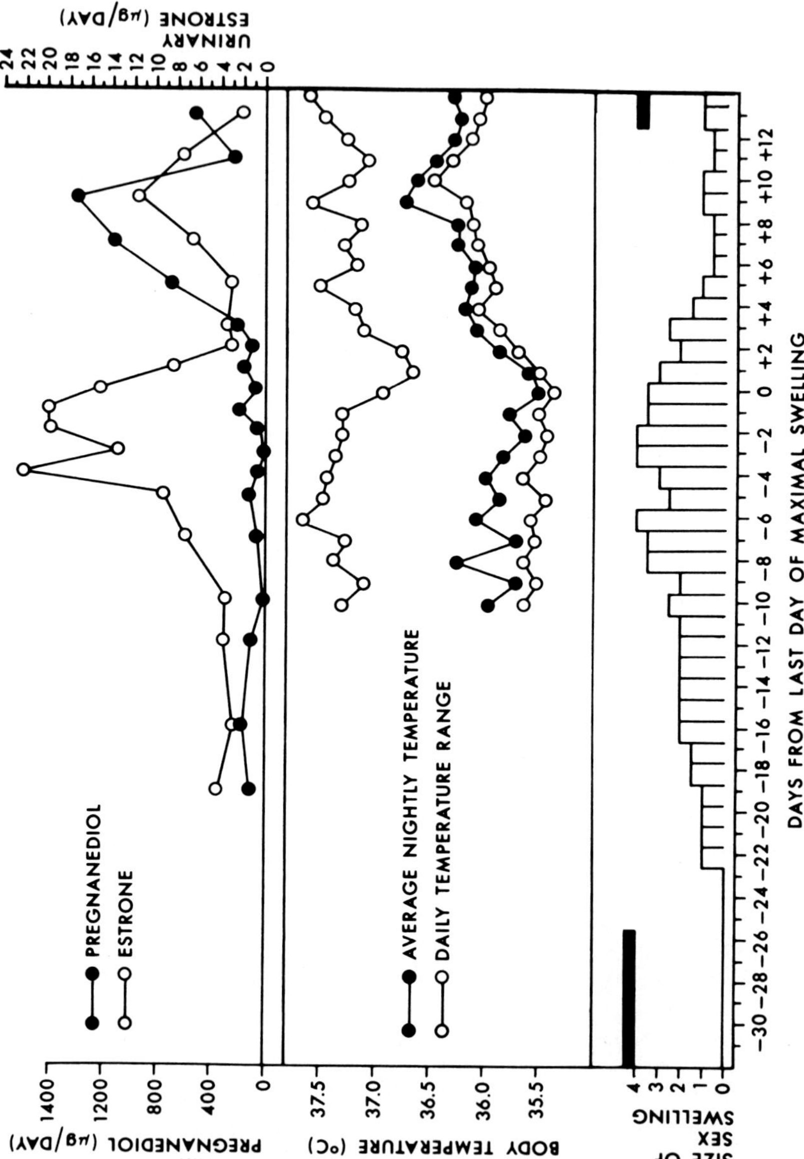

Fig. 16. Correlations of body temperature, urinary esterone and pregnanediol excretion, and sexual swelling during the menstrual cycle of a chimpanzee. Repeated sampling of nocturnal temperature was used for determination of nadirs (comparison with the daily temperature range shows nocturnal temperature is basal). Note correlation between falling estrogen levels, temperature nadir, and Day 0. (Although swelling showed a slight reduction in size on Day −1, it was not regarded as significant, as the criterion for initiation of detumescence requires a 1-point decrement in swelling size.)

observations of ovarian morphology available in the chimpanzee, and supports the conclusion that ovulation occurs most often in the chimpanzee on Day 0 or very early on the day of detumescence, Day +1. However, this generalization must be tempered with the caution that there is likely to be some variability in the peri-ovulatory temporal relationships so that individual variations from the norm may be frequent.

A similar correlation between ovulation and the beginning of detumescence has been reported by Blakely *et al.* (1977) in *Macaca nemestrina,* based upon laparoscopic studies. In this species, mean progesterone levels rose to 1 ng/ml on the day of ovulation, with a rapid increase thereafter in nine cycles studied (Blaine *et al.,* 1975).

Stevens *et al.* (1970), reporting on endocrine correlations in 22 cycles of 12 baboons (*Papis anubis*), noted that deturgescence of the swelling occurred in all animals three or four days following the midcycle urinary estrogen peak. An LH peak was detected in plasma and urine on the same day as the urinary estrogen peak. Therefore, ovulation might be expected approximately two to three days before swelling detumescence.

Kling and Westfahl (1978) made a careful comparison of endocrine changes during five menstrual cycles of *P. cyanocephalus* and man. They noted a similar correlation between LH peak and swelling deturgescence to that described in *P. anubis* (deturgescence occurred three to six days after the LH peak; personal communication). In their paper, Kling and Westfahl drew attention to the fact that progesterone levels rise relatively slowly in the baboon (significant rise three days after the LH peak), compared with the human (same day as LH peak). The chimpanzee, as mentioned in Section III,A,1,e, showed elevated LH levels on the last two days of maximal sexual swelling, suggesting that the rise of progesterone levels after the LH peak is rapid as in man and *M. nemestrina.*

Therefore, in species such as chimpanzee and *M. nemestrina,* ovulation is closely correlated with the time of swelling detumescence because of rapid elevation of progesterone levels after the LH peak. In the baboon, on the other hand, ovulation precedes swelling detumescence by several days because progesterone levels rise much more slowly.

2. Gorilla

The concentration of copulatory activity during the period of maximal labial swelling reported by Nadler (1975b), and the fact that LH peaks always occurred on the second day of labial swelling in four cycles analyzed (Nadler *et al.*, 1979), strongly suggests that ovulation occurs on the third day of labial swelling, just before or during labial detumescence. This conclusion is supported by the observation that secretory changes occurred on the first and second days of labial detumescence in two cycles studied (Section III,C,2).

3. Orangutan

The lack of sexual swelling in the orangutan makes a comparable analysis impossible.

V. REPRODUCTIVE SENESCENCE

This subject has been studied only in the female chimpanzee, as records of aged gorillas and orangutans are lacking. Menstrual cycle duration and frequency and conception rate were compared in 10 chimpanzees aged 35–48 with the corresponding data of the same animals when aged 15–25 (Graham, 1979). All of the animals continued cycling until the end of the study or until death from natural causes. The oldest chimpanzee in the study was aged 48 and was still cycling; this animal is the longest-living female chimpanzee recorded. These data suggest that menopause may not be a characteristic of the chimpanzee.

Mean cycle frequency was slightly reduced at age 35–48, but cycle length was shorter in some aged individuals, longer in others, compared with their younger years. Three conceptions occurred in animals aged 35 or more; only one of these (at age 38) produced a living term infant. The age of the oldest female to conceive was estimated to be 40 years, but the pregnancy ended in a premature stillbirth. The conception rate (% mated cycles resulting in conception) was greatly reduced during age 35–48 (3.85%) compared with age 15–25 (25%).

All of the chimpanzees in this study showed evidence of aging, and there is little reason to believe that female chimpanzees will often live to beyond age 50 in captivity. Increased irregularity of cycles and decreased conception rate suggest a decline of fertility in the aged chimpanzees, although menopause did not occur. More recent study of these animals and a single *Pan paniscus* provide some indications of impending or actual menopause (Gould *et al.*, in press). The only nonhuman primate species in which menopause has been unequivocably described is the rhesus monkey (Hodgen *et al.*, 1977). Menopause in the laboratory rhesus monkey and human being may be the consequence of a prolonged lifespan resulting from artificial environmental conditions.

VI. CONCLUSIONS

This analysis shows that the best-studied of the great apes, the chimpanzee, closely resembles the human in most aspects of its menstrual cycle physiology. To the extent they have been studied, this is also true of the orangutan and gorilla. Some of these resemblances are unique to the great apes, such as a long midluteal estrogen peak (documented in chimpanzee, gorilla, and orangutan) and spontaneous uterine decidualization (in the chimpanzee).

Some differences between apes and man also exist; for example, although steroid excretion during the menstrual cycle generally resembles that of man in many respects more closely than monkeys, there are quantative differences evidenced best by the low excretion of estriol. The sexual swelling of the chimpanzee, and the labial swelling of the gorilla are not present in humans, but in this instance the difference enhances the value of these apes in experimental studies, because of the value of these changes as an external indicator of changing hormone levels and the approximate time of ovulation.

REFERENCES

Aulmann, G. (1932). *Zool. Gart. Leipzig* **5,** 81–90.
Allen, E., Diddle, A. W., and Elder, J. H. (1935). *Am. J. Physiol.* **110,** 593–596.
Blaine, C. R., White, R. J., Blakley, G. A., and Ross, W. F. (1975). *In* "Contemporary Primatology" (S. Knodo, M. Kawai, and A. Ehara, eds.), pp. 141–151. Karger, Basel.
Blakley, G. A., Blaine, C. R., and Morton, W. R. (1977). *Lab. Anim. Sci.* **27,** 352–355.
Burger, H. G., Catt, K. J., and Brown, J. B. (1968). *J. Clin. Endocrinol.* **28,** 1508–1512.
Clark, G. (1945). *Fed. Proc.* **4,** 14.
Clark, G. (1947). *Endocrinology* **41,** 327–329.
Clark, G. (1949). *Yale J. Biol. Med.* **21,** 245–247.
Clark, G., and Birch, H. G. (1946). *Anat. Rec.* **94,** 453–454.
Clark, G., and Birch, H. G. (1948). *Endocrinology* **43,** 218–231.
Clarke, R. W. (1940). *Am. J. Physiol.* **131,** 325–330.
Collins, D. C., Graham, C. E., and Preedy, J. R. K. (1975). *Endocrinology* **96,** 93–101.
Delaforge, J. P., Thomas, K., and Ferin, J. (1970). *Acta Eur. Fertil.* **2,** 141–166.
Elder, J. H., and Yerkes, R. M. (1936). *Anat. Rec.* **67,** 119–143.
Erikson, L. B. (1963). *Fertil. Steril.* **14,** 273–283.
Fish, W. R., Young, W. C. and Dorfman, R. I. (1941). *Endocrinology* **28,** 585–592.
Gilbert, C. and Gillman, J. (1945). *S. Afr. J. Med. Sci. (Biol. Suppl.)* **10,** 15–18.

Gillman, J., and Gilbert, C. (1956a). *Am. J. Anat.* **98,** 231–271.

Gillman, J., and Gilbert, C. (1956b). *Exp. Med. Surg.* **14,** 31–38.

Gillman, J., and Gilbert, C. (1956c). *Exp. Med. Surg.* **14,** 39–47.

Ginetzinsky, A. G. (1958). *Nature (London),* **182,** 1218–1219.

Gosselin, R. E., Hobson, W. C., and Vale, W. (1979a). *Ann. Meeting Soc. Study Reproduction, August 21-26, Quebec, Canada* Abstract No. 195.

Gosselin, R. E., Fuller, G. B., Coy, D. H., Schally, A. V., and Hobson, W. C. (1979b).
Proc. Soc. Exp. Biol. Med. **161,** 21–24.

Gould, K. G., Flint, M., and Graham, C. E. *Maturitas.* (in press).

Gould, K. G., Martin, D. E. and Graham, C. E. (1976). "Scanning Electron Microscopy/
1976," Vol. VI, pp. 335–342. IITRI, Chicago, Illinois.

Graham, C. E. (1970). *In* "The Chimpanzee" (G. H. Bourne, ed.), Vol. III, pp. 183–220.
Karger, New York.

Graham, C. E. (1973). *Folia. Primat.* **19,** 458–468.

Graham, C. E. (1976). *In* "The Laboratory Animal in the Study of Reproduction" (T.
Antikatzides, S. Ericksen, and A. Spiegel, eds.). Gustav Fischer Verlag, Stuttgart and
New York.

Graham, C. E. (1979). *Am. J. Phys. Anthrop.* **50,** 291–300.

Graham, C. E., Collins, D. C., Robinson, H., and Preedy, J. R. K. (1972). *Endocrinology*
91, 13–24.

Graham, C. E., Keeling, M., Chapman, C., Cummins, L. B., and Haynie, J. (1973). *Am.
J. Phys. Anthrop.* **38,** 211–215.

Graham, C. E., Wright, K., Collins, D. C., and Preedy, J. R. K. (1974). *IRCS (Res.
Endocrine Syst.; Physiol.; Reprod., Obstet. Gynecol.)* **2,** 1697.

Graham, C. E., Warner, H., Collins, D. C., and Preedy, J. R. K. (1977). *J. Reprod. Fert.*
50, 23–28.

Graham, C. E., Gould, K. G., Wright, K., and Collins, D. C. (1978). *In* "Recent Advances
in Primatology" (D. J. Chives and E. H. R. Ford, eds.). Academic Press, New York.

Graham, C. E., Gould, K. G., Collins, D. C., and Preedy, J. R. K. (1979). *Endocrinology*
105, 269–275.

Guerrero, R. *et al.* (1976). *Acta Endocrinol.* **81,** 133–149.

Hess, D. L., and Resko, J. A. (1973). *Endocrinology* **92,** 446–453.

Hobson, W., and Fuller, G. B. (1977). *Biol. Reprod.* **17,** 294–297.

Hobson, W. C., Gosselin, R., and Fuller, G. B. (1978). *Proc. Ann. Meeting, Endocrine Soc.,*
Abstract 765, p. 460.

Hodgen, G. D., Goodman, A. L., O'Connor, A., and Johnson, D. K. (1977). *Am. J.
Obstet. Gynecol.* **127,** 581–584.

Hoff, J. D., Lasley, B. L., Wang, C. R., and Yen, S. S. C. (1977). *J. Clin. Endocrinol.
Metab.* **44,** 302–312.

Hopper, B., and Tullner, W. W. (1970). *Endocrinology* **86,** 1225–1230.

Howland, B. E., Faiman, C., and Butler, T. M. (1971). *Biol. Reprod.* **4,** 101–105.

Jewett, D. A., and Dukelow, W. R. (1971). *Folia Primatol.* **16,** 216–220.

Johanssen, E. D. B., and Wide, L. (1969). *Acta Endocrinol.* **62,** 82–88.

Johanssen, E. D. B., Wide, L. and Gemzell, C. (1971). *Acta Endocrinol.* **68,** 502–512.

Kirschner, M. A., and Lipsett, M. B. (1964). *Steroids,* **3,** 277–294.

Kling, O. R., and Westfahl, P. K. (1978). *Biol. Reprod.* **18,** 392–400.

Klopper, A. (1957). *J. Obstet. Gynaecol. Br. Emp.* **64,** 504–511.

Krohn, P. L., and Zuckerman, S. (1937). *J. Physiol. (London)* **88,** 369–387.

Ludwig, A. W., Boas, N. F., and Soffer, L. J. (1950). *Proc. Soc. Exp. Biol. Med.* **73,** 137–140.

Martin, D. E., Graham, C. E., and Gould, K. G. (1978). *Symp. Zool. Soc. London* **43**, 249–260.

Morris, N. M., Underwood, L. E., and Easterling, W. (1976). *Fertil. Steril.* **27**, 780–783.

Nadler, R. D. (1975a). *Anat. Rec.* **181**, 791–797.

Nadler, R. D. (1975b). *Science* **191**, 813–814.

Nadler, R. D. (1977). *Arch. Sex. Behav.* **6**, 457–475.

Nadler, R. D., Graham, C. E., Collins, D. C., and Gould, K. G. (1979). *Endocrinology* **105**, 290–296.

Naidoo, S. S. (1971). *Proc. Soc. Exp. Biol. Med.* **136**, 1014–1019.

Naidoo, S. S., and Quereshi, A. M. (1974). *Can. J. Physiol. Pharmacol.* **52**, 206–214.

Newton, J., Joyce, D., Pearce, B., Revell, C., and Tyler, J. (1971). *J. Reprod. Fertil.* **27**, 481–484.

Nillius, S. J., and Wide, L. (1972). *J. Obstet. Gynaec. Brit. Comm.* **10**, 865–873.

Nissen, H. W., and Yerkes, R. M. (1943). *Anat. Rec.* **86**, 567–578.

Noback, C. V. (1939). *Anat. Rec.* **73**, 209–221.

Noyes, R. W., Hertig, A. T., and Rock, J. (1950). *Fertil. Steril.* **1**, 3–25.

Parsons, L., and Sommers, S. C. (1962). *In* "Gynecology." Saunders, Philadelphia, Pennsylvania.

Pearce, R. H., and Watson, E. M. (1949). *Can. J. Res. Sect. E* **27**, 43.

Plant, T. M. (1974). *J. Reprod. Fertil.* **41**, 205–209.

Plant, T. M., James, Y. H. T., and Michael, R. P. (1969). *J. Endocrinol.* **43**, 493–494.

Ragan, C., and Meyer, K. (1949). *Clin. Invest.* **28**, 56–59.

Reyes, F. I., Winter, J. S. D., Faiman, C., and Hobson, W. C. (1975). *Endocrinology* **96**, 1447–1455.

Savage–Rumbaugh, E. S., and Wilkerson, B. J. (1978). *J. Hum. Evol.* **7**, 327–344.

Solleveld, H. A., and van Zwieten, M. J. (1978). *J. Med. Primatol.* **7**, 19–25.

Stevens, V. C., Sparks, S. J. and Powell, J. E. (1970). *Endocrinology* **87**, 658–666.

Thomas, K., Cardon, M., Donnez, J., and Ferin, J. (1973). *Contraception* **7**, 289–297.

Tinklepaugh, O. L. (1930). *Am. J. Phys. Anthropol.* **14**, 487–489.

Yen, S. S. C., VandenBerg, G., Rebar, R., and Ehara, Y. (1972). *J. Clin. Endocrinol. Metab.* **35**, 931–937.

Yerkes, R. M. and Elder, J. H. (1936a). The sexual and reproductive cycles of chimpanzee, *Proc. Nat. Acad. Sci.* **22**, No. 5, 276–283.

Yerkes, R. M., and Elder, J. H. (1936b). *Comp. Psychol. Monogr.* **13**, 1–39.

Young, W. C., and Yerkes, R. M. (1943). *Endocrinology* **33**, 121–154.

Yussman, M. A., and Taymour, M. L. (1970). *J. Clin. Endocrinol.* **30**, 396–399.

Yussman, M. A., Taymour, M. L., Miyata, J., and Pheteplace, C. (1970). *Fertil. Steril.* **21**, 119–125.

Zuckerman, S., and Fulton, J. F. (1934). *J. Anat.* **69**, 38–46.

Chapter 2

ENDOCRINOLOGY OF PREGNANCY IN APES

Charles Faiman
Francisco I. Reyes
Jeremy S. D. Winter
William C. Hobson

I. INTRODUCTION

Although interest in reproductive endocrinology of the great apes dates back some five decades to the first demonstration of the presence of chorionic gonadotropin (CG) in the urine of a pregnant orangutan (Aschheim and Zondek, 1928) and pregnant chimpanzees (Zuckerman,

45

Copyright © 1981 by Academic Press, Inc.
All rights of reproduction in any form reserved.
ISBN 0-12-295020-8

1935), only fragmentary information had appeared in the literature by the early 1970s. This fact was clearly underscored by Graham (1970) and Solomon and Leung (1972).

The advent of sensitive and specific assay techniques for the measurement of both peptide and steroid hormones, coupled with the collaborative efforts of reproductive physiologists and primatologists over the past decade, have helped to bridge some of the gaps in our knowledge of ape reproduction. However, most major advances in human reproduction utilizing primate models have been made in the rhesus monkey and baboon rather than in the great apes. Not surprisingly, the restricted availability of breeding colonies of great apes, coupled with the difficulties (both physical and fiscal) surrounding their use as experimental subjects, in large measure have been responsible for slow progress. The chimpanzee, fortunately, has been the exception in this regard. Enough data have accumulated so that the conclusion seems well-founded that this primate species, unlike others, possesses an endocrinologic environment during pregnancy that closely mimics that of the human female.

It is the purpose of this chapter to provide a current overview of the endocrine changes during pregnancy in the great apes. Of necessity, this review will have its major emphasis on hormonal events in the chimpanzee. Comparisons with the endocrinology of pregnancy in humans and other primate species will be made throughout.

The older literature has been the subject of a number of reviews (Graham, 1970; Hobson, 1971; Solomon and Leung, 1972; Tullner, 1972), so that only the more recent contributions will be reviewed comprehensively. The interested reader is also referred to recent reviews on the endocrinology of pregnancy in both the human (Fuchs and Klopper, 1977; Goodwin *et al.*, 1976; Klopper and Diczfalusy, 1969) and the rhesus monkey (Kaplan and Grumbach, 1978; Knight and O'Connor, 1977). Physical, behavioral, and clinical aspects of pregnancy in chimpanzees have been described by Keeling and Roberts (1972). Breeding in apes is discussed by Martin (Chapter 14, this volume).

II. GESTATION IN APES

A. Length of Gestation

The duration of pregnancy in a number of primate species as estimated from the time of conception is summarized in Table 1. It is important to note that the values given for the orangutan and gorilla

TABLE 1

Length of gestation in some simian primates.

Family	Common Name	Species	Average Length of Gestation (days)[a]	References[b]
Pongidae	Chimpanzee	*Pan troglodytes*	228	Peacock and Rogers (1959)
	Orangutan	*Pongo pygmaeus*	245–275	Aulmann (1932)
	Gorilla	*Gorilla gorilla gorilla*	255–265	Noback (1939)
Hominidae	Man	*Homo sapiens*	265	Gibson and McKeown (1950)
Cercopithicidae	Rhesus monkey	*Macaca mulatta*	154–168	Hartman (1932); van Wagenen and Asling (1964)
	Baboon	*Papio cynocephalus*	189	Hobson (1970)

[a] Calculations based upon time from conception.

[b] See also Martin, Chapter 14, this volume.

should be considered as reasonable approximations, whereas those for the other species are based upon sound data (see also Martin, Chapter 14, this volume).

B. Sexual Skin during Gestation

Periods of sexual skin edema, which mimic those of the menstrual cycle, occur in some pregnant chimpanzees (Clark and Birch, 1948). These are of variable degree and tend to occur at irregular intervals most frequently during the first trimester of pregnancy. Slight to marked swelling of the genital region has been observed during pregnancy in the orangutan (Noback, 1939; Schultz, 1938).

III. HORMONAL CHANGES DURING PREGNANCY

A. Chorionic Gonadotropin

1. Great Apes

Substantive data only are available in the chimpanzee, chorionic gonadotropin (CG) having been studied only in a handful of orangutans and gorillas. Chimpanzee CG has been measured serially in the serum of four chimpanzees throughout pregnancy (Fig. 1, Reyes *et al.*, 1975) by a radioimmunoassay technique. Levels became detectable nine to fifteen days following the midcycle pre-ovulatory LH peak and rose exponentially to reach maximal values of 56–164 IU/ml. The mean doubling time was 1.55 days. Following a plateau from 35–65 days gestation, levels declined to reach a nadir at 140–160 days; a secondary, smaller rise was noted towards term but values remained below 1 IU/ml. A similar pattern has been noted for urinary CG excretion patterns (Boorman *et al.*, 1974; Clegg and Weaver, 1972; Nixon *et al.*, 1972).

Only two pregnant gorillas have been studied in detail. Tullner and Gray (1968) measured urinary CG by bioassay from 57-days' gestation (first sample) to term in a single gorilla. Martin (1976) described the urinary CG pattern determined by radioimmunoassay in another; CG became detectable at about 10 days, levels rose to reach a peak at 80–150 days, then declined following which there was a smaller, secondary peak prior to term.

Urinary CG has been studied serially in three orangutan pregnancies. Hodgen *et al.* (1977) observed positive responses using the subhuman primate pregnancy test in a single animal throughout an eight-month

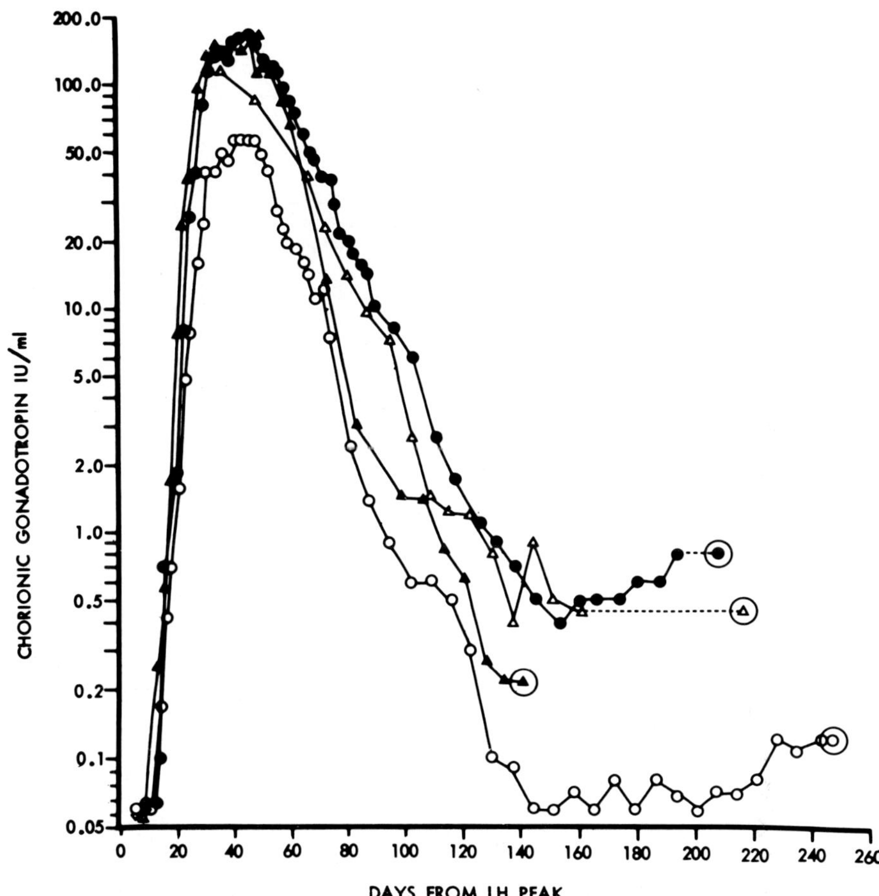

Fig. 1. Semilogarithmic plot of serum chorionic gonadotropin levels in I.U. of hCG standard during pregnancy in four chimpanzees. The large circles indicate termination of pregnancy (by hysterotomy in ▲ and spontaneous delivery in the others). Animal ● had a twin pregnancy. (Reyes *et al.*, 1975; reproduced with permission of the J. B. Lippincott Co.)

interval until term, whereas Davis (1977) using the same test procedure, failed to detect CG during the last three months of gestation in another animal. Tullner (1974) detected hormonal activity by bioassay in a third animal up to one month, but not at 17 days, prior to delivery.

2. *Comparison with Human, Rhesus Monkey, and Baboon*

Table 2 shows a comparison of CG patterns in a number of primate species. Depending upon the sensitivity of the assay technique em-

TABLE 2

Chorionic gonadotropin (CG) patterns in blood or urine of some simian primates.

Species	Earliest Detection (days)[a]	Time of First Trimester Peak (days)[a]	Presence until Term	Presence of Third Trimester Peak	References
Chimpanzee	9	35–65	+	+	Reyes *et al.* (1975)
Orangutan	30	?	+	?	Hodgen *et al.* (1977); Tullner (1974)
Gorilla	10	80–150(?)	+	+	Martin (1976); Tullner and Gray (1968)
Human	8–9	35–65	+	+	Boroditsky *et al.* (1975); Faiman (1976); Kosasa *et al.* (1973)
Rhesus monkey	9	20–25	−	−	Hobson *et al.* (1975); Hodgen *et al.* (1975)
Baboon	12	20–40	±	±	Hobson (1970); Stevens *et al.* (1976); Tullner and Hertz (1971)

[a] Time from ovulatory LH peak.

ployed and the numbers of subjects studied from the onset of pregnancy, CG is detectable in body fluids as early as the eighth to ninth day following the ovulatory LH peak (presumably the seventh or eighth day following fertilization), and as early as one to two days following implantation of the blastocyst in humans (Faiman, 1976; Kosasa *et al.*, 1973), the chimpanzee (Howland *et al.*, 1971; Reyes *et al.*, 1975), the gorilla (Martin, 1976), the rhesus monkey (Meyer, 1972; Hobson *et al.*, 1975; Hodgen *et al.*, 1975), and the baboon (Hobson, 1970). Sampling has not been performed in the orangutan prior to the first positive tests at about 30 days (Tullner, 1974; Hodgen *et al.*, 1977).

Highest CG levels are seen during the first trimester in all species studied with the possible exception of the gorilla in which the peak may occur somewhat later (80–150 days in one animal). The doubling time to reach the peak, the timing of the peak, and the levels attained during the peak are virtually identical in the chimpanzee and human (see Reyes *et al.*, 1975). As in humans, CG remains detectable throughout pregnancy in the chimpanzee and gorilla, albeit at lower levels than in man, and there appears to be a small third trimester peak in all three species (Reyes *et al.*, 1975; Boroditsky *et al.*, 1975; Martin, 1976). No comparative data are available for the orangutan. In contrast, although data are sparse, peak levels appear somewhat earlier (18–40 days) in the baboon, and values are low or undetectable until term (Hobson, 1970; Tullner and Hertz, 1971; Stevens *et al.*, 1976). In the rhesus monkey, peak levels occur between Days 20 and 25; thereafter, CG levels decline to become undetectable after 40 days (Hobson *et al.*, 1975; Hodgen *et al.*, 1975).

A comparison of CG patterns during pregnancy in humans, the chimpanzee and the rhesus monkey is shown in Fig. 2.

3. Physiologic Role of CG during Pregnancy

The functions of CG during pregnancy, outlined in Table 3, remain speculative. Its best documented role is in maintenance of early pregnancy by prolonging the functional life span of the corpus luteum (Faiman, 1976; Meyer, 1972). Pituitary gonadotropins are not essential for early pregnancy maintenance, at least in humans, as is evident from the many normal pregnancies in hypopituitary women after ovulation induction with exogenous gonadotropins.

Less clear is the role CG may play in sexual differentiation of the male fetus. In humans, CG is a well-known stimulator of testicular androgen production. In the human fetus, there is a close temporal relationship between the presence of CG in the circulation and the initiation of Leydig cell testosterone secretion. Both precede the ap-

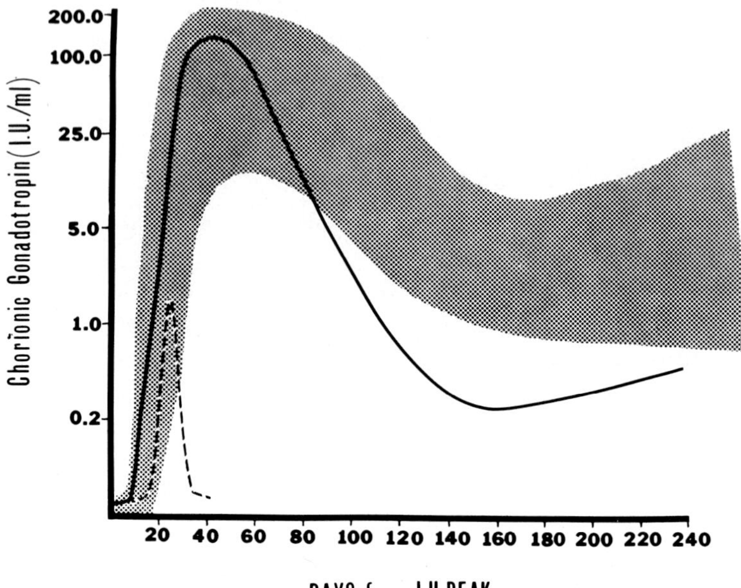

Fig. 2. Serum concentrations of chorionic gonadotropin during pregnancy in chimpanzees (—) and rhesus monkeys (- - -), as compared to the normal range in humans (shaded area). (Hobson *et al.*, 1976; reproduced with permission of the Hemisphere Publishing Corp.)

pearance of pituitary LH. Furthermore, sexual differentiation appears to be normal in anencephalic infants. Taken together, these findings have led us to suggest that the primary stimulus for testicular testosterone production in the human fetus is CG (Clements *et al.*, 1976). The presence of hypotrophic gonads and genitalia in anencephalic infants, however, supports the contention that pituitary gonadotropins subserve an ancillary role in fetal gonadal development and function. In the rhesus monkey, where CG disappears early in gestation (see Section A,2), pituitary gonadotropins may play a more decisive role than in humans (Gulyas *et al.*, 1977).

TABLE 3

Possible physiologic roles for chorionic gonadotropin during pregnancy.

1. Early maintenance of corpus luteum.
2. Sexual differentiation of the male fetus.
3. Effect on placental metabolism and steroidogenesis.
4. Prevention of maternal rejection of the products of conception.

CG also appears to influence placental metabolism and steroid production (Brody, 1969). Recent evidence has suggested, as well, that the failure of the immunologic system of the mother to reject the fetus may be due to a reversible, noncytotoxic alteration of maternal lymphocyte immunocompetence by CG present on the trophoblastic surface (Adcock *et al.*, 1973). This observation has not been substantiated, however (Gundert *et al.*, 1975; Pattilo *et al.*, 1976).

B. Lactogenic Hormones

1. *Placental Lactogen*

Placental lactogen, also called chorionic somatomammotropin, is a peptide hormone produced by the syncytiotrophoblast that shares structural homologies with both pituitary growth hormone and prolactin. Although it has been the subject of intensive investigation in both humans (Friesen and Singer, 1976) and the rhesus monkey (Belanger *et al.*, 1971), we are not aware of any studies in the great apes.

In man, placental lactogen is largely confined to the maternal compartment. It is present in the maternal circulation by five to six weeks gestation at levels in serum of 10–15 ng/ml; levels rise, most notably between 20 and 30 weeks, to reach values of 6–10 µg/ml at term.

A physiological role for placental lactogen remains unclear. It is unlikely that it plays a direct role in fetal growth. However, it may participate with steroid hormones in mobilizing fat for maternal caloric requirements while inhibiting maternal glucose utilization, thus sparing glucose for fetal use. Although placental lactogen clearly possesses lactogenic activity in numerous test systems, its role in stimulating maternal mammary development during pregnancy remains undefined.

In the rhesus monkey, Belanger *et al.* (1971) demonstrated a similar pattern of circulating placental lactogen concentrations to that observed in humans; its rapid disappearance in the postpartum period is also similar in both species.

2. *Pituitary Prolactin*

In humans, pituitary prolactin secretion increases during pregnancy, presumably due to the stimulatory influence of enhanced estrogen production (Guyda, 1976; see Section C,1). Elevations of serum prolactin levels above the nongravid range are evident as early as eight weeks of gestation with the first trimester level averaging 30 ng/ml; thereafter, there is a progressive rise to term with mean levels approximating 200

54   *Charles Faiman* et al.

ng/ml. Variations both between and within individuals tend to be large, and values may range from 40 to 600 ng/ml during the last trimester.

The only report of prolactin in the great apes is that by Reyes *et al.* (1975), who studied serum prolactin levels in two chimpanzees prior to and following the inception of pregnancy (Fig. 3). Blood samples were obtained without anesthesia. As in humans, values during pregnancy are higher than those observed in the nongravid state and rise with advancing gestation; however, values tend to be somewhat lower than those observed during human pregnancy.

In contrast, serum prolactin levels do not rise above nongravid values throughout pregnancy in the rhesus monkey (Friesen *et al.*, 1972), except during the last week prior to parturition when an approximate three-fold increment is observed (Weiss *et al.*, 1976). The lack of a substantial rise in prolactin levels throughout gestation in the monkey may be related to the comparatively much lower increments in estrogen production during pregnancy in this species (see Section C,1).

The role of prolactin in stimulating maternal mammary growth during pregnancy is not well-defined. Presumably it acts in concert with placental lactogen and the high levels of estrogen and progesterone found during pregnancy. An additional role of prolactin during pregnancy has been proffered recently; it may be involved in the regulation of amniotic fluid volume (Tyson and Pinto, 1978).

Following delivery and the rapid clearance of placental steroids, which are thought to antagonize its lactogenic properties, prolactin plays a principle role in stimulating the production of breast milk.

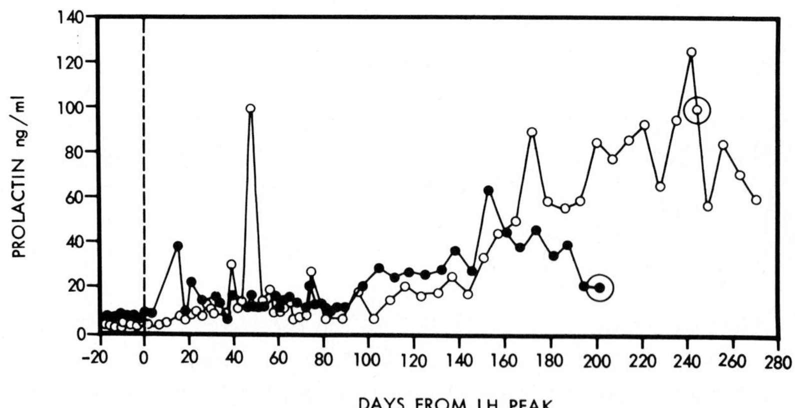

Fig. 3. Serum prolactin levels just prior to and during pregnancy in two chimpanzees. The large circles indicate the day of delivery. (Reyes *et al.*, 1975; reproduced with permission of the J. B. Lippincott Co.)

C. Steroid Hormones

1. Source Of Steroid Hormones During Pregnancy

A number of concepts have evolved that aid in understanding the production and metabolism of steroid hormones during pregnancy in primates. The source of estrogen and progesterone in the maternal compartment during early pregnancy is predominently ovarian, consisting of both luteal and extraluteal components (Williams *et al.*, 1978). Notwithstanding the numerous observations in both humans and the rhesus monkey that the corpus luteum of pregnancy may continue to be functionally active throughout pregnancy, the ovary becomes dispensible by five to six weeks of gestation; it is no longer required for maintenance of maternal estrogen and progesterone levels and of pregnancy itself (Csapo *et al.*, 1974; Diczfalusy and Borell, 1961; Hodgen and Tullner, 1975; Yoshimi *et al.*, 1969). Thereafter, the placenta takes over these functions, and there is no dependence upon the maternal ovary, adrenal, or pituitary. Since neither the placenta, the fetus, nor the maternal components are capable of completely synthesizing the bulk of estrogen and progesterone alone, the concept of a *materno–feto–placental unit* has emerged (Beling, 1977; Ryan and Hopper, 1974). Although there appear to be numerous quantitative differences in progesterone and estrogen metabolism throughout pregnancy among the various primate species (Kaplan and Grumbach, 1978; Ryan and Hopper, 1974; and see below), the presence of a materno–feto–placental unit in which steroid precursors from both the maternal and fetal compartments are transformed into progesterone and estrogens by the placenta appears to be universal.

The bulk of progesterone produced during pregnancy after early gestation is placental in origin. Since the placenta is virtually incapable of *de novo* synthesis of progesterone precursors from acetate, the major precursor for placental progesterone production appears to be cholesterol and pregnenolone of maternal origin (Beling, 1977; Klopper and Fuchs, 1977; see Fig. 4). The major urinary metabolite of progesterone in humans is pregnanediol.

The production of enormous quantities of estrogen during pregnancy in humans is brought about in the placenta by aromatization of C-19 precursors produced mainly in the fetal adrenal glands (Beling, 1977; see Fig. 4). Fetal dehydroepiandrosterone sulfate (DHAS) is converted to androstenedione and estrone by the placenta; placental estradiol (E_2) is derived from placental testosterone and estrone (E_1). Fetal 16αOH-DHAS, an hepatic product of adrenal DHAS, serves as the major precursor for placental estriol (E_3) synthesis. At term, 50% of the precursors

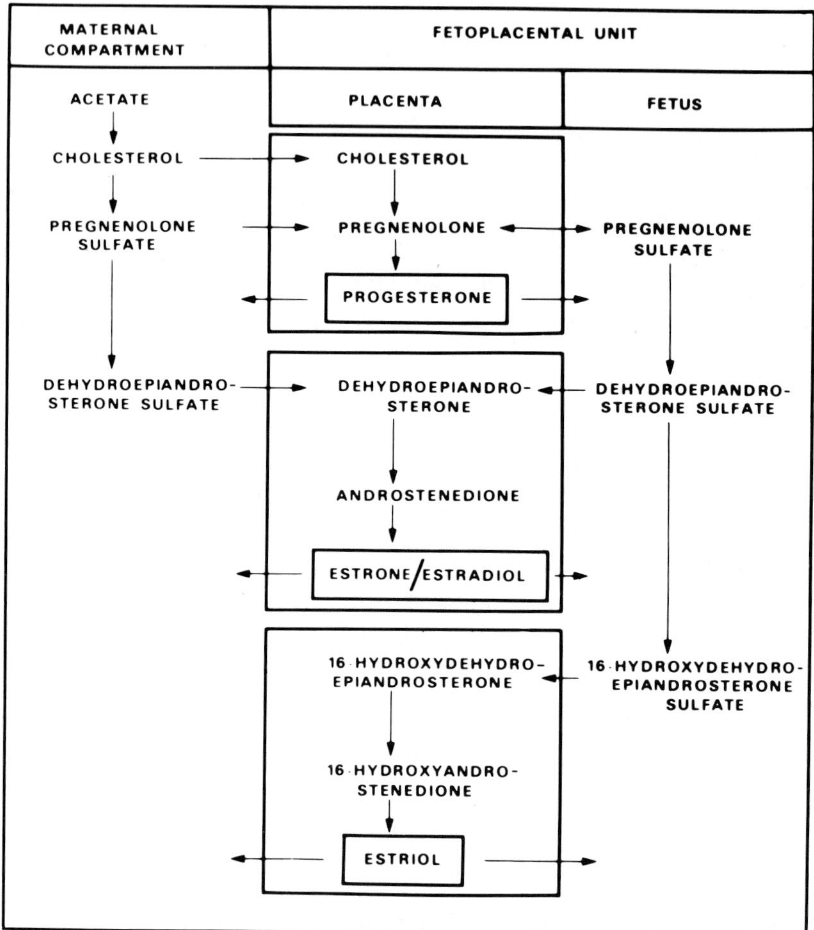

Fig. 4. Major steroid biosynthetic pathways in the materno–feto–placental unit. (Ryan and Hopper, 1974; reproduced with permission of S. Karger AG.)

for E_1 and E_2, and 90% for E_3 are of fetal origin; the source of the remainder is the maternal adrenal. Considerable amounts of synthesized estrogens are transferred to both fetal and maternal compartments. In the mother these undergo hepatic conjugation as sulfates and glucosiduronates prior to urinary excretion.

Thus, vast amounts of estrogen precursors originate from the fetal adrenal that possesses a unique, large fetal zone during intrauterine life and atrophies shortly following delivery. This fetal zone structure is found in humans and the great apes; it is also observed in the rhesus monkey and baboon but its growth during gestation in these species appears less marked than in man and the great apes (Bloch, 1968; Challis *et al.*, 1974; Lanman, 1957; Pepe *et al.*, 1977).

Because estriol production is predominantly dependent upon an intact feto-placental unit, urinary and serum estriol determinations have been used to monitor and predict the outcome of high-risk pregnancies in humans. Although a great deal of debate still rages over the use of these tests as prognostic indicators, they have been used with success in conjunction with tests that assess fetal maturity to intervene in the termination of pregnancy prior to term in individual cases. No such pregnancy surveillance as yet has been attempted for the salvage of nonhuman primate fetuses at risk, although the high estrogen values in apes suggests this approach to improve breeding efficiency might be worth exploring.

2.   *Progesterone*

a.   Urinary Progesterone Metabolites.   Pregnanediol was first reported to be present during the fifth or sixth months of pregnancy in pools of urine from two chimpanzees by Fish *et al.* (1942). Values approximated 2 mg/day in this study, somewhat lower than the 6.5 mg/day reported by Elmadjian and Forchielli (1965) in a presumptively pregnant chimpanzee; considerably higher values in the range of 5–25 mg/day occur at five to six months gestation in women (Ryan and Hopper, 1974). No data are available in either the orangutan or gorilla. The comparatively low values for urinary pregnanediol in chimpanzees may be a reflection of incomplete urine collection. Furthermore, the extensive purification steps by Fish *et al.* (1942) were not corrected for recovery so that the estimated values are minimal ones.

In contrast, the pregnant rhesus monkey and baboon excrete minimal amounts of pregnanediol (Liskowski and Wolf, 1972; Merkatz and Beling, 1969), the major urinary metabolite of progesterone in these species being androsterone.

b.   Serum Progesterone Levels.   Serum progesterone concentrations during pregnancy in the chimpanzee appear to resemble values during pregnancy in humans more closely than does urinary pregnanediol excretion. This is illustrated in Fig. 5 and Table 4. In the study by Reyes *et al.* (1975), serum progesterone levels rose during the luteal phase of the fertile cycle; this was followed by a modest fall at 15–18 days of gestation, and then a secondary rise or plateau was noted. Levels further increased from 140 days until term when maximal values were attained (49–120 ng/ml). No data are available in the orangutan or gorilla.

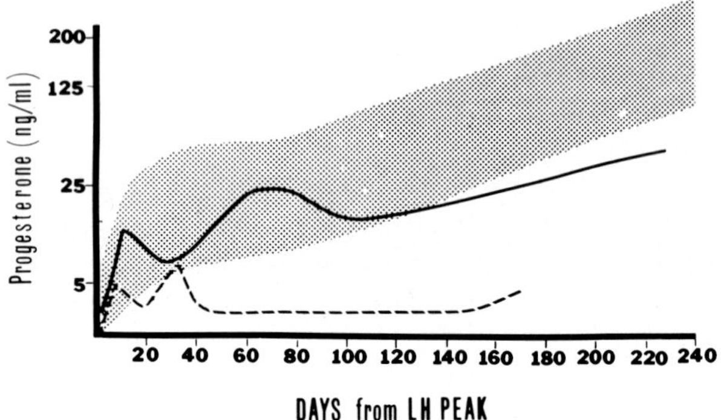

Fig. 5. Serum levels of progesterone during pregnancy in chimpanzees (—) and rhesus monkeys (- - -), as compared to the normal range in humans (shaded area). (Hobson *et al.*, 1976; reproduced with permission of the Hemisphere Publishing Corp.)

Progesterone levels in both the chimpanzee and human are considerably higher during pregnancy than values observed during the luteal phase of the nonfertile menstrual cycle (5–25 ng/ml). In contrast, values in the rhesus monkey and baboon during pregnancy are considerably lower than those in the chimpanzee and human and are not appreciably higher than those observed during the luteal phase of the menstrual cycle.

TABLE 4

Comparison of serum progesterone levels at term in various primate species.

Species	Serum Progesterone (ng/ml)	References
Chimpanzee	49–120[a]	Reyes *et al.* (1975)
Human	45–210[a]	Boroditsky *et al.* (1975, 1978); Buster *et al.* (1979a); Johansson (1969); Tulchinsky and Hobel (1973)
Rhesus monkey	2–10[a]	Bosu *et al.* (1973) Hodgen *et al.* (1972) Neill *et al.* (1969) Weiss *et al.* (1976)
Baboon	7, 12[b]	Albrecht and Townsley (1976, 1978)

[a] Range of values reported.
[b] Mean values in the two studies.

3. Estrogens

a. Urinary Estrogen Metabolites. Allen *et al.* (1935) first reported the presence of bioassayable estrogen in the urine of the pregnant chimpanzee. This finding has been confirmed by chemical assays; no reliable quantitative data are available, although the major urinary estrogen appears to be estriol (see Graham, 1970).

Two reports regarding urinary estrogen in the pregnant gorilla have appeared. Hopper *et al.* (1968) estimated fractionated urinary estrogens in a pregnant gorilla sampled serially from the ninth week of gestation until just prior to term. Values rose progressively to term, as in the human, with estriol being the major metabolite (see Table 5). Total estrogen excretion at term approximated 4 mg/day (expressed as per gm creatinine because of uncertainty as to completeness of collection), a value approximately one-tenth that observed at term in humans (Ryan and Hopper, 1974). Seaton (1978) measured total urinary estrogen excretion in serial samples prior to the inception of pregnancy and throughout gestation in another gorilla. Values rose from 25 μg/gm creatinine in the nongravid state (similar to values in women) to 2 mg/gm creatinine in the 50 days prior to term. This value is probably an underestimate since estriol is the predominant urinary metabolite but is recognized only one-third as well as the estradiol standard employed in the radioimmunoassay.

No reports on urinary estrogen excretion in the orangutan are available.

Table 5 compares the urinary estrogen excretion at term in a number of primate species. Note that, as in humans, the predominant estrogen metabolite in the gorilla (and chimpanzee, see above) is estriol; it is estrone in both the rhesus monkey and baboon. Furthermore, the total amounts of estrogen excreted are considerably lower in the latter two species. These findings may reflect the relative immaturity of the fetal adrenal zone and the inability of the fetal liver to 16α-hydroxylate DHAS in these species (see Fig. 4).

b. Serum Estrogen Levels. Of the great apes, the only species in which serum estrogen levels have been examined during pregnancy has been the chimpanzee. Three animals were studied serially during pregnancy from its inception (Reyes *et al.*, 1975). Early pregnancy was heralded by a rise in serum estradiol (E_2) levels (unconjugated) to values higher than those seen during nonfertile cycles (>30 ng/dl) by two to three weeks following conception, coincident with the ascending limb of the first trimester CG peak (see Section III,A,1). A similar, but smaller increment in serum estrone (E_1) levels was observed; estriol (E_3) was

Table 5

Mean values of urinary estrogen excretion in primate pregnancy at term.

Species	Estrogen Metabolites(μg/day)			References
	Estrone	Estradiol	Estriol	
Gorilla	1300	120	2900	Hopper *et al.* (1968)
Human	1400	500	29,000	Brown (1956)
Rhesus monkey	30	3	1.5	Hopper and Tullner (1967)
Baboon	131	12	13	Merkatz and Beling (1969);
	840	—	—	Townsley (1974)

intermittently detectable in two of the three animals until 45 days gestation. After rising at the inception of pregnancy, E_1 and E_2 concentrations showed a transient decline between 30 and 60 days and rose thereafter to term when maximum values were reached (E_1, 180–300 ng/dl; and E_2, 500–800 ng/dl). E_3 rose after 60 to 80 days to term when levels of 400–1000 ng/dl were reached.

Figure 6 illustrates the serum patterns of these three estrogens throughout pregnancy. Table 6 summarizes estrogen levels attained in a number of primate species at term. Note that the pattern of serum estrogen levels in the chimpanzee are remarkably similar to those observed in man; levels are within or slightly below the ranges seen during human pregnancy.

In contrast, E_2 levels in the rhesus monkey do not rise throughout gestation; values plateau near 50 ng/dl from 70 days gestation until a week before term at which time there is a further increment to near 70 ng/dl at term (Challis *et al.*, 1975; Hodgen *et al.*, 1972; Weiss *et al.*, 1976). Mean serum concentrations of both E_2 and E_1 are 25- to 40-fold lower than those observed in humans at term. In the baboon, serum estradiol levels also plateau from 60 to 155 days gestation at approximately 300 ng/dl and rise during the last month to concentrations of 750 ng/dl at term. Thus, although the pattern of estradiol in the baboon does not resemble that seen in humans and in the chimpanzee, the levels attained at term are similar. Estriol has not been reported to be present in the circulation during pregnancy in either the rhesus monkey or baboon.

4. Role of Steroid Hormones During Pregnancy

The exact role of the enormous circulating concentrations of biologically active estrogens and progesterone during pregnancy remains

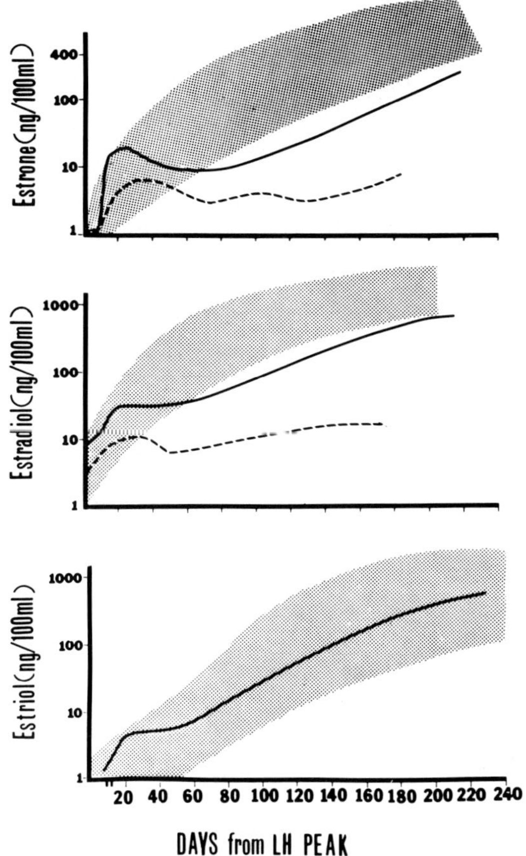

Fig. 6. Serum estrogen levels (estrone, estradiol, and estriol) during pregnancy in chimpanzees (—) and rhesus monkeys (- - -), as compared to the normal ranges in humans (shaded areas). Rhesus monkeys do not have measurable levels of estriol during pregnancy. (Hobson *et al.*, 1976; reproduced with permission of the Hemisphere Publishing Corp.)

speculative. Although preparation of the endometrium for implantation is a well-recognized effect of estrogen and progesterone, the function of these steroid hormones during pregnancy is largely unknown. It has been postulated that progesterone functions to diminish uterine contractility and to allow for pregnancy maintenance. Enormous levels of progesterone seem not to be required, however, as is evident from circulating progesterone concentrations during pregnancy in the rhesus monkey; levels are not substantially above nonfertile luteal values throughout gestation (see Table 4 and Fig. 5).

TABLE 6

Serum estrogen levels in primate pregnancy at term.

Species	Serum Estrogen Level (ng/dl)[a]			References
	Estrone	Estradiol	Estriol	
Chimpanzee	180–300	500–800	400–1000	Reyes *et al.* (1975)
Human	300–2000	550–3500	550–2600	Boroditsky *et al.* (1978); Buster *et al.* (1979); Lindberg *et al.* (1974); Loriaux *et al.* (1972); Tulchinsky and Abraham (1971); Tulchinsky *et al.* (1972)
Rhesus monkey	10–50	20–100	Not detectable	Bosu *et al.* (1973); Challis *et al.* (1975); Hodgen *et al.* (1972); Resko *et al.* (1975); Weiss *et al.* (1976);
Baboon	—	750 ± 170	—	Albrecht and Townsley (1978)

[a] Values reported are ranges, except for baboon which are mean ± SE.

The steroid hormones of pregnancy stimulate, presumably in concert with the lactogenic hormones, acinar growth and development of the breast. They also appear to function as inhibitors of lactogenesis; it is not until the rapid steroid hormone decline in the puerperium that active milk production takes place.

The elevated steroid levels during pregnancy coupled with generous levels of placental lactogen in humans appear to result in a glucose-sparing effect for fetal use, a role presumably mediated by peripheral insulin antagonism (Friesen and Singer, 1976). Progesterone may also act to enhance protein catabolism (Klopper and Fuchs, 1977) and thus provide for more glucose substrate.

From a comparative standpoint, there appears to be a correlation between estrogen and progesterone production during pregnancy and the length of gestation among the various primate species (see Tables 1, 4, 5, and 6). Whether this signifies some underlying role for steroids in prolonging gestation or simply represents enhanced steroid production by the larger feto-placental unit found in pregnancies of longer duration remains speculative (Ryan and Hopper, 1974).

IV. ENDOCRINOLOGY OF LABOR

Unlike the ruminant, in which a clear role for involvement of the fetal hypothalamo–pituitary–adrenal axis in initiating labor is well-established, the endocrine mechanisms involved in the onset of labor in humans and nonhuman primates remain unclear (Challis and Manning, 1978; Knight and O'Connor, 1977; MacDonald *et al.*, 1978; Nathanielsz, 1978). It seems apparent from experiments in sheep, rhesus monkey, baboon, and man that there are critical interrelationships among the fetal adrenal, placenta, and fetal membranes that ultimately lead to enhanced prostaglandin production and the initiation of labor. Although much work has been done in nonhuman primate models, most notably the rhesus monkey and baboon, no studies have been performed utilizing the great apes as models for the initiation of labor in humans. It seems evident from the vast differences in function of the feto-placental unit among the various primate and nonprimate species so far studied (see Section C,1 and references above) that extrapolation of findings in these species to man is of dubious value. The greater similarity of endocrine patterns of apes during pregnancy qualifies them as ideal surrogates for studies aimed at elucidating the endocrine mechanisms involved in the initiation and maintenance of labor in humans.

V.   SUMMARY AND CONCLUSIONS

Largely because of availability, two primate species have been utilized in the past decade as models for human pregnancy: namely, the rhesus monkey and the baboon. It seems quite clear that in many respects they are not ideal. In particular, the feto-placental unit and steroid metabolic pathways are very different in these species from those in humans and the great apes (Leung and Solomon, 1972; Ryan and Hopper, 1974; Solomon and Leung, 1972; Townsley, 1974). Thus, their utility as models of human pregnancy is limited.

In contrast, the chimpanzee and gorilla both appear to possess hormonal patterns and levels during pregnancy that closely resemble the situation in humans (Figs. 2, 5, and 6). However, considerable gaps in our knowledge remain, most notably with regard to the orangutan; descriptive data of hormonal changes during pregnancy of apes are fragmentary, and detailed studies of the formation and metabolism of steroid hormones during pregnancy are lacking.

Nevertheless, the available data suggest that a well-organized effort to exploit the great apes, particularly the chimpanzee, as an endocrine surrogate of human pregnancy would pay generous scientific dividends. Because there are so few chimpanzees available and because the expense and technical difficulties encountered in their use are so great, a well-funded, multilocation, and multidisciplinary approach is indicated. Initially, effort should be directed toward further defining hormone secretion, excretion, and metabolism in the chimpanzee and comparing the results with corresponding information in humans and other animal models. As this information refines our estimates of the areas in which use of the great apes will be the most fruitful, experiments can be undertaken that apply directly to important human problems.

Some of these problems include: (1) elucidation of the control of placental function; (2) assessment of the flux of hormones in the amniotic fluid compartment and the utility of amniotic fluid sampling as an index of the hormonal milieu of the fetus; (3) fetal–placental–maternal interactions leading to parturition; (4) the effects of exogenous hormone exposure on fetal well-being and development; (5) nutritional factors influencing fetal endocrine development; and (6) the relative roles of maternal, placental, and fetal hormones on the progress and outcome of fetal hypothalamic–pituitary–gonadal development.

Obviously, some of these experiments await the development of more sophisticated surgical and manipulative techniques in order to prevent wastage of extremely valuable research animals. Other experiments depend on the development of a proper milieu in which large breeding colonies, capable of producing adequate numbers of pregnancies, are available and are staffed by multidisciplinary teams of primate physiologists, endocrinologists, and steroid chemists. Finally, these efforts will depend on an adequate and stable source of funding.

Chimpanzees present a unique opportunity to make significant advances in our understanding of human pregnancy and fetal development. However, in view of their status as an endangered species, chimpanzees should be used selectively for experiments that cannot be performed in other species, using protocols that will avoid fetal wastage.

ACKNOWLEDGMENTS

Supported in part by the Medical Research Council of Canada (Grant PG-5) and by the Federal Department of Research and Technology of the Federal Republic of Germany under a coordinated research program, Ecologic-Toxicologic Effects of Foreign Compounds in Nonhuman Primates and Other Laboratory Animals.

The authors wish to thank Mrs. J. A. McDougall for typing the manuscript.

REFERENCES

Adcock, E. W., III *et al.* (1973). *Science* **181**, 845–847.

Albrecht, E. D., and Townsley, J. D. (1976). *Biol. Reprod.* **14**, 610–612.

Albrecht, E. D., and Townsley, J. D. (1978). *Biol. Reprod.* **18**, 247–250.

Allen, E., Diddle, A. W., and Elder, J. H. (1935). *Am. J. Physiol.* **110**, 593–596.

Aschheim, S., and Zondek, B. (1928). *Klin. Wochenschr.* **7**, 1453–1457.

Aulmann, G. (1932). *Zool. Gart. (Leipzig)* **5**, 81–90.

Belanger, C., Shome, B., Friesen, H., and Myers, R. E. (1971). *J. Clin. Invest.* **50**, 2660–2667.

Beling, C. (1977). *In* "Endocrinology of Pregnancy" (F. Fuchs and A. Klopper, eds.), pp. 76–98. Harper, New York.

Bloch, E. (1968). *In* "Functions of the Adrenal Cortex" (K. W. McKerns, ed.), Vol. III, pp. 721–772. North-Holland Publ., Amsterdam.

Boorman, G. A., Speltie, T. M., and Fitzgerald, G. H. (1974). *J. Med. Primatol.* **3**, 269–275.

Boroditsky, R. S., Reyes, F. I., Winter, J. S. D., and Faiman, C. (1975). *Am. J. Obstet. Gynecol.* **121**, 238–241.

Boroditsky, R. S., Reyes, F. I., Winter, J. S. D., and Faiman, C. (1978). *Obstet. Gynecol.* **51**, 686–691.

Bosu, W. T. K., Johansson, E. D. B., and Gemzell, C. (1973). *Acta Endocrinol. (Kbh.).* **74**, 743–755.

Brody, S. (1969). *In* "Foetus and Placenta" (A. Klopper and E. Diczfalusy, eds.), pp. 299–411. Blackwell, Oxford.

Brown, J. B. (1956). *Lancet* **i**, 704–707.

Buster, J. E. *et al.* (1979a). *J. Clin. Endocrinol. Metab.* **48**, 133–138.

Buster, J. E. *et al.* (1979b). *J. Clin. Endocrinol. Metab.* **48,** 139–142.

Challis, J. R. G., and Manning, F. A. (1978). *Semin. Perinatol.* **2,** 247–260.

Challis, J. R. G., Davies, I. J., Benirschke, K., Hendrickx, A. G., and Ryan, K. J. (1974). *Endocrinology* **95,** 1300–1305.

Challis, J. R. G., Davies, I. J., Benirschke, K., Hendrickx, A. G., and Ryan, K. J. (1975). *Endocrinology* **96,** 185–192.

Clark, G., and Birch, H. G. (1948). *Endocrinology* **43,** 218–231.

Clegg, M. T., and Weaver, M. (1972). *Proc. Soc. Exp. Biol. Med.* **139,** 1170–1174.

Clements, J. A., Reyes, F. I., Winter, J. S. D., and Faiman, C. (1976). *J. Clin. Endocrinol. Metab.* **42,** 9–19.

Csapo, A. I., Pulkkinen, M. O., and Kaihola, H. L. (1974). *Am. J. Obstet. Gynecol.* **118,** 985–989.

Davis, R. R. (1977). *J. Med. Primatol.* **6,** 315–318.

Diczfalusy, E., and Borell, U. (1961). *J. Clin. Endocrinol. Metab.* **21,** 1119–1126.

Elmadjian, G., and Forchielli, D. (1965). *Proc. Int. Congr. Hormonal Steroids, 1st* Vol. 2, pp. 535–544. Academic Press, New York.

Faiman, C. (1976). *In* "Perinatal Medicine" (J. W. Goodwin, J. O. Godden, and G. W. Chance, eds.), pp. 330–340. Williams and Wilkins, Baltimore, Maryland.

Fish, W. R., Dorfman, R. I., and Young, W. C. (1942). *J. Biol. Chem.* **143,** 715–720.

Friesen, H. G., and Singer, W. (1976). *In* "Perinatal Medicine" (J. W. Goodwin, J. O. Godden, and G. W. Chance, eds.), pp. 341–353. Williams and Wilkins, Baltimore, Maryland.

Friesen, H., Hwang, P., Guyda, H., Tolis, G., Tyson, J., and Myers, R. (1972). *In* "Prolactin and Carcinogenesis" (A. R. Boyns and K. Griffiths, eds.), pp. 64–80. Alpha Omega Alpha Publ., Cardiff, Wales.

Fuchs, F., and Klopper, A. (1977). *In* "Endocrinology of Pregnancy." Harper, New York.

Gibson, J. R., and McKeown, T. (1950). *Brit. J. Soc. Med.* **4,** 221–233.

Goodwin, J. W., Godden, J. O., and Chance, G. W. (1976). *In* "Perinatal Medicine," pp. 321–382. Williams and Wilkins, Baltimore, Maryland.

Graham, C. E. (1970). *In* "The Chimpanzee" (G. H. Bourne, ed.), Vol. III, pp. 183–220. Karger, Basel.

Gulyas, B. J., Tullner, W. W., and Hodgen, G. D. (1977). *Biol. Reprod.* **17,** 650–660.

Gundert, D., Merz, W. E., Hilgenfeldt, U., and Brossmer, R. (1975). *FEBS Lett.* **53,** 309–312.

Guyda, H. J. (1976). *In* "Perinatal Medicine" (J. W. Goodwin, J. O. Godden, and G. W. Chance, eds.), pp. 354–362. Williams and Wilkins, Baltimore, Maryland.

Hartman, C. G. (1932). *Contr. Embryol. Carnegie Inst.* **23,** 1–161.

Hobson, B. M. (1970). *Folia Primatol.* **12,** 111–115.

Hobson, B. M. (1971). *Adv. Reprod. Physiol.* **5,** 67–102.

Hobson, W., Faiman, C., Dougherty, W. J., Reyes, F. I., and Winter, J. S. D. (1975). *Fertil. Steril.* **26,** 93–97.

Hobson, W., Coulston, F., Faiman, C., Winter, J. S. D., and Reyes, F. (1976). *J. Tox. Environ. Health* **1,** 657–668.

Hodgen, G. D., and Tullner, W. W. (1975). *Steroids* **25,** 275–282.

Hodgen, G. D., Dufau, M. L., Catt, K. J., and Tullner, W. W. (1972). *Endocrinology* **91,** 896–900.

Hodgen, G. D., Niemann, W. H., and Tullner, W. W. (1975). *Endocrinology* **96,** 789–791.

Hodgen, G. D., Turner, C. K., Smith, E. E., and Bush, R. M. (1977). *Lab Anim. Sci.* **27,** 99–101.

Hopper, B. R., and Tullner, W. W. (1967). *Steroids* **9,** 517–527.

Hopper, B. R., Tullner, W. W., and Gray, C. W. (1968). *Proc. Soc. Exp. Biol. Med.* **129,** 213–214.

Howland, B. E., Faiman, C., and Butler, T. M. (1971). *Biol. Reprod.* **4,** 101–105.

Johansson, E. D. B. (1969). *Acta Endocrinol. (Kbh.).* **61,** 607–617.

Kaplan, S. L., and Grumbach, M. M. (1978). *Clin. Endocrinol. Metab.* **7,** 487–511.

Keeling, M. E., and Roberts, J. R. (1972). *In* "The Chimpanzee" (G. H. Bourne, ed.), Vol. V, pp. 127–152. Karger, Basel.

Klopper, A., and Diczfalusy, E. (1969). "Foetus and Placenta". Blackwell, Oxford.

Klopper, A., and Fuchs, F. (1977). *In* "Endocrinology of Pregnancy" (F. Fuchs and A. Klopper, eds.), pp. 99–122. Harper, New York.

Knight, J., and O'Connor, M. (1977). "The Fetus and Birth," *Ciba Found. Symp. No. 47* (new series). Elsevier, New York.

Kosasa, T., Levesque, L., Goldstein, D. P., and Taymor, M. L. (1973). *J. Clin. Endocrinol. Metab.* **36,** 622–624.

Lanman, J. T. (1957). *Endocrinology* **61,** 684–691.

Leung, K., and Solomon, S. (1972). *Endocrinology* **91,** 341–349.

Lindberg, B. S., Johansson, E. D. B., and Nilsson, B. A. (1974). *Acta Obstet. Gynecol. Scand. Suppl.* **32,** 21–36.

Liskowski, L., and Wolf, R. C. (1972). *Proc. Soc. Exp. Biol. Med.* **139,** 1123–1126.

Loriaux, D. L., Ruder, J. H., Knab, D. R., and Lipsett, M. B. (1972). *J. Clin. Endocrinol. Metab.* **35,** 887–891.

MacDonald, P. C., Porter, J. C., Schwartz, B. E., and Johnston, J. M. (1978). *Semin. Perinatol.* **2,** 273–286.

Martin, R. (1976). *New Sci.* **72,** 100–102.

Merkatz, I. R., and Beling, C. G. (1969). *J. Reprod. Fertil. Suppl.* **6,** 129–135.

Meyer, R. K. (1972). *Acta Endocrinol. (Kbh.) Suppl.* **166,** 214–217.

Nathanielsz, P. W. (1978). *Ann. Rev. Physiol.* **40,** 411–445.

Neill, J. D., Johansson, E. D. B., and Knobil, E. (1969). *Endocrinology* **84,** 45–48.

Nixon, W. E., Hodgen, G. D., Niemann, W. H., Ross, G. T., and Tullner, W. W. (1972). *Endocrinology* **90,** 1105–1109.

Noback, C. V. (1939). *Anat. Rec.* **73,** 209–225.

Pattilo, R. A., Shalaby, M. R., Hussa, R. O., Bahl, O. P., and Mattingly, R. F. (1976). *Obstet. Gynecol.* **47,** 557–561.

Peacock, L. J., and Rogers, C. M. (1959). *Science* **129,** 959.

Pepe, G. J., Titus, J. A., and Townsley, J. D. (1977). *Biol. Reprod.* **17,** 701–705.

Resko, J. A., Ploem, J. G., and Stadelman, H. L. (1975). *Endocrinology* **97,** 425–430.

Reyes, F. I., Winter, J. S. D., Faiman, C., and Hobson, W. C. (1975). *Endocrinology* **96,** 1447–1455.

Ryan, K. J., and Hopper, B. R. (1974). *In* "Contributions to Primatology" (H. Kuhn *et al.*, eds.), Vol. III, pp. 258–283. Karger, Basel.

Schultz, A. H. (1938). *J. Mammol.* **19,** 363–366.

Seaton, B. (1978). *J. Reprod. Fertil.* **53,** 231–236.

Solomon, S., and Leung, K., (1972). *Acta Endocrinol. (Kbh.) Suppl.* **166,** 178–190.

Stevens, V. C., Bohn, H., and Powell, J. E. (1976). *Am. J. Obstet. Gynecol.* **124,** 51–54.

Townsley, J. D. (1974). *Endocrinology* **95,** 1759–1762.

Tulchinsky, D., and Abraham, G. E. (1971). *J. Clin. Endocrinol. Metab.* **33,** 775–782.

Tulchinsky, D., and Hobel, C. J. (1973). *Am. J. Obstet. Gynecol.* **117,** 884–893.

Tulchinsky, D., Hobel, C. J., Yeager, E., and Marshall, J. R. (1972). *Am. J. Obstet. Gynecol.* **112,** 1095–1100.

Tullner, W. W. (1972). *Acta Endocrinol. (Kbh.) Suppl.* **166,** 200–213.

Tullner, W. W. (1974). *In* "Contributions to Primatology" (H. Kuhn *et al.*, eds.), Vol. III, pp. 235–257. Karger, Basel.

Tullner, W. W., and Gray, C. W. (1968). *Proc. Soc. Exp. Biol. Med.* **128,** 954–956.

Tullner, W. W., and Hertz, R. (1971). *In* "Medical Primatology" (J. Moor-Jankowski and K. I. Goldsmith, eds.), pp. 553–561. Karger, Basel.

Tyson, J. E., and Pinto, H. (1978). *Clin. Obstet. Gynecol.* **5,** 411–434.

van Wagenen, G., and Asling, C. W. (1964). *Am. J. Anat.* **114,** 107–132.

Weiss, G., Butler, W. R., Hotchkiss, J., Dierschke, D. J., and Knobil, E. (1976). *Proc. Soc. Exp. Biol. Med.* **151,** 113–116.

Williams, R. F., Johnson, D. K., and Hodgen, G. D. (1978). *Steroids* **32,** 539–545.

Yoshimi, T., Strott, C. A., Marshall, J. R., and Lipsett, M. B. (1969). *J. Clin. Endocrinol. Metab.* **29,** 225–230.

Zuckerman, S. (1935). *Am. J. Physiol.* **110,** 597–601.

Chapter 3

POSTPARTUM AMENORRHEA AND BEHAVIOR OF APES

Ronald D. Nadler
Charles E. Graham
Delwood C. Collins
O. Ray Kling

I. INTRODUCTION

Following parturition, mammals experience a period of variable duration in which reproductive cyclicity is inhibited. In menstruating species, this condition is termed *postpartum amenorrhea*. Lactation is associated with postpartum amenorrhea, but lactation may cease before cyclicity resumes. The duration of lactation, suckling, and amenorrhea has not been studied extensively in apes, but reports indicate that several years is common (see Chapters 9–12 on sexuality in the wild).

II. DURATION OF POSTPARTUM AMENORRHEA

Evidence indicates that proximity of an infant strongly influences the duration of amenorrhea and lactation, and that effect is mediated by

69

Copyright © 1981 by Academic Press, Inc.
All rights of reproduction in any form reserved.
ISBN 0-12-295020-8

the frequency and intensity of suckling. If the infant ape is removed from the mother at birth or dies early in life, cyclicity resumes within a few weeks or months. Little information exists regarding the effect of infant contact on great apes, since infants are usually removed quickly to facilitate repeated breeding. We conducted a study of the duration of amenorrhea in chimpanzees at the Yerkes Regional Primate Research Center of Emory University. The length of amenorrhea of mothers that had their viable infants removed within 48 hours of birth was compared with that of mothers that were permitted to retain their infants for three to eight months or more than one year (Table 1). Although there was considerable individual variability, postpartum amenorrhea was significantly longer when infant chimpanzees were left with the mother and allowed to nurse (p < 0.005; Student's *t*-test), compared to those whose infants were removed at birth. A trend towards increased duration of postpartum amenorrhea existed with increased duration of infant retention, although no statistical difference was shown between animals retaining their infants for three to eight months and those retaining them more than one year (Graham *et al.*, 1977). The > 1-year retention group, however, included five animals that recommenced cyclicity before infant removal. If these animals are eliminated, the mean (± SE) duration of amenorrhea increases to 477 (± 60), which is significantly longer than that for the third to eighth month group.

III.　ENDOCRINOLOGY OF POSTPARTUM AMENORRHEA

The endocrinological basis for postpartum inhibition of cyclicity, and the prolongation of the inhibition by suckling in mammals is unknown. We have studied this problem in the chimpanzee, comparing gonad-

TABLE 1

Duration of postpartum amenorrhea in chimpanzees retaining infants for various intervals of time.

Duration of Infant Retention	Duration of Amenorrhea		Interval between Infant Removal and Cycle Resumption	
	N	$\bar{x} \pm$ SE	N	$\bar{x} \pm$ SE
< 2 days	27	118 ± 22.7	27	118 ± 22.7
3–8 months	8	277 ± 36	8	104 ± 39.4
> 1 year	14	370 ± 48	9	96 ± 26

otropin levels in eight animals that had their infants removed at birth with five that retained their infants for 2.5–14 months. Blood samples were drawn every two weeks, beginning at parturition and continuing until after cyclicity resumed. LH, FSH, estradiol-17β, and progesterone were measured by radioimmunoassay, as described in Graham *et al.* (1979).

Since postpartum amenorrhea is variable, data were pooled and normalized to the date of parturition for the first 10 weeks of postpartum amenorrhea. In a similar manner, data were normalized for the eight weeks preceding the last day of maximal sexual swelling of the first postpartum cycle (Fig. 1).

Progesterone levels were low throughout the entire period of postpartum amenorrhea, usually less than 0.2 ng/ml. Estrogen levels were generally below 100 pg/ml throughout postpartum amenorrhea, although transient elevations were noted. The last sample from each animal shown in Fig. 1 was obtained shortly after sexual-swelling detumescence associated with the first postpartum menstrual cycle. Mean progesterone and estradiol-17β levels were elevated at this time, but standard errors were large; this variability reflects the fact that some cycles did not show a luteal phase elevation of ovarian steroids, suggesting that the first postpartum cycle is frequently anovulatory.

LH levels were initially high (4–16 mU/ml, one week postpartum), presumably reflecting a cross-reaction in the assay with chorionic gonadotropin not yet cleared from the circulation. Thereafter, mean LH levels fell to < 4 mU/ml for the duration of postpartum amenorrhea. On the other hand, mean FSH levels, initially low (1–4 mU/ml, one week postpartum), rose by three weeks postpartum to a level in excess of 6 mU/ml, with a range of 4–13 mU/ml; FSH declined shortly before the first postpartum sexual-swelling detumescence. Figure 2 illustrates the levels of LH, FSH, estradiol-17β, and progesterone for a female chimpanzee whose infant was removed shortly after birth.

Although the period of amenorrhea was longer in the mothers that retained their infants, their hormone profiles prior to the onset of cyclicity were similar, with two exceptions, to those of the mothers whose infants were removed at birth. The former group showed significantly lower levels of estradiol-17β two to six weeks after parturition, and higher FSH levels eight weeks before the first postpartum sex-swelling detumescence. These data suggest that proximity of the infant tends to suppress ovarian steroid secretion and maintain elevation of FSH levels.

The postpartum endocrine changes in four orangutans that retained their infants for 66–77 weeks after birth were determined (Fig. 3). Twice-

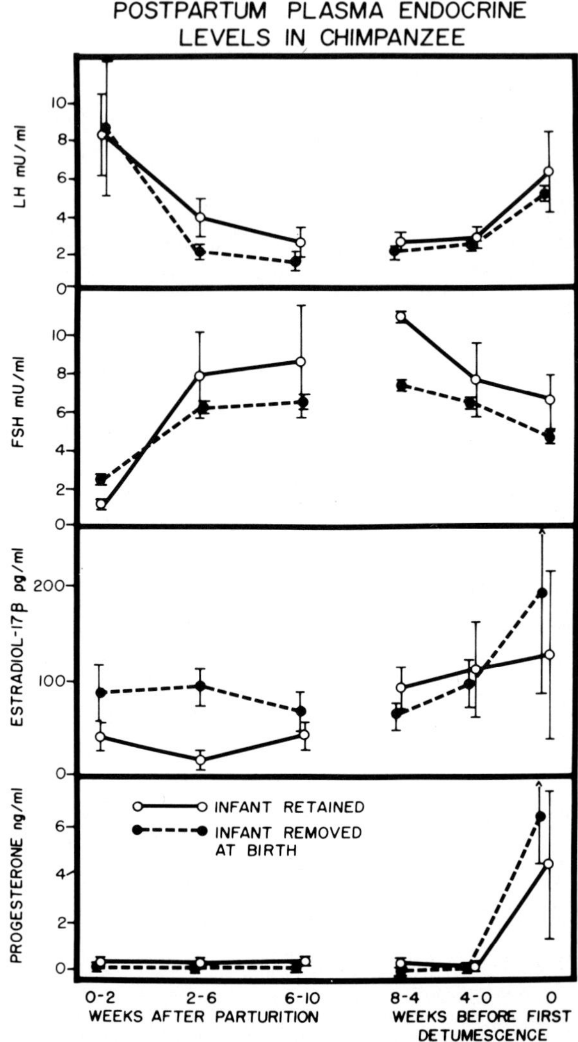

Fig. 1. Mean postpartum endocrine levels in eight chimpanzees whose infants were separated within 48 hours of birth, and five animals that retained infants for 2.5–14 months. Each point represents the mean and standard error of samples obtained from each animal every two weeks during the sampling period.

monthly sampling commenced one to two weeks after delivery in three subjects, and 31 weeks postpartum in a fourth. The endocrine pattern was identical to that we observed in the chimpanzee, with the characteristic elevation of FSH, depression of LH, and generally low estradiol levels with occasional random elevations. Progesterone levels were

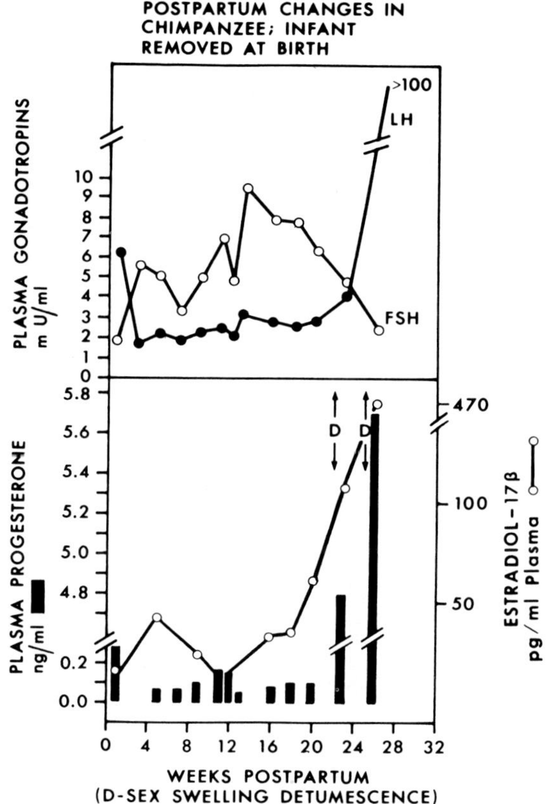

Fig. 2. Postpartum endocrine levels in a single chimpanzee whose infant was removed at birth. D indicates periods of sexual-swelling detumescence.

less than 0.4 ng/ml with rare exceptions. Two animals showed elevated (> 5 ng/ml) progesterone levels indicative of returned cyclicity four and 16 weeks after infant removal.

Several studies show that postpartum endocrine changes in the human female are very similar to those found in the apes (Canales *et al.*, 1974; Reyes *et al.*, 1972; Rolland *et al.*, 1975; Tolis *et al.*, 1974). A striking characteristic of the puerperium in these species, paralleled to a lesser extent only by the early follicular phase of the menstrual cycle, is the relatively high levels of FSH and low levels of LH (Table 2). Similar results were reported also for monkeys (Goodman and Hodgen, 1978). Since secretion of these two hormones appears to be controlled by a single releasing factor, the relatively greater amount of FSH in the postpartum period deserves consideration. We postulate that the absence of ovulation in the postpartum condition is related to the unusual

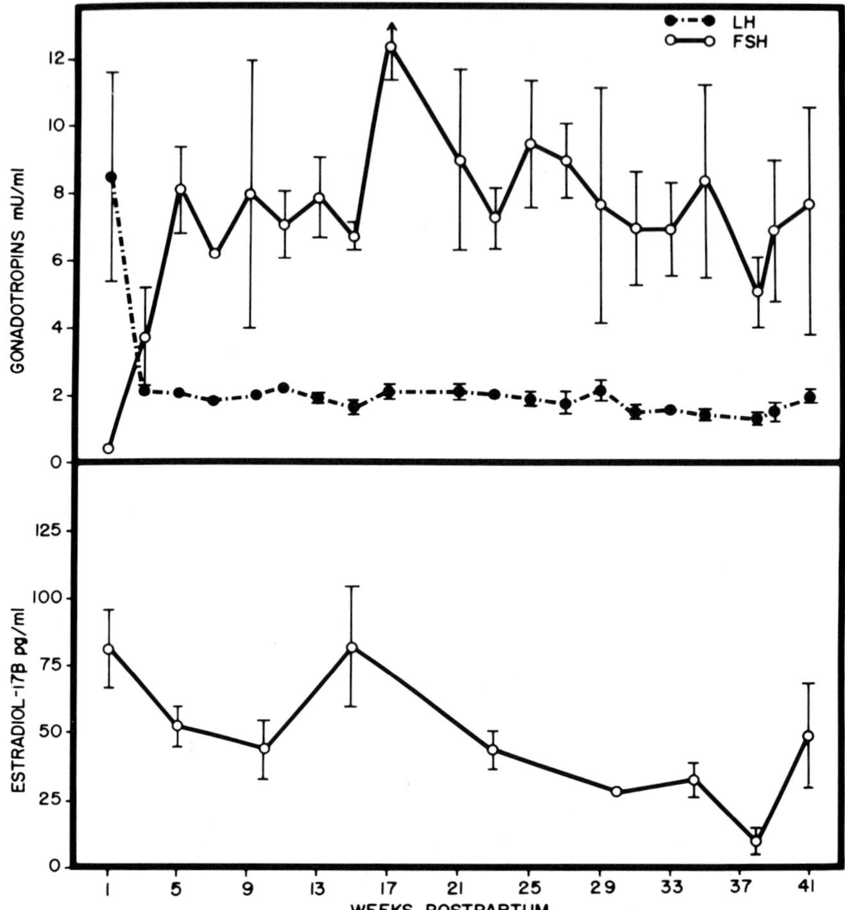

Fig. 3. Mean (± standard error) postpartum endocrine levels in three orangutans that retained their infants in excess of 41 weeks without resumption of cyclicity. Progesterone levels (not shown) were less than 0.4 ng/ml plasma throughout.

balance (one might say imbalance) of gonadotropins. In the human, prolactin levels are considerably elevated during the puerperium, and this may play a key role in inhibiting ovulation.

In Chapter 1 (Graham, this volume), we present evidence that estrogens regulate the secretion of LH and FSH. The effect of estrogen is one of predominantly negative feedback with respect to FSH, in contrast to its effect on LH. The elevated FSH levels in the puerperium may result from a loss of the negative feedback effect of estrogen on FSH gonadotrophs due to the general depression of estrogen levels at this time to early follicular phase values. The data in Table 2 suggest

TABLE 2

Mean gonadotropin levels in female chimpanzees: mU/ml plasma (n).

	Midfollicular	Midluteal	Pregnant	Oophorectomized	Postpartum
LH	4.8 (14)	6.76 (16)	>25.0 (12) (CG)[a]	>41 (4)	2.7 (16)
FSH	6.7 (14)	5.66 (14)	1.2 (12)	>55 (4)	7.86 (16)
LH/FSH ratio	0.7	1.2	—	—	0.34

[a] CG—This value presumably reflects cross-reaction of the assay with chorionic gonadotropin.

that LH/FSH ratios are low in both the postpartum and midfollicular phase animals, compared to midluteal phase animals. This appears to result from a reduction in LH levels and an elevation in FSH levels, suggesting a similar differential feedback of estrogen on the secretion of the gonadotropins in both conditions.

If this interpretation is correct, administration of exogenous estrogen should inhibit endogenous FSH levels and the response to LHRH of FSH, but not LH. Treatment of postpartum chimpanzees with 10 mg or 15 mg mestranol/day for seven days progressively inhibited the capacity of the pituitary to secrete FSH in response to an LHRH challenge, whereas the secretion of LH was unchanged, or actually augmented at the higher dose level. A similar effect was noted in early follicular phase animals (Graham *et al.*, 1979). These data clearly suggest that a selective inhibitory effect of estrogen on FSH secretion exists at the level of the pituitary. Thus, the increased levels of FSH during the puerperium may result from the low circulating levels of estrogen. Significant estrogen feedback remains, however, since estrogen levels do not fall to oophorectomized levels, and gonadotropins do not rise to the high levels typical of castrates (Tables 2 and 3).

The generally low levels of circulating estrogen in the postpartum chimpanzee indicate that follicular development is partially inhibited, although random high values (Fig. 2) suggest that occasional, partial follicular development occurs. The lack of follicular development in the face of high levels of FSH is puzzling. Perhaps higher levels of LH are required to synergize with the FSH. Alternatively, a separate mechanism, perhaps linked to suckling-related prolactin elevation (Jaffe *et al.*, 1973), may play a role in the inhibition of follicular development.

Circulating adrenal and ovarian steroids were measured in several physiological states utilizing chromatographic separation and radioimmunoassay (Kling and Westfahl, 1978). The high $E_1:E_2$ ratio in the puerperium suggests that a major source of estrogen originates from the adrenal gland. In conditions where the ovarian contribution is im-

TABLE 3

Steroid hormone levels in chimpanzees: Mean $(\overline{X})$ ± standard error (SE); pg/ml plasma.

Status	n		T	DHT	Δ^4-A	E_1	E_2	P_4	17-OHP$_4$	DHEA	E_1/E_2
Postpartum[a]	4	$\overline{x}$	141	565	550	83	49	237	412	6342	1.69
		SE	52	540	63	7	8	22	129	2242	
Oophorectomized	6[e]	$\overline{x}$	116	1552[b]	1260[b]	37	26	379	614	9008	1.43
				(943)	(627)						
		SE	8	401[b]	404[b]	8	7	102	150	1416	
				(143)	(314)						
Early follicular	3	$\overline{x}$	94	479	723	37	13[d]	680	232	6087	
		SE	14	179	124	10		271	91	892	
Midfollicular	4	$\overline{x}$	331	1859[c]	2425	72	125	470	622	12567	.57
		SE	67	1240	360	8	21	193	147	3385	
Luteal	4	$\overline{x}$	273	889	1893	75	108	8040	1261	10470[f]	.69
		SE	36	312	301	7	18	3471	39	962	

[a] 4–11 months postpartum.
[b] Elimination of suspect high values from two samples in one assay yields the values in parentheses.
[c] High mean due to one high value ($n = 3$).
[d] $n = 1$.
[e] Two animals each sampled three times at three-month intervals.
[f] $n = 2$.

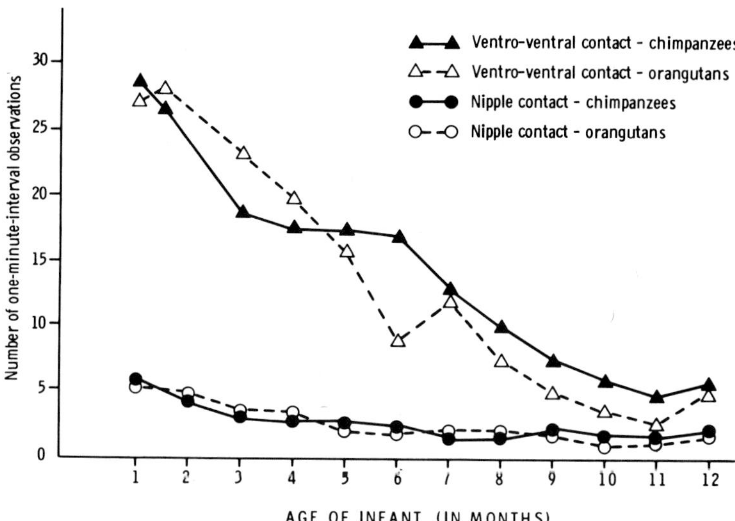

Fig. 4. Mother–infant relations of chimpanzees and orangutans during the infants' first year of life.

paired, the E_1:E_2 ratio is elevated, due to the reduction of ovarian estradiol secretion, and the resulting predominance of estrone, originating from peripheral aromatization of adrenal Δ^4-androstenedione. This phenomenon is well-described in the menopausal woman (Grodin *et al.*, 1971) and is clearly exhibited by the ovariectomized chimpanzees when compared with the three phases of the menstrual cycle (Table 3). However, when ovarian secretion is impaired, one would expect the net secretion to be reduced, as is seen in the ovariectomized animal; the increase in estrone is only relative. However, in the postpartum animals, the levels of both estrogens are twice the ovariectomized levels, indicating either altered ovarian estrogen production or increased adrenal production. Values for circulating adrenal steroids such as cortisol suggest no elevation of adrenal secretion. Therefore, the difference between ovariectomized and postpartum levels probably reflects altered secretion by the ovary. This could be the result of Δ^4-androstenedione production by the ovarian stroma followed by peripheral conversion to estrone.

IV. MOTHER–INFANT RELATIONS DURING POSTPARTUM AMENORRHEA

The patterns of behavioral interaction between the female chimpanzees and their infants showed considerable consistency during the period of observation for three of the four pairs studied. The remaining female began abusing her infant about two months postpartum, necessitating its early separation at 80 days. The behavior patterns for this female and its infant are considered separately. For the three mother–infant pairs that remained together for approximately one year, there was a gradual and continuous decline in suckling, from about 20% of the time observed during the first month to four percent toward the end of the first year (Fig. 4). Although the relative amount of time spent suckling varied somewhat, the relationship between the pairs in terms of the frequency of suckling was the same for eight of the 11 months the animals were studied. This indicates that there was a consistent individual pattern between each mother and her offspring, and similar patterns between pairs with respect to the frequency of suckling. For the pair from which the infant was separated early because of infant abuse, suckling was considerably more frequent during the first month than in the other pairs, declined sharply by about 50% during the second month, and declined another 50% during the third month prior to separation.

Ventro-ventral contact between the mothers and infants was observed more than 90% of the time during the first two months. For three of the pairs, there followed a continuous decline such that the frequency of ventro-ventral contact at one year was reduced to about 17% of the time observed. For the pair in which the infant was separated early, ventro-ventral contact declined about 10% during the second month, and then a sharp reduction occurred to only 22% of the observed time during the month of separation.

With the exception of the latter pair, total contact between mother and infant remained at approximately 100% observed time through the first six months, despite the decline in ventro-ventral contact. This resulted from increased forms of contact other than ventro-ventral, i.e., touching, restraint or retrieval of the infant by the mother, social play, and social examination and grooming. These other forms of contact increased from about seven percent observed time to 50% during the first six months, and remained at approximately 50% of observed time for the remainder of the year. The reduction in total contact during the second half of the year is due to the continued decline in ventro-ventral contact and the plateau in other types of contact at 50% of observed time. For the early separated pair, total contact remained at about 100% for the first two months and then dropped to 75% on the month of separation. Analysis indicated that total contact was not reduced further because other types of contact offset the sharp decline in ventro-ventral contact. For this pair, however, the other types of contact were frequently forms of abuse by the mother and constituted the basis for separation.

A study of the behavioral interactions between female orangutans and their infants was carried out on four mother–infant pairs (Fig. 4). The data on ventro-ventral contact and nipple contact showed a pattern of decline during the first year that was quite similar to that of the chimpanzees. Nipple contact was at its highest levels during the first three months postpartum, occupying approximately 19% of the total time observed. Nipple contact then decreased sharply to 10% of the total time during the fourth month and seven percent during the fifth month. Nipple contact remained fairly stable at about seven percent observed time from the sixth month postpartum onward. Ventro-ventral contact remained at a high level during the first four months, ranging from 66% to 95% of total observation time. From the fifth to eighth months, ventro-ventral contact fluctuated between 25% and 50%, and then declined to approximately 17%, where it remained relatively constant. While intimate forms of contact were decreasing, other forms of contact were increasing during the first year. Other forms of contact were relatively low during the first four months (eight percent of total time) but increased during the next five months to approximately 17% and reached a level of 50% total time by the end of the first year postpartum.

The data on relevant aspects of the mother–infant relationship in these two species indicate that there are substantial behavioral changes during the first year, especially in such measures as nipple contact and ventro-ventral contact. On the other hand, the plasma levels of the hormones FSH, LH, progesterone, and estradiol-17β are relatively constant and not responsive to major reductions in the intensity of the mother–infant relationship, including suckling, as long as some minimal degree of nurturing persists. It is possible that a closer relationship exists between prolactin concentrations and behavior. However, we were unable to obtain undisturbed levels of prolactin in the present study because of the confounding effect of the anesthetic used to obtain the blood samples.

V. DISCUSSION AND CONCLUSIONS

Short (1976) has discussed the comparative aspects of postpartum amenorrhea in feral apes, man at the hunter–gatherer level of existence, and modern man. All groups except modern man show a prolonged interval between successive births, 3.8 years in the gorilla (Harcourt *et al.*, Chapter 10, this volume), 5.7 years in the chimpanzee (Tutin and McGinnis, Chapter 9, this volume), and 4.1 years in the !Kung hunter–gatherers of the Kalahari Desert, resulting in the production of two to five offspring per female. The principal factor holding birthrate in check is the contraceptive effect of lactation, as evidenced by the fact that when an infant dies or is removed from the mother shortly after birth, the birth interval is markedly reduced. It is not uncommon for captive chimpanzees, whose infants are often removed at birth or soon after, to experience a dozen pregnancies. Our own work in the laboratory showed a marked effect of infant proximity on the duration of postpartum amenorrhea when the infant was left with the mother for one year. Generally, laboratory-born ape infants are removed from the mother when they are a year or two old at most, for study and facilitation of repeated breeding. Short, however, reported on one group of captive gorillas in which infants remained with their mothers, and birth intervals were comparable to the wild state. He noted that the difference between the apes, the !Kung, and modern man appears to relate primarily to the pattern and frequency of suckling. The suckling frequency in the captive gorillas mentioned previously, in wild gorillas and chimpanzees, and among the !Kung is two to four times per hour, and in the !Kung at least, suckling continues during the night. Our laboratory studies indicate that although there was a considerable reduction in nipple contact in the first year postpartum in chimpanzees and orangutans, nipple contact was maintained for at least four to seven percent of observed time, comparable to data from wild animals.

Among Western women, breast-feeding is usually restricted to a maximum frequency of four-hour intervals during the daytime only. Even this reduced frequency of breast-feeding prolongs postpartum acyclicity compared to women who do not breast-feed at all. Rolland *et al.* (1975) demonstrated a close relationship between frequency of suckling and elevated prolactin levels in women, suggesting that suckling-induced prolactin elevation might in some way inhibit cyclicity in higher primates; if so, the mechanism of inhibition is uncertain. Our work with the apes provided no hint as to how the profile of circulating gonadotropins and steroid hormones characteristic of postpartum amenorrhea is maintained.

It is clear that breast-feeding is a significant natural means of contraception in some human populations and in apes. Although it may not be possible to regain the benefits of this method of birth-spacing in Western countries by resorting to more frequent breast-feeding, it may be that discovery of the mechanism of menstrual-cycle inhibition related to suckling could lead to development of a pharmacological contraceptive approach that would activate the natural inhibitory mechanism. Because of the similarity in postpartum endocrine patterns to man, apes, particularly chimpanzees, are excellent subjects for such research.

REFERENCES

Canales, E. S., Zarate, A., Garrido, J., Leon, C., Soria, J., and Schally, A. V. (1974). *J. Clin. Endocrinol. Metab.* **38,** 1140–1142.

Goodman, A. L., and Hodgen, G. D. (1978). *Steroids* **31,** 731–744.

Graham, C. E., Nadler, R. D., Neill, J. D., and Collins, D. C. (1977). Paper presented at the American Society of Primatologists, April 16–19, Seattle, Washington.

Graham, C. E., Gould, K. G., Collins, D. C., and Preedy, J. R. K. (1979). *Endocrinology* **105,** 269–275.

Grodin, J. M., Siiteri, P. K., and McDonald, P. C. (1971). *In* "Menopause and Aging" (K. J. Ryan and D. C. Gibson, eds.), pp. 15–35. U. S. Dept. of Health, Education and Welfare publication #(NIH) 73-319, Washington, D. C.

Jaffe, R. B., Yuen, B. H., Keye, W. R., and Midgley, A. R. (1973). *Am. J. Obstet. Gynecol.* **117,** 757–773.

Kling, O. R., and Westfahl, P. K. (1978). *Biol. Reprod.* **18,** 392–400.

Reyes, F. I., Winter, J. S. D., and Faiman, C. (1972). *Am. J. Obstet. Gynecol.* **114,** 589–594.

Rolland, R., Lequin, R. M., Schellerens, L. A., and DeJong, F. H. (1975). *Clin. Endocrinol.* **4,** 15–25.

Short, R. V. (1976). *Proc. R. Soc. London Ser. B.* **195,** 3–24.

Tolis, G., Guyda, H., Pillorger, R., and Friesen, H. G. (1974). *Endocrinol. Res. Commun.* **1,** 293.

Chapter 4

REPRODUCTIVE AND ENDOCRINE DEVELOPMENT IN THE GREAT APES

William C. Hobson
Gene B. Fuller
Jeremy S. D. Winter
Charles Faiman
Francisco I. Reyes

INTRODUCTION

The processes which direct or initiate the maturation of the reproductive–endocrine system from early fetal life through puberty remain
incompletely understood. Our ignorance of the physiology of sexual

83

Copyright © 1981 by Academic Press, Inc.
All rights of reproduction in any form reserved.
ISBN 0-12-295020-8

maturation in humans is in large part due to the lack of an appropriate animal model. Extrapolation of rodent studies to man is hindered both by the real differences between their adult reproductive processes and by their different rates of maturation. Cause-and-effect relationships among the endocrine hormones during development are difficult to establish in the fast-maturing rodents since the exact relationship of events is not apparent from simple inspection of baseline growth and endocrine patterns.

Not only do studies in the great apes present an opportunity to understand human reproductive endocrine development, but they are also important for the husbandry of the great apes themselves. An adequate understanding of the relationship of sexual development to adult reproductive physiology and behavior is necessary to realize the maximum reproductive potential of the apes, to ensure preservation of the species in captivity, and to provide adequate supplies of experimental subjects.

The discussion of reproductive endocrine development in the great apes in this chapter is divided into four categories: fetal development, infancy, the juvenile period, and puberty. The fetal and infant periods are included in this discussion of sexual development because recent data suggest that several important maturational events take place during these times. For purposes of this discussion infancy is defined as the period extending from birth through age two, and the juvenile (or prepubertal) period is defined as the quiescent interval between infancy and the first endocrine changes leading to puberty.

The objective of this chapter is to compare the somewhat scanty data on sexual endocrine development of great apes with the more abundant information on man and lower nonhuman primates, in order to point out the areas in which great apes can best serve as endocrine surrogates for man.

I. FETAL DEVELOPMENT

Studies of fetal sexual development in nonhuman primates have recently become possible by the simultaneous development of more sensitive radioimmunoassay techniques and specialized surgical procedures for sampling and manipulating the fetus. An excellent review of this work, mostly performed in rhesus monkeys, has been published by Resko (1977). Limited availability of aborted normal human fetuses and the finding that hormone concentrations in the amniotic fluid parallel those in the fetal circulation have provided some data on maturational processes occurring in humans (Reyes *et al.*, 1974).

No studies of fetal sexual development in the great apes have been published. This is not surprising in view of the limited number of pregnancies available and the high cost of experimentation with the great apes. There is considerable circumstantial evidence, however, that suggests that the great apes may be the only appropriate model of certain human maternal-fetal endocrine relationships. The importance of these areas may in time lead to limited usage of the chimpanzee for studies of sexual development of the primate fetus.

As discussed by Faiman *et al.* (Chapter 2, this volume), the great apes alone among the primate species studied have a maternal-fetal-placental unit which approximates that of man. For instance, only in humans and apes are steroid precursors known to be supplied by the mother, 16-hydroxylated by the fetal liver, and aromatized by the placenta (Faiman *et al.*, Chapter 2, this volume). This relationship is utilized by clinicians to follow the well-being of the human fetus by measurements of estriol in maternal serum or urine. In view of the important effects of steroid exposure on subsequent sex-development and the potential for disrupting this development by exposure to "hormonal toxins," our present knowledge of fetal endocrine physiology appears to be grossly inadequate. Resko (1977) in a series of experiments in rhesus monkeys confirmed observations in the human that exposure to testosterone during fetal life has profound effects on fetal sexual development. These changes include anatomical masculinization of female fetuses, abnormal adolescent sex behavior, and altered feedback responses of the adult pituitary-hypothalamic unit to estrogen. Similarly, the well-publicized studies of Herbst *et al.* (1971) and Noller and Kurland (1976) demonstrated that prenatal exposure to diethylstilbestrol may predispose human female children to vaginal adenosis and cervical cancer after puberty. The potential for an abnormal fetal endocrine milieu to adversely affect subsequent sexual development in humans is clearly pointed out by the anatomical masculinization induced by excess androgen production in girls with adrenogenital syndrome. Girls with this disease are reported to have masculinized sex behavior patterns (Ehrhardt and Baker, 1974).

Obviously, further studies of the effects of fetal exposure to hormones are needed in order to understand how to correct inborn defects and also to protect the fetus from toxic effects of maternal hormone therapy. Since experimental manipulation of human fetuses is not ethically possible and the physiology of steroid secretion and steroid metabolism in lower nonhuman primate species differs from man, some experiments in the great apes are anticipated. Such studies will probably take the form of repeated amniocentesis and/or nondestructive surgical procedures since it is not easy to justify any procedure likely to jeopardize the life or health of great apes. Data from studies in great apes could be used to evaluate the applicability of similar studies in lower nonhuman primates to man, to determine the detailed time-course of hypothalamic-pituitary-gonadal development *in utero*, to elucidate the acquisition of feedback relationships in the fetus, and to study the influence of maternal endocrine secretions on the fetus. By judicious use of great apes, particularly of the chimpanzee, the problems which are not soluble by studies of rodents or lower nonhuman primates may be resolved.

II. INFANCY

Considerable circumstantial evidence is accumulating which suggests that important maturational changes in hypothalamic-pituitary-gonadal relationships occur during infancy in primates. Term maternal levels of estradiol, estrone, estriol, and progesterone are quite high in chimpanzees (Reyes *et al.*, 1975). Therefore, it is not surprising to find that serum gonadotropins in the infant chimpanzee are low at birth (Winter *et al.*, 1975; Fuller *et al.*, 1981), presumably due to steroid feedback suppression. Responding to removal of the maternal steroid source, both FSH and LH rise rapidly in the infant chimpanzee during the first week of life. In a comparative study, Winter *et al.*, (1975) found a striking sex difference in LH and FSH levels in infant humans and chimpanzees. In female chimpanzee infants (Fig. 1), serum LH and FSH concentrations continued to rise until about three months of age, subsequently declining to a nadir around one year of age. In infant male chimpanzees the FSH rise is lower and shorter, peaking at about one month and declining to the prepubertal range by four months. Thus, serum levels of FSH are significantly higher in female chimpanzees throughout infancy than those in male infants. These observations have been confirmed in chimpanzees (Fuller *et al.*, 1981) and in humans (Garnier *et al.*, 1974). A similar sex difference has been reported in rhesus monkeys (Fuller *et al.*, 1981). Longitudinal studies in infant female chimpanzees show fluctuations which are reminiscent of the cyclic gonadotropin release seen in early human female adolescence (Faiman *et al.*, 1973).

While the early postnatal rise of serum gonadotropins can be ascribed

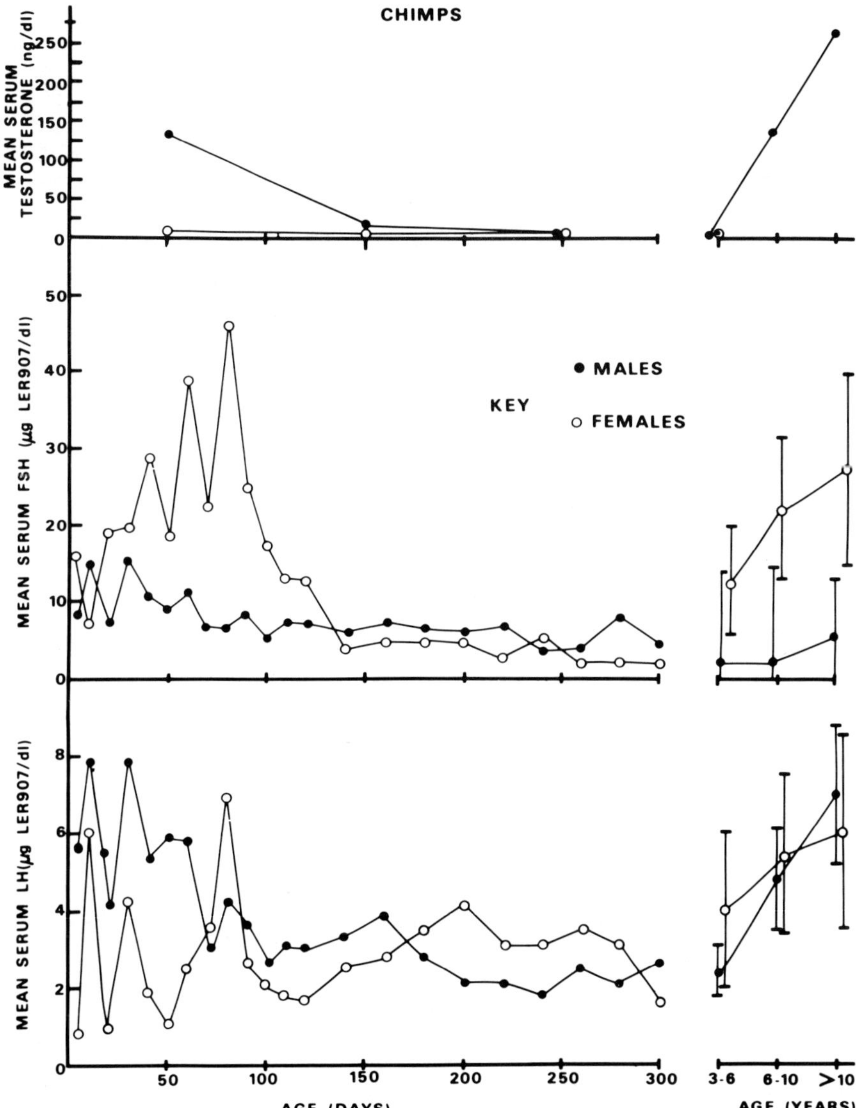

Fig. 1. Serum FSH, LH, and testosterone concentrations in representative infant male and female chimpanzees. These are compared to the mean (± 1 SD) hormone levels at 3–6, 6–10, and 10–15 years of age. (Winter *et al.* 1978; reproduced with permission of the W. B. Saunders Co. Ltd.)

to withdrawal of placental steroids, the cause of the subsequent decline is not as clear. In human male infants, testicular androgen secretion may lead to suppression of the infant hypothalamic-pituitary unit. During the first few weeks of life, serum testosterone levels rise to pre-pubertal concentrations in infant human males and rhesus monkeys (Winter *et al.*, 1976; Robinson and Bridson, 1978). Unfortunately the only studies of age-related changes of testosterone in chimpanzees did not include many samples from infants less than a year old (McCormack, 1971; Martin *et al.*, 1977; Fuller *et al.*, 1980). In human females, evidence for a period of early postnatal gonadal activity is not so clear. The estradiol rise in infant girls is neither as marked nor as sustained as the testosterone rise in male infants (Winter *et al.*, 1976), perhaps accounting for the greater FSH secretion in females than in males during infancy.

At birth the human adrenal gland is proportionally 10–20 times larger by weight than the adult gland. A similar relationship exists at birth in the chimpanzee (Lanman, 1957). The large size is due to a histologically distinct reticular zone, commonly known as the fetal zone. Regression of the zone is accompanied by a decrease in serum dehydroepiandrosterone sulphate (DHAS) and in the 3β-hydroxy-δ^5 steroids found in the urine of newborn humans, chimpanzees, and orangutans (Shackleton and Gustafson, 1971; Shackleton, 1973). These steroids are not present in large quantities in the newborn rhesus monkey, presumably due to the early disappearance of the fetal zone in this species (Shackleton, 1973). Therefore, it would appear that the great apes, and possibly the baboon (Townsley and Pepe, 1977), are more suitable for study of the endocrine adaptation to extra-uterine life.

There are few studies of hormone feedback dynamics during infancy in any primate. Garnier *et al.* (1974) reported that administration of luteinizing hormone-releasing hormone (LHRH) to human infants elicited a greater FSH response in girls than in boys, and that sensitivity to LHRH diminished between infancy and the juvenile period. Our attempts to duplicate these findings in rhesus monkeys failed due to the lack of a consistent response to LHRH by rhesus monkeys of any age (Hobson and Fuller, 1977). Elucidation of the developmental role of LHRH in primates has been hindered by the inconsistent responses of all the macaques to bolus injections of LHRH (Ehara *et al.*, 1972; Arimura *et al.*, 1973; Shah *et al.*, 1976; Hobson and Fuller, 1977). Whether or not these observations are due to the mode of administration or to inherent species differences in pituitary function is not resolved. In any case the chimpanzee presents an ideal model to study LHRH responses during sexual maturation since, like man, it responds

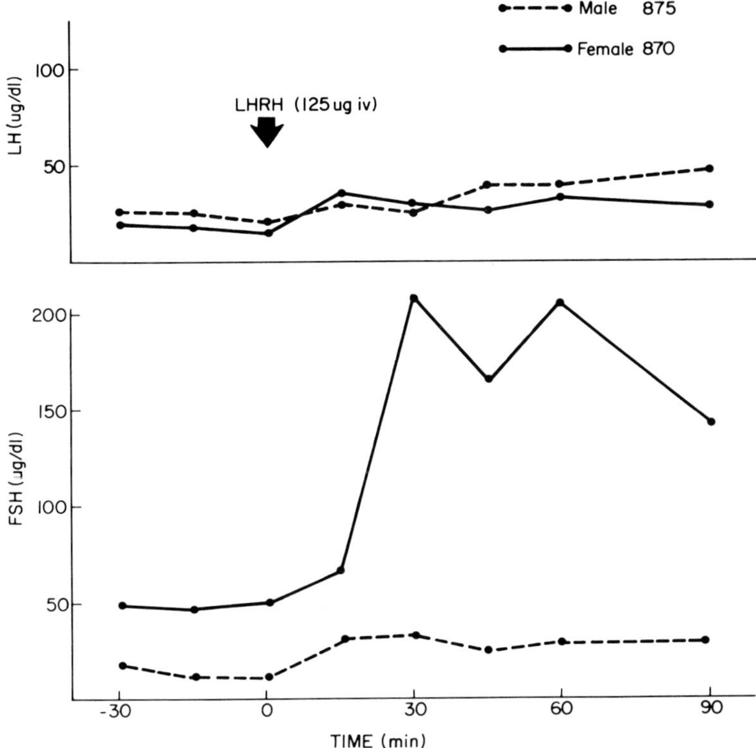

Fig. 2. Serum gonadotropin responses to LHRH injection in representative male and female prepubertal chimpanzees.

to bolus injections of LHRH both before (Hobson and Fuller, 1981 and Fig. 2) and after puberty (Hobson and Fuller, 1977; Graham *et al.*, 1979; Graham, Chapter 1, this volume). To date, no studies of LHRH responsivity during infancy in the great apes have been published.

In a series of experiments designed to examine the reasons for the fall in LH and FSH during late infancy, we studied the ability of exogenously administered estrogen to suppress gonadotropin concentrations in chimpanzees between six months and nine years of age (Hobson and Fuller, 1981). In adults, administration of estrogen to female chimpanzees during the follicular phase of the menstrual cycle is followed by an immediate decline in LH and FSH serum concentrations, and then by a brisk "preovulatory-like" LH surge (Hobson and Fuller, 1981; Fig. 3). Similar patterns are observed in women (Monroe *et al.*, 1972) and adult female rhesus monkeys (Dierschke *et al.*, 1973). When similar doses of estradiol were given to juvenile or pubertal chimpanzees,

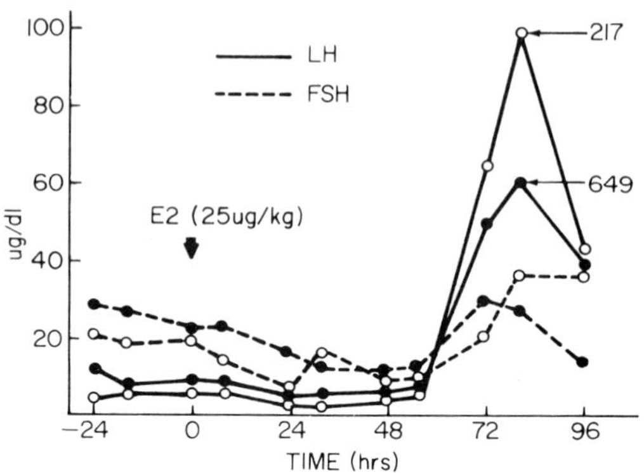

Fig. 3. Serum gonadotropin responses to estradiol (E₂) injection (s.c. in oil) in adult female chimpanzees treated between Days 2–4 of the menstrual cycle.

neither a consistent gonadotropin suppression nor a gonadotropin surge was observed (Fig. 4). (See Section III for a more detailed discussion of this experiment.) In the infants, however, the relatively high levels of FSH were readily suppressed by administration of exogenous estrogen (Fig. 5; positive feedback effects were not observed). These preliminary experiments indicate that the hypothalamic-pituitary response to exogenous steroids changes between infancy and the juvenile period. Taken together with the observations that sensitivity to LHRH is lost at this time and that endogenous steroid secretion seems to diminish, these results suggest that a profound change in hypothalamic-pituitary-gonadal function occurs during late infancy. These changes, which we call the "infant endocrine transition," could almost be termed a "puberty in reverse." These data also point out that chimpanzees are ideally suited to endocrine studies of infancy—studies which could not be accomplished in humans for ethical reasons, and which are more difficult to perform in smaller nonhuman primates, in part because of their small blood volume.

III.   THE JUVENILE PERIOD

The juvenile period, here defined as the quiescent interval between infancy and puberty, is characterized in man by low and stable levels of both gonadotropins and sex steroids. The endocrine profile of this period has not been defined in gorillas and orangutans, but has been described in some detail in chimpanzees (Hobson *et al.*, 1981; Martin *et al.*, 1977). In chimpanzees the period begins with the last endocrine

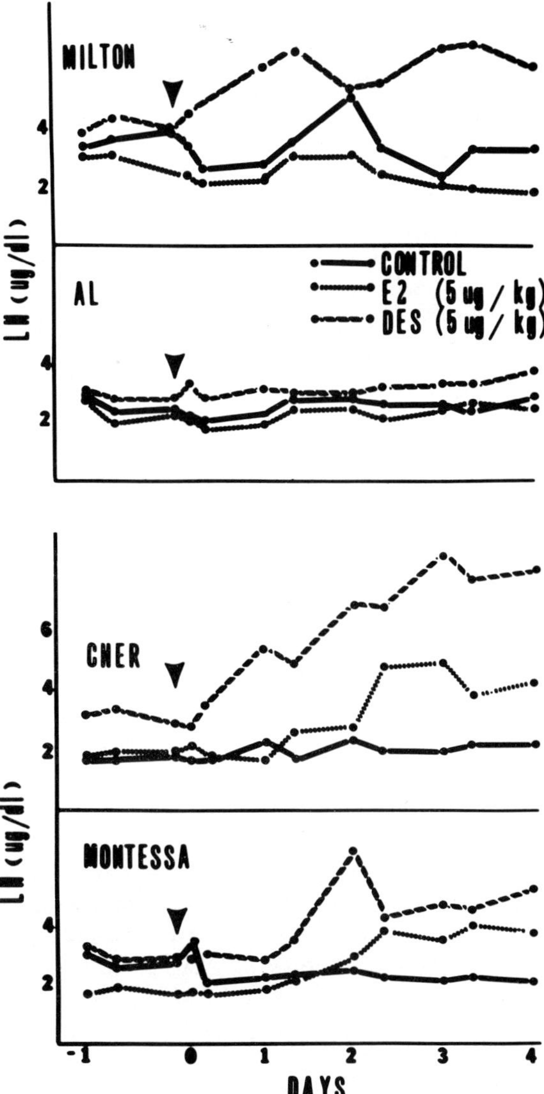

Fig. 4. Serum LH concentrations in two male (Milton and Al) and two female (Cher and Montessa) prepubertal chimpanzees given a single injection of estradiol (E₂) or diethylstilbestrol (DES) (s.c. in oil).

                                                        William C. Hobson et al.

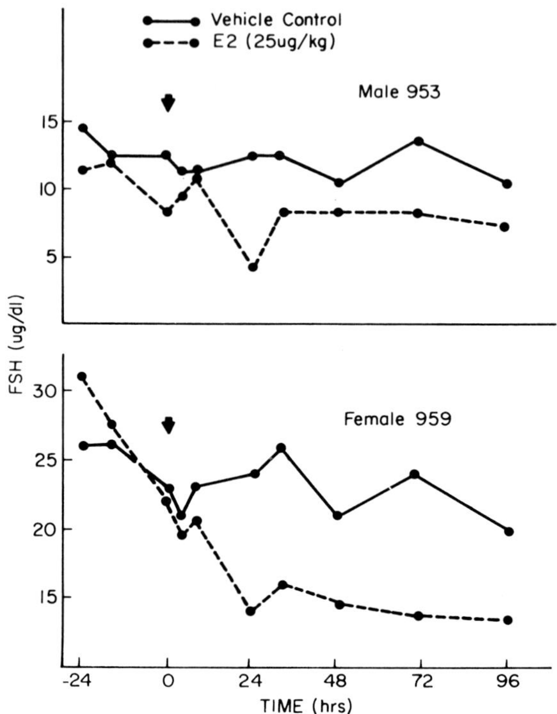

Fig. 5. Representative FSH responses of infant (7–9 months) chimpanzees to estradiol (E_2) injection (s.c. in oil).

changes of infancy around two years of age and lasts until the first hormonal changes at about age six, which lead to puberty.

Serum gonadotropin concentrations in chimpanzees decline during late infancy and remain low throughout the juvenile period (Hobson *et al.*, 1980; Fig. 1). Analysis of serial serum samples from younger female chimpanzees in this age range reveals cyclic fluctuations of gonadotropin levels which are similar to those described in early human adolescence (Faiman *et al.*, 1973).

The mechanism by which reproductive function in the higher primates is inhibited between infancy and puberty remains obscure in spite of considerable scientific investigation. It is generally recognized that the quiescent interval is necessary to prevent reproductive function until it can be accommodated by body size. Several mechanisms have been proposed. They are generally divided into three categories:

1. Greater sensitivity by one or more components of the hypothalamic-pituitary-gonadal (HPG) system to negative hormone feedback.

2. Immaturity of a component of the HPG system (often linked to body growth).

3. Production of an active inhibitor of the HPG system which disappears at puberty.

An acceptable theory not only must account for the hormonal quiescence during the juvenile period, but also must explain the observed changes which occur at puberty and the recent findings of dynamic changes occurring late in infancy.

Currently, the most widely accepted hypothesis accounting for the low and stable gonadotropin and sex-steroid secretion during the juvenile period is that the pituitary and/or hypothalamus is exquisitely sensitive to negative steroid feedback prior to puberty, and that this threshold is elevated at the time of puberty (Winter *et al.*, 1978). This hypothesis is primarily based on data from experiments with rodents and ruminants in which prepubertal castration results in an immediate and sustained increase in gonadotropin levels. The amounts of exogenous steroids required to prevent this postcastration rise are smaller before than after puberty. These data are supported by the demonstration that very small amounts of exogenous estrogen can suppress urinary levels of LH in prepubertal girls (Kelch *et al.*, 1973; Reiter *et al.*, 1974). Effects of exogenous estrogen on serum gonadotropins in prepubertal children have not been reported in detail.

The role of the adrenal in suppressing gonadotropin secretion prior to puberty has been the subject of considerable speculation. In children, adrenal maturation (termed *adrenarche*) precedes the onset of puberty by one to two years (Cutler *et al.*, 1978). This temporal association has led to speculation of a cause-and-effect relationship. Adrenarche begins in children around six to eight years of age and is characterized by dramatic increase in the circulating levels of dehydroepiandrosterone (DHEA) and dehydroepiandrosterone sulfate (DHAS), and by less dramatic rises in androstenedione. During this period, adrenal cortisol secretion remains relatively constant (Ducharme *et al.*, 1976), suggesting that ACTH levels are stable and that the increases in DHEA and DHAS are not ACTH-mediated. There have been several attempts to establish the existence of a "new," non-ACTH pituitary hormone responsible for this adrenal maturation. One attempt in dogs (Parker and Odell, 1977) provided some preliminary evidence for such a hormone which was tentatively called cortical adrenal-stimulating hormone. A persistent difficulty with this and other attempts to identify an adrenarchal hormone or to associate adrenarche with puberty has been the lack of an adequate animal model; until recently, adrenarche had been demonstrated to occur only in children.

Recently, Cutler *et al.*, (1978) completed a detailed comparative study of time-related adrenal function in several species. Using circulating levels of DHEA, DHAS, and androstenedione as markers, they looked for adrenarche in rats, guinea pigs, hamsters, goats, sheep, cattle, dogs, rhesus monkeys, and chimpanzees. Of the nonprimate species, only the rabbit and dog showed significant age-related increases in adrenal steroids. However, the low basal levels of DHAS, DHEA, and androstenedione preclude the use of these species as a model of man. Rhesus monkeys had very high levels of adrenal androgens, but no increase associated with puberty or age was demonstrated. In contrast to these species, but similar to man, androstenedione, DHAS, and DHA increased dramatically with age in chimpanzees (Fig. 6). The increase was statistically significant by five years of age and preceded the first signs of puberty.

These findings are supported by results from our own laboratory (Hobson *et al.*, 1981). Longitudinal blood samples were collected from unanesthetized male and female chimpanzees between three and six years of age. Serum concentrations of estrone, estradiol, androstenedione, progesterone, 170H-progesterone, and testosterone were measured by radioimmunoassay (RIA) in samples collected at monthly in-

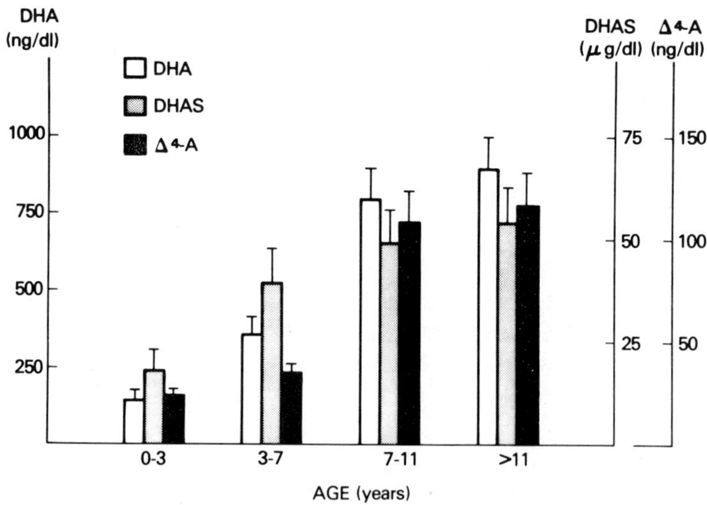

Fig. 6. Plasma concentrations (mean ± SE) of DHA, DHAS, and androstenedione (δ^4-A) in chimpanzees with increasing age. The number of animals in each group were 10, 10, 6, and 6 for the 0–3, 3–7, 7–11, and >11 groups, respectively. The numbers in each group were divided equally between males and females. (Cutler *et al.*, 1978; reproduced with permission of the Williams & Wilkins Co.)

tervals, and also in samples collected at two-hour intervals for 24-hour periods. DHEA levels in these chimpanzees showed a 15–20-fold increase between four and six years of age. The other steroids remained relatively low and stable during this time (Fig. 7). The finding that chimpanzees undergo a humanlike adrenarche will certainly lead to further studies designed to understand both the physiological significance of these events and their relation to the initiation of sexual maturation in the higher primates.

In other studies in this laboratory, adolescent chimpanzees were used to investigate the control of gonadotropin secretion before puberty. The hypothesis that gonadotropin secretion prior to puberty is maintained at low levels by enhanced sensitivity to steroid feedback (probably estrogen) was tested by exposing infant, juvenile, and pubertal chimpanzees to varying amounts of estrogen and measuring its effect on gonadotropin levels. Kelch *et al.*, (1973) reported that 10 ug/kg estradiol benzoate suppressed both urinary and plasma FSH in two prepubertal girls, and also cited preliminary evidence that the feedback sensitivity to ethinyl estradiol decreased at puberty. In contrast to Kelch's study of only two subjects, we found that although estradiol benzoate (at doses of 5 ug/kg or higher) clearly suppressed serum LH and FSH during infancy and late puberty. It was not able to consistently modify levels of either gonadotropin in 12 chimpanzees between two and seven years of age (Hobson and Fuller, 1981). While suppression of serum LH and/or FSH concentrations was observed in some animals (irrespective of sex or age), gonadotropin elevations occurred in an equal number, and most commonly there was no change at all (Fig. 4). In contrast, adult chimpanzees of both sexes respond to an injection of estradiol benzoate (25 ug/kg) with an immediate reduction of both LH and FSH serum concentrations (Fig. 3). As in women, this initial phase of gonadotropin suppression is followed by pre-ovulatorylike LH and FSH surges. Similar surges were never observed in estrogen-treated prepubertal chimpanzees of either sex. Likewise, prepubertal girls (Reiter *et al.*, 1974) and rhesus monkeys (Dierschke *et al.*, 1974) do not release LH in response to estrogen administration.

The proposal that gonadotropin levels are controlled by extreme negative feedback sensitivity to gonadal steroid must be re-examined in light of these data. In addition to our observation that administration of estrogen prior to puberty does not consistently suppress LH and FSH levels, we have also found that removal of gonadal steroid feedback by castration in prepubertal rhesus monkeys is not followed by pronounced gonadotropin changes until about the expected time of puberty (Winter *et al.*, 1977). In combination these data imply that control of

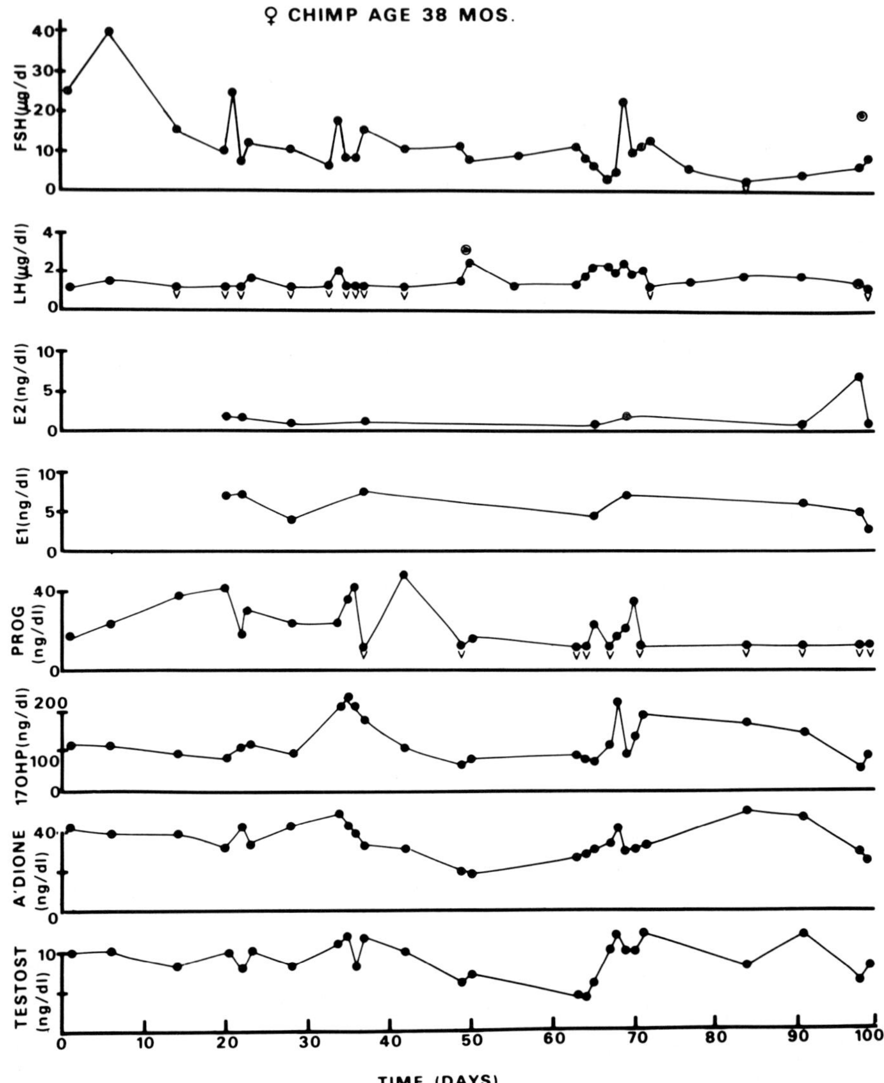

Fig. 7. Serial serum concentrations of FSH, LH, estradiol (E₂), estrone (E₁), proges-terone, 17-hydroxyprogesterone (17OHP), androstenedione, and testosterone in a three-year-old prepubertal female chimpanzee. The symbol v̇ refers to values below the limit of assay sensitivity. (Winter *et al.*, 1978; reproduced with permission of the W. B. Saunders Co. Ltd.)

FSH and LH before puberty may be due to a nonestrogen, nongonadal source.

Although we did not find that estrogen suppressed basal LH or FSH levels prior to puberty, we subsequently demonstrated the "potential" for negative estrogen feedback on one component of prepubertal gonadotropin secretion. When LHRH is given to juvenile chimpanzees or girls, FSH is released in preference to LH (Hobson and Fuller, 1981 and Fig. 2; Garnier *et al.*, 1974). However, we found that if estradiol (5 ug/kg or more) is given prior to LHRH administration, both the LH and FSH responses to LHRH are blocked in prepubertal chimpanzees (Fig. 8). These results imply that LHRH secretion is minimal prior to puberty since estrogen clearly blocks gonadotropin release by exogenous LHRH, and exogenous estrogen does not affect basal gonadotropin levels at this time. Additionally, it is difficult to assume that LHRH secretion is maximally suppressed by endogenous gonadal steroids because we found that castration causes little or no immediate change of basal gonadotropin levels in prepubertal rhesus monkeys. While it is tempting to speculate that acquisition of estrogen-suppressable gonadotropin secretion is simply a function of some maturational process within the hypothalamus, this theory is not consistent with our observation that the higher levels of LH and FSH observed during infancy are readily suppressed by estrogen; estrogen-suppressable gonadotro-

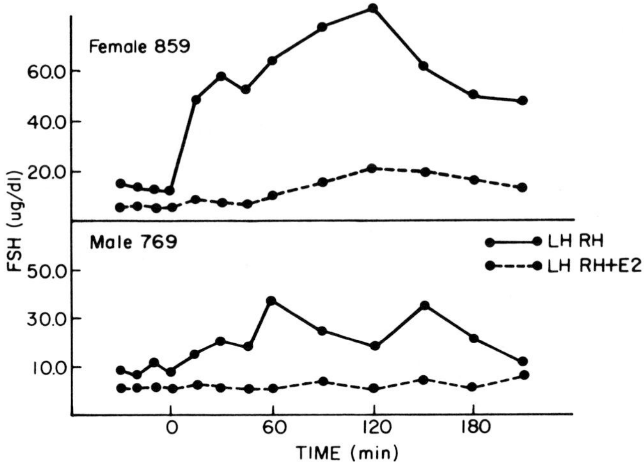

Fig. 8. Estradiol (E_2) block of the FSH response to LHRH. LHRH (125 ug i.v.) was given at time 0. Estradiol (5 ug/kg s.c. in oil) or vehicle was given 24 and 48 hr prior to the LHRH injection. The vehicle control and estradiol experiments were separated by a 10-day interval.

pin secretion is present during infancy but is lost during the juvenile years. Speculation aside, these results do point out that the adolescent chimpanzee may be fruitfully used to elucidate the mechanisms which control hypothalamic-pituitary function prior to puberty.

IV. PUBERTY

While the exact nature of the mechanism controlling gonadotropin secretion during childhood remains obscure, there is considerable evidence that the timing of the eventual onset of puberty is determined by a process of neuronal maturation. This has been attributed to an ultimate progressive reduction of hypothalamic sensitivity to the negative feedback effect of sex steroids (Grumbach *et al.*, 1974; Winter *et al.*, 1978). Levels of LHRH cannot be reliably measured in the peripheral circulation. However, a recent report described a puberty-related increase in urinary excretion of a material which is immunologically related to LHRH and may be a breakdown product of it (Root *et al.*, 1977). If this, in fact, represents increased LHRH secretion, it should be followed, in order, by enhanced pituitary secretion of FSH and LH, an increased rate and degree of gonadal maturation, rising levels of estrogen or androgen in the peripheral circulation, and the appearance of secondary sex characteristics.

While the primary hormonal step in this process is probably enhanced LHRH secretion, the signal undergoes secondary amplification at the level of the pituitary and ovary as puberty progresses, thereby accelerating the rate of pubertal development. Thus, puberty is accompanied by increased gonadotropin responsiveness to LHRH in humans (Dickerman *et al.*, 1976) and in chimpanzees (Hobson and Fuller, 1981). The gonads also show increased sensitivity to gonadotropic stimulation (Winter and Faiman, 1972). These amplifications are probably related to changes in the affinity or number of hormone receptors in the various organs. Elsner *et al.* (1977) reported puberty-related changes in the cytoplasmic and nuclear receptors of genital target tissues that rendered the organs more responsive to the circulating sex steroids.

Detailed descriptions of endocrine patterns during puberty in the great apes have not been reported. The available endocrine data are difficult to correlate with pubertal changes since the great apes do not present the easily observed external signs by which puberty can be staged. In humans, axillary hair growth, breast development, gonadal size, and time of menarche are used as indices of pubertal stages, whereas in the great apes not all of these features are readily apparent.

The time of first perineal swelling and time of menarche are useful markers, but they require careful observation initiated well in advance of these events. Young and Yerkes (1943) found in 18 chimpanzees that the average age at menarche was 8 years, 11 months, and that the range extended from 7 years, 4 months to 10 years, 2 months.

Similar to secretory patterns of other hormones, gonadotropins and gonadal steroids are secreted episodically; changes in the mean levels of circulating hormones result primarily from changes in frequency of these secretory episodes. The earliest endocrine sign of impending puberty in humans has been the detection of episodic rises in serum LH (Penny *et al.*, 1977). These secretory episodes are initially detected only during sleep and are well-correlated with episodes of rapid-eye-movement (REM) sleep (Boyar *et al.*, 1972; Judd *et al.*, 1977). By the time sexual development is complete, these sleep-related elevations of gonadotropin secretion are no longer apparent. Preliminary evidence from our work suggests that a similar phenomenon occurs in prepubertal chimpanzees. When gonadotropins were measured in samples collected at two-hour intervals from two 6-year-old chimpanzee females, significant elevations in serum FSH occurred between 0–0400 hr on each night of sampling (unpublished observations). If subsequent experimentation confirms this finding, the chimpanzee will be the first non-human primate model of puberty-associated nocturnal gonadotropin rises. Whether or not the elevation in FSH rather than LH will be a consistent finding remains to be seen.

Interestingly enough, evidence is beginning to accumulate that both the nocturnal gonadotropin surges and the first significant elevations of daytime serum LH and FSH levels in chimpanzees are more closely correlated with body size than with chronologic age. Wilen and Naftolin (1976) found an excellent correlation in pubertal rhesus monkeys between body size and the time of menarche; they postulated that the onset of puberty is somehow dependent on body size. Systematic studies in the great apes correlating body size with either hormonal or external signs of puberty have not been reported. Circumstantial evidence from our breeding colony suggests that in fact this relationship may be true in the great apes. Whereas the nocturnal gonadotropin spikes occurred in the two 6-year-old females who weighed 29–31 kg, no signs of impending puberty (including nocturnal FSH spikes) have been observed in other older chimpanzees of a smaller body size (20–25 kg). For purposes of classification we use an arbitrary weight limit of 25 kg as the threshold above which chimpanzees are no longer considered to be prepubertal, but further documentation is needed.

In a study of testosterone concentrations and plasma binding in male chimpanzees, McCormack (1971) found that the increase in serum testosterone concentrations paralleled the increase in body weight during puberty. She also reported that both the capacity and affinity of the testosterone-binding globulin (TBG) decreased during puberty.

In contrast to humans, the growth spurt in rhesus monkeys occurs during or after puberty (Wilen and Naftolin, 1976). Smith *et al.* (1975) reported that the prepubertal growth spurt in female chimpanzees was poorly defined. However, careful inspection of his data as well as those of Young and Yerkes (1943) and Gavan (1971) would suggest that the question is still open. Careful study of chimpanzees under optimal nutritional conditions might reveal a true increase in the rate of weight gain prior to puberty. There is some evidence for a prepubertal growth spurt in gorillas (Gijzen and Tijskens, 1971). Seven of eight female gorillas studied at Yerkes Primate Research Center showed an increased rate of weight gain (approximately two-fold) occurring between estimated ages of 4 and 5.5 years (C. Graham, personal communication).

In most girls, ovulation does not begin until six months or more after menarche. Similarly, estrogen is not able to induce gonadotropin surges until well along in puberty in either humans or rhesus monkeys (Reiter *et al.*, 1974; Dierschke *et al.*, 1974). Even after ovulation has begun, it is not unusual to observe anovular or short luteal-phase menstrual cycle in both man and monkeys. Similar studies have not been reported in chimpanzees. However, the reports of postmenarchal infertility in chimpanzees (Young and Yerkes, 1943) correspond with the well-known relative infertility of adolescent girls.

At best, pubertal endocrine data in the great apes is sporadic. However, the limited data available is intriguing in that it suggests that the great apes may represent the only known surrogate for study of some parameters of human sexual development during puberty. For instance, only the chimpanzee has been shown to have nocturnal gonadotropin rises associated with the onset of puberty. In addition, the chimpanzee may be the best primate model for investigating the role of LHRH secretion in pubertal onset since the macaques do not appear to respond uniformly to LHRH. When adequate descriptions of longitudinal endocrine changes in the great apes become available, no doubt other appropriate uses of these animals may be suggested.

V. CONCLUSIONS

In almost no other area of reproductive physiology has the lack of an adequate animal model impeded the acquisition of knowledge as it has in the field of human prepubertal endocrine and sexual development. An overriding question in this area, and the subject of much current research, is the nature of the neuroendocrine processes which normally act to block the acquisition of reproductive capacity in man and other primates until the period of maternal dependence and growth has ended. The logical advantages of such a delay are apparent to the casual observer and are underscored by the severe clinical problems associated with precocious sexual development. In the past, studies of sexual development in the higher primates have been chiefly descriptive. This, of course, is in large part due to the fact that most studies have been done in children where manipulative experiments are not possible. Perhaps it is now time to suggest that we move beyond this descriptive phase of research and consider manipulative studies in non-human primates. The available information strongly indicates that many of these studies will have to be performed on great apes. It may be only in these animal models that we can determine the extent to which aberrations of prepubertal endocrine function are associated with problems in fertility and sex-related adult behavior.

Appropriate experimentation in the great apes must recognize both the scarcity of these animals and their unique research potential. Obviously, experiments that would significantly reduce the fertility of our captive chimpanzee, orangutan, or gorilla population would not be appropriate. Fortunately, new experimental techniques utilizing non-invasive or noncompromising procedures have been developed which promise to make a large amount of experimentation in the great apes possible without reducing their ultimate value.

ACKNOWLEDGMENTS

This work was supported in part by the Medical Research Council of Canada (Grant PG-5) and in part by the Federal Department of Research and Technology of the Federal Republic of Germany under a coordinated research program, Ecologic–Toxicologic Effects of Foreign Compounds in Nonhuman Primates and other Laboratory Animals.
The authors also wish to thank Becky Hobson for preparation and typing of this manuscript.

REFERENCES

Arimura, A., Spies, H. G., and Schally, A. V. (1973). *J. Clin. Endocrinol. Metab.* **36,** 372–374.

Boyar, R., Finkelstein, J., Roffwarg, H., Kapen, S., Weitzman, E., and Hellman, L. (1972). *N. Engl. J. Med.* **287,** 582–586.

Cutler, G. B., Jr., Glenn, M., Bush, M., Hodgen, G. D., Graham, C. E., and Loriaux, D. L. (1978). *Endocrinology* **103,** 2112–2118.

Dickerman, Z., Prager-Lewin, R., and Laron, Z. (1976). *Am. J. Dis. Child.* **130,** 634–638.

Dierschke, D. J., Yamiji, T., Karsch, F. J., Weick, R. F., Weiss, G., and Knobil, E. (1973). *Endocrinology* **92,** 1496–1501.

Dierschke, D. J., Weiss, G., and Knobil, E. (1974). *Endocrinology* **94,** 198–206.

Ducharme, J. R., Forest, M. G., DePeretti, E., Sempe, M., Collu, R., and Bertrand, J. (1976). *J. Clin. Endocrinol. Metab.* **42,** 468–476.

Ehara, Y., Ryan, K. S. and Yen, S. S. C. (1972). *Contraception* **6,** 465–478.

Ehrhardt, A. A., and Baker, S. W. (1974). *In* "Sex Differences in Behavior" (R. C. Freidman, R. M. Richart, and R. L. Vandewiele, eds.), pp. 33–51. Wiley, New York.

Elsner, C. W., Illingworth, D. V., de Groot de la Cruz, K., Flickinger, G. L., and Mikhail, G. (1977). *J. Steroid Biochem.* **8,** 151–155.

Faiman, C., Winter, J. S. D., and Grotts, D. (1973). *Proc. Soc. Exp. Biol. Med.* **144,** 952–955.

Fuller, G. B., Winter, J. S. D., Faiman, C., Reyes, F. I., and Hobson, W. C. (1981). *Biol. Reprod.* (in press).

Garnier, P. E., Chaussain, J.-L., Binet, E., Schlumberger, A., and Job, J.-C. (1974). *Acta Endocrinol.* **77,** 422–434.

Gavan, J. A. (1971). *In* "The Chimpanzee," Vol. 4, pp. 46–102. Karger, Basel.

Gijzen, A., and Tijskens, J. (1971). *Int. Zoo Yearb.* **11,** 183–193.

Graham, C. E., Gould, K. G., Collins, D. C., and Preedy, J. R. K. (1979). *Endocrinology* **105,** 269–275.

Grumbach, M. M., Roth, J. C., Kaplan, S. L., and Kelch, R. P. (1974). *In* "Control of Onset of Puberty" (M. M. Grumbach, G. D. Grave, and F. E. Mayer, eds.), pp. 115–166. Wiley, New York.

Herbst, A. L., Ulfelder, H., and Poskanzer, D. C. (1971). *N. Engl. J. Med.* **284,** 878–881.

Hobson, W., and Fuller, G. B. (1977). *Biol. Reprod.* **17,** 294–297.

Hobson, W. C., and Fuller, G. B. (1981). *J. Clin. Endocrinol. Metab.* (in press).

Hobson, W. C., Winter, J. S. D., Reyes, F. I., Faiman, C., and Fuller, G. B. (1981). *Biol. Reprod.* (in press).

Judd, H. L., Parker, D. C., and Yen, S. S. C. (1977). *J. Clin. Endocrinol. Metab.* **44,** 865–869.

Kelch, R. P., Kaplan, S. L., and Grumbach, M. M. (1973). *J. Clin. Invest.* **52,** 1122–1128.

Lanman, J. T. (1957). *Endocrinology* **61,** 684–691.

Martin, D. E., Swenson, R. B., and Collins, D. C. (1977). *Steroids* **29,** 471–481.

McCormack, S. A. (1971). *Endocrinology* **89,** 1171–1177.

Monroe, S. E., Jaffe, R. B., and Midgley, A. R., Jr. (1972). *J. Clin. Endocrinol.* **34,** 343–347.

Noller, K. L. and Kurland, L. T. (1976). *J. Toxicol. Environ. Health Suppl.* **1,** 1–11.

Parker, L., and Odell, W. (1977). *Clin. Res.* **25,** 299A (Abstract).

Penny, R., Olambiwonu, N. O., and Frasier, S. D. (1977). *J. Clin. Endocrinol. Metab.* **45,** 307–311.

Reiter, E. O., Kulin, E. H., and Hamwood, S. T. (1974). *Pediatr. Res.* **8,** 740–745.

Resko, J. A. (1977). *Adv. Sex Hormone Res.* **3,** 139–168.

Reyes, F. I., Boroditsky, R. S., Winter, J. S. D., and Faiman, C. (1974). *J. Clin. Endocrinol. Metab.* **38,** 612–617.

Reyes, F. I., Winter, J. S. D., Faiman, C., and Hobson, W. C. (1975). *Endocrinology* **96,** 1447–1455.

Robinson, J. A., and Bridson, W. E. (1978). *Biol. Reprod.* **19,** 773–778.

Root, A. W., Reiter, E. O., and Duckett, G. E. (1977). *J. Clin. Endocrinol. Metab.* **44,** 909–914.

Shackleton, C. H. L., and Gustafsson, J. A. (1971). *Steroids* **18,** 175–186.

Shackleton, C. H. L. (1973). *J. Steroid Biochem.* **5,** 113–118.

Shah, G. V., Sheth, A. R., Godgil, B. A., and Swamy, X. R. (1976). *Indian J. Exp. Biol.* **14,** 171–174.

Smith, A. H., Butler, T. M., and Pace, N. (1975). *Folia Primatol.* **24,** 29–59.

Townsley, J. D., and Pepe, G. J. (1977). *Acta Endocrinol.* **85,** 415–421.

Wilen, R., and Naftolin, F. (1976). *Biol. Reprod.* **15,** 356–360.

Winter, J. S. D., and Faiman, C. (1972). *J. Clin. Endocrinol. Metab.* **35,** 561–564.

Winter, J. S. D., Faiman, C., Hobson, W. C., Prasad, A. V., and Reyes, F. I. (1975). *J. Clin. Endocrinol. Metab.* **40,** 545–551.

Winter, J. S. D., Hughes, I. A., Reyes, F. I., and Faiman, C. (1976). *J. Clin. Endocrinol. Metab.* **42,** 679–686.

Winter, J. S. D., Ellsworth, L. R., Faiman, C., Reyes, F. I., and Hobson, W. C. (1977). Endocrine Society (Abstract).

Winter, J. S. D., Faiman, C., Reyes, F. I., and Hobson, W. C. (1978). *Clin. Endocrinol. Metab.* **7,** 513–530.

Young, W. C., and Yerkes, R. M. (1943). *Endocrinology* **33,** 121–154.

Chapter 5

THE FEMALE APE GENITAL TRACT AND ITS SECRETIONS

Kenneth G. Gould
David E. Martin

I. INTRODUCTION

The great apes are physiologically and anatomically closely related to human beings. In common with human females and almost all other primate species, the great apes possess a simple unilocular uterus (uterus simplex) located within the pelvis, and two oviducts (or Fallopian tubes), which lie between the uterus and the ovaries. The general

105

Copyright © 1981 by Academic Press, Inc.
All rights of reproduction in any form reserved.
ISBN 0-12-295020-8

relations of pelvic organs in the apes are similar to those in humans (Fig. 1a & b).

Our information on the detailed structure and physiology of the female reproductive organs of the great apes is incomplete, and the comparisons that can be directly made are limited. More data is available in this area on the chimpanzee than the other apes, and for this reason the chimpanzee has been selected as the reference for the purpose of this discussion. Certain specific areas of reproductive tract function, *viz.*, ovarian secretions and menstrual cycle physiology, are discussed elsewhere in this volume and will receive only cursory mention here.

II.   GROSS ANATOMY

The usable literature on gross anatomy of the reproductive organs of the great apes is sparse. Although a fairly large list of publications can be obtained, one of the more impressive features of their content is the degree of repetition among them. This situation is probably a direct consequence of the restricted availability of material for study. The most complete information available on the anatomy of internal reproductive organs is that of Wislocki (1932), whose publication has been much quoted since. The greatest volume of literature was produced in the late nineteenth and early twentieth centuries (e.g., von Bischoff, 1880; Ehlers, 1881; Horiuchi, 1893; Bolk, 1907), but much is of dubious value, with one report (Gratiolet and Alix, 1865) incorrectly describing the uterus of the chimpanzee as bicornuate.

The general arrangement of the internal genitalia of *Pan, Pongo,* and *Gorilla* is similar to that in *Homo* (Figs. 1 and 2), with a unilocular uterus situated dorsal to the bladder, and a cervix of relatively moderate size (compared to lower primates), communicating to the exterior via a dilatable vagina, which is lined with a moderately rugose membrane.

Such information as is available suggests that the uterus of *Gorilla* is larger than that of *Homo*, that the uterus of *Pan* is approximately the same size as *Homo*, and that *Pongo* possesses a uterus considerably smaller than the other species. The increased size of the gorilla uterus has been attributed to extreme development of all the muscular elements. In both *Pan* and *Gorilla* the long axis of the uterus barely deviates from that of the vagina and differs from *Homo* in this regard. In all the apes, as in *Homo*, the cervical canal is relatively straight and unobstructed.

The difference in overall size of the uterus, together with the different orientation of the pelvic skeleton, relative to the spinal column, combine

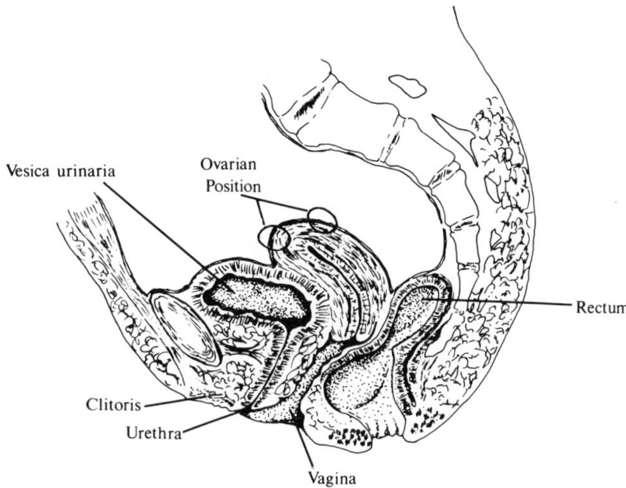

Fig. 1a. Diagram of pelvic organs in *Homo*. Note variations in ovarian position relative to the uterine corpus.

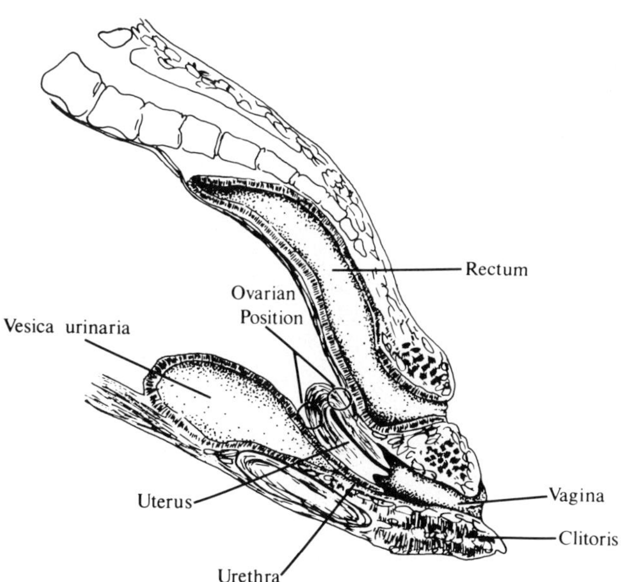

Fig. 1b. Diagram of pelvic organs in *Pan*. Compare the variable ovarian position with that of *Homo*.

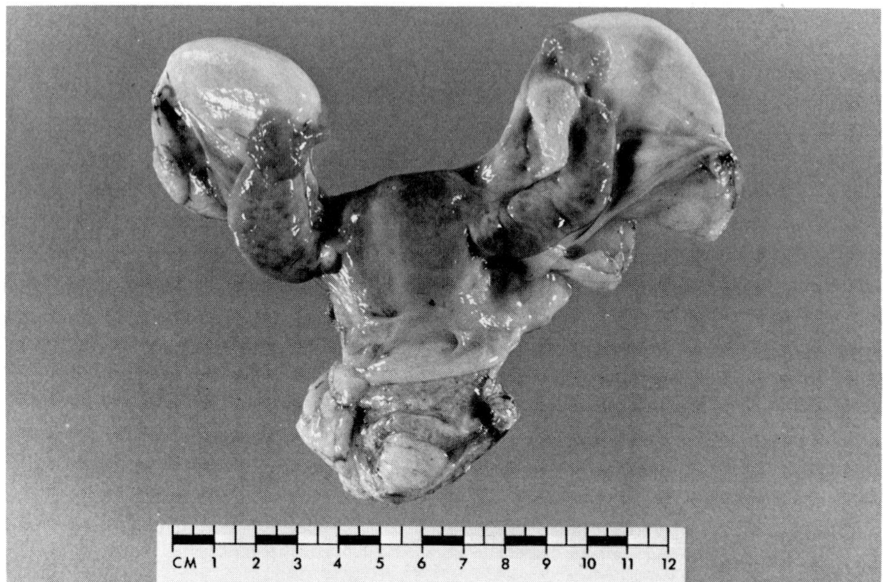

Fig. 2. Genital tract of *Gorilla*. This tract is from a 16-year-old animal.

to place the uterus of *Pan* and *Pongo* deeper in the pelvis, posterior to the fundus of the partially distended bladder rather than dorsal as in the human. This altered relationship is significant when access at laparotomy, or visualization during laparoscopy, are considered. In *Gorilla* the relative size and shape of uterus and pelvis result in a reduction in size of the vesico-uterine and recto-uterine pockets and relatively exposed uterine position.

Differences between the anthropoids and man with respect to gross anatomy of the oviducts and ovaries have been inadequately documented. Recorded dimensions of the ovary of *Pan* and *Gorilla* range from $1.5 \times 0.5 \times 0.2$ cm to $4.5 \times 2.5 \times 2.0$ cm and are within the accepted normal range for *Homo*. Some older reports are based on immature specimens and are unreliable when extrapolated to provide information on the adult condition.

It appears that the oviducts are more flexed in *Pan* than in *Homo* and are approximately 7 cm long. In a single *Gorilla* specimen the extramural portion of the oviducts measured 4.7 and 4.5 cm. The ovaries in *Pan* are positioned dorsocaudal to the uterine corpus, which provides some interference with their ready observation at laparoscopy. This is also the case in *Pongo* and *Gorilla*.

The mesotubarium ovarica, a ligamentous structure between one pole of the ovary and the tip of the oviductal fimbrium, is present and well-developed in *Homo*, but is much reduced or absent in *Pan*. This structure has been the subject of physiological and morphological investigation using scanning electron microscopy and conventional light microscopy in the human (Okamura *et al.*, 1977a, b). Specific functions for this structure have not been thoroughly documented, but a role in movement of the fimbriae during oocyte pickup is postulated (Okamura *et al.*, 1977c). The existence and microscopic structure of any homologous structure in *Pongo* and *Gorilla* has not been documented.

The vagina of an adult female chimpanzee is approximately 12.5 cm long, and increases in diameter in the posterior direction. Dorsal and ventral fornices are present and well-developed, the latter being larger. The anterior vagina, in general, has transverse mucosal folds, and the posterior portion demonstrates shallow longitudinal folding. Marked foldings and local protrusions of the vaginal wall are frequently observed in the anterior portion of the vagina. It is a frequent observation that the cervical canal is eccentrically placed within the ectocervix when visualized with aid of a vaginal speculum. The endocervical mucosa varies in its extent, and the junction with ectocervical mucosa may or may not be visible. There has been no systematic documentation of changes in the position of this junction with menstrual cycle, age, or parity, such as have been recorded in *Homo* (Graham, 1973b).

The vagina does not open directly to the exterior in *Pan*, but via a vestibule formed by the sexual swelling that surrounds and involves the perineal area. The vestibule fluctuates in size corresponding with fluctuations in size of the sexual swelling. There has been debate on the homology of the sexual swelling, and Hill equated it with the labia minora (Ehlers, 1881; Hill, 1958). If this homology is correct, the labia majora are absent in the adult.

The urethra opens ventrally into the vaginal-vestibular canal approximately at the junction of these components. Thus, it can only be visualized by means of a speculum.

The vagina of *Pongo* and *Gorilla* is smaller in all dimensions than in *Pan*, but the general anatomy is similar. There is, however, a less detectable tumescence in *Gorilla* than *Pan* (Nadler, 1975), and no swelling is detectable during the menstrual cycle in *Pongo*.

III. MICROANATOMY

A comprehensive review of the microanatomy of the chimpanzee genital system has been provided by Graham and Bradley (1972). The current discussion will refer to points of comparison between the apes, with illustration drawn from a somewhat restricted data base with respect to *Gorilla* and *Pongo.*

A. Ovary

The adult ovary possesses a smooth surface, with a well-defined germinal epithelium. The ovary of the gorilla demonstrates small papillae with a connective tissue core. These papillae apparently are not the same as the papillae described on the surface of the human ovary during pregnancy (Eckstein, 1958; Saglic, 1938). In an active female many large developing and atretic follicles are present. The incidence of follicular atresia is high, a characteristic shared with other primate species. With increasing maturation a progressive reduction in number of follicles is evident, with concomitant increase in aniline blue-staining fibrous material in the cortex. The germinal epithelium in *Pan* is 5–8 μm high (Fig. 3), and in *Gorilla* 10–12 μm. In *Gorilla* the germinal epithelium is comprised of columnar cells; in *Pan* and *Homo* the cells

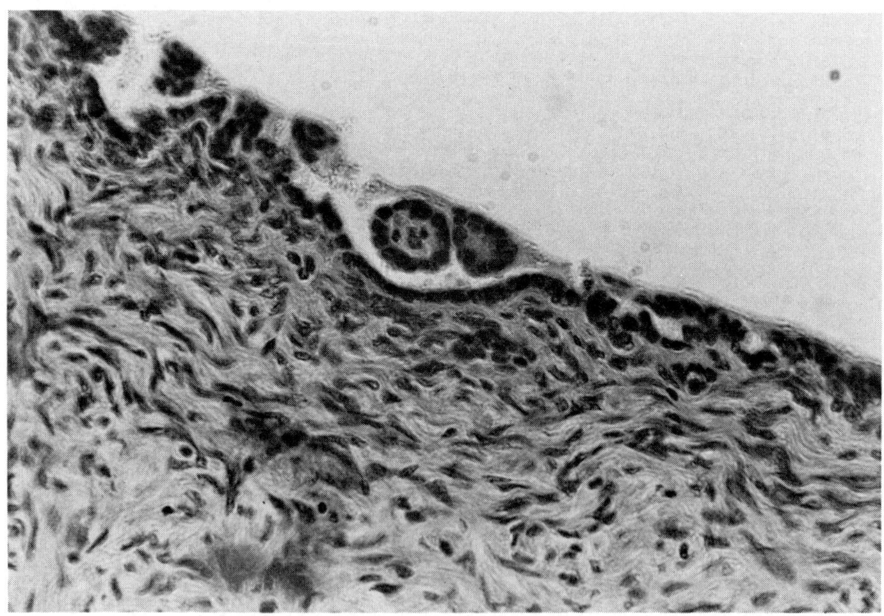

Fig. 3. Germinal epithelium of *Pan.* Note the cuboidal nature of the cells. Picture width 450 μm. H and E.

are cuboidal, while in the specimens of *Pongo* available for study it is a very low cuboidal, approaching the squamous state (Fig. 4).

Graham and Bradley (1972) have provided a discussion on the amount and organization of thecal tissue in the chimpanzee ovary. There is a distinct division between theca interna and externa in the adult, although this is not clearly seen in younger animals.

It has been maintained that there are two generations of interstitial cells in some species of mammal with a primary type derived from the granulosa and germinal cells, the secondary type from the theca interna of developing follicles. Graham and Bradley (1972) concluded that the enlarged interstitial cells seen most commonly in association with atretic follicles of adult *Pan* can be regarded as of thecal origin, in accordance with the study of Mossman *et al.* (1964). A similar conclusion was reached with regard to *Homo* (literature cited by Graham and Bradley, 1972).

Saglic (1938) did not recognize interstitial cells in the orangutan ovary, but hypertrophied thecal cells, apparently associated with atretic follicles, have been identified in our material (Fig. 5).

It appears that hilus cells are found routinely only in *Gorilla* and *Homo*. In *Pan* primary follicles can be detected some distance onto the

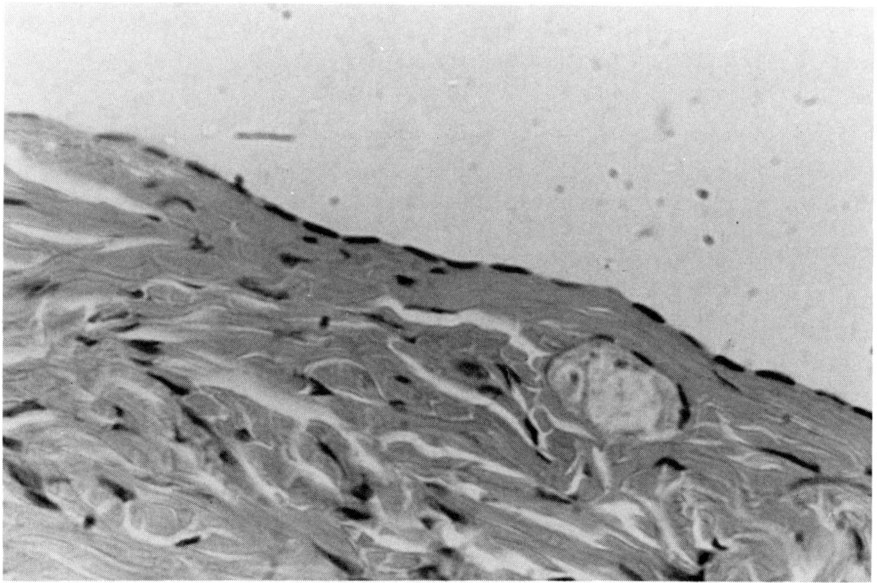

Fig. 4. Germinal epithelium of *Pongo*. Note how low the cells of the epithelium are in this specimen. Picture width 450 μm. H and E.

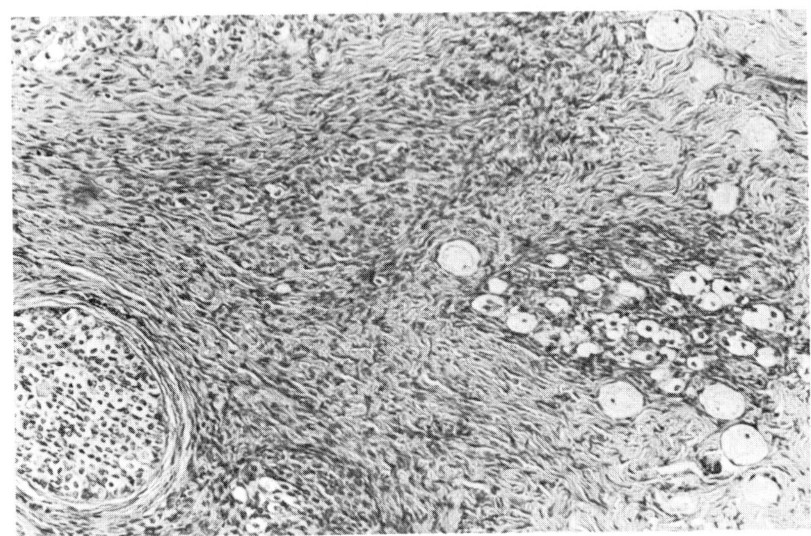

Fig. 5. Ovary of *Pongo,* hypertrophied thecal cells are present in the lower-right quadrant of the micrograph. Picture width 450 μm. H and E.

surface of the ovarian pedicle, and regeneration of ovarian tissue after "oophorectomy" has been reported by Young (1944). This redevelopment is attributable to such remaining tissue.

B. Oviduct

The oviduct provides a means for passage of the ovum from the site of ovulation to the site of implantation. It is far from being a passive conduit, however (Gould, 1974; Dukelow and Riegle, 1974). It provides an active mechanism for oocyte collection and a source of nutrition to the mature ovum, ensures the maintenance of an environment appropriate for fertilization (Gould, 1974), and carries and nourishes the developing embryo prior to its delivery into the uterus.

The basic structure of the oviduct is similar in all hominoids, with a trumpet-shaped fimbriated development at the ovarian end (Fig. 6), and a muscular walled tube leading to the uterus. The histological structure of the oviductal lining demonstrates the presence of peg cells and secretory and ciliated cells. The fimbriated end of the oviduct is most developed in *Pan* and *Pongo,* less so in *Gorilla,* and least in *Homo.* The mesotubarium ovarica is most clearly developed in *Homo.*

In all species there is a preponderance of ciliated cells in the fimbria

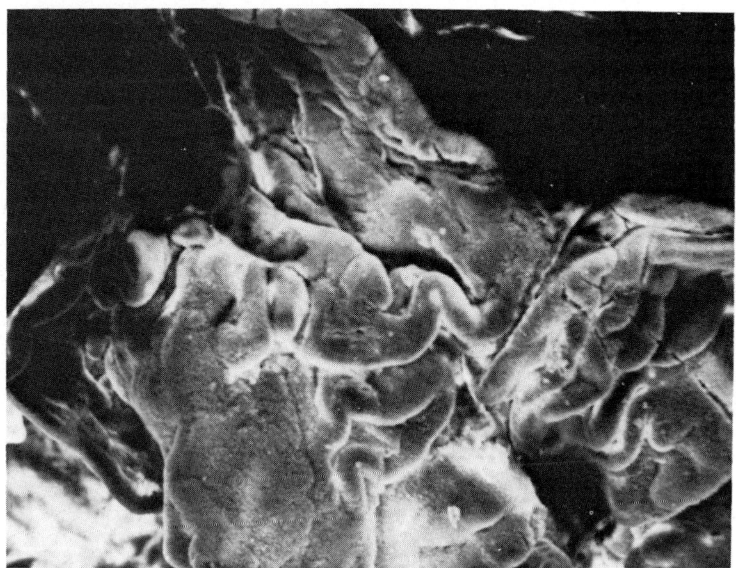

Fig. 6. Oviductal fimbrium of *Pongo*. This scanning electron micrograph shows the complexity of the trumpetlike ostium. Picture width 4 mm.

(Figs. 7 and 8). These cells are assumed to be important in ovum pick-up from the follicle, but 75% of women with Kartagener's syndrome (Afzelius *et al.*, 1975; immotility of cilia as a result of absence of the dynein arms on the ciliary root), are fertile, so ciliary action is not a *sine qua non* in this regard.

The lumen of the oviduct is thrown into multiple longitudinal folds that are gradually reduced toward the uterine end. The muscularis increases in thickness toward the uterus and blends into the myometrium. There is, in the apes, a significant portion of the oviduct within the myometrium; this is the isthmic portion.

There is an extensive adrenergic innervation, as demonstrated by histochemical techniques, which has given rise to the assumption of a sympathetic control of oviduct activity. Ovum transport cannot be influenced, however, by application of sympathomimetic drugs, which raises questions concerning the assumption (Pauerstein and Eddy, 1979).

C. Uterus

The serosal layer of the uterus overlies a narrow longitudinal muscle layer. This layer is subdivided by connective tissue bundles and be-

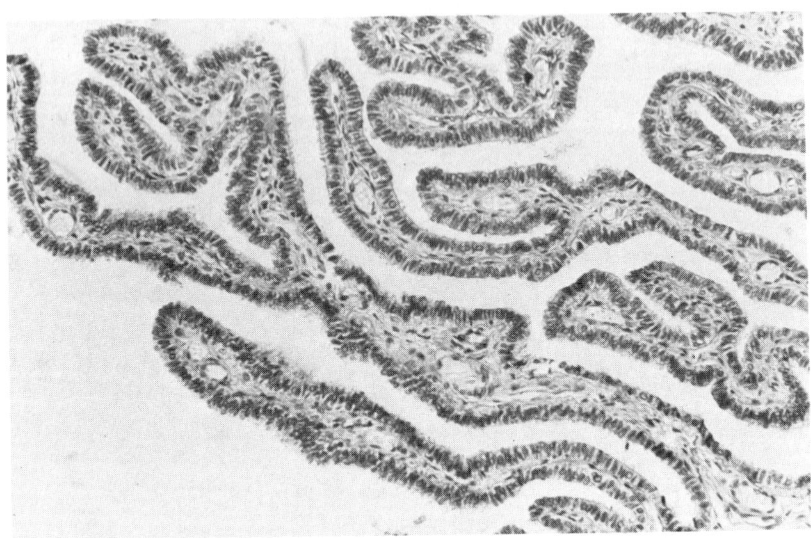

Fig. 7. Fimbrium of *Pan.* Note the complexity of the folding of tissue in this region. Picture width 800 μm. H and E.

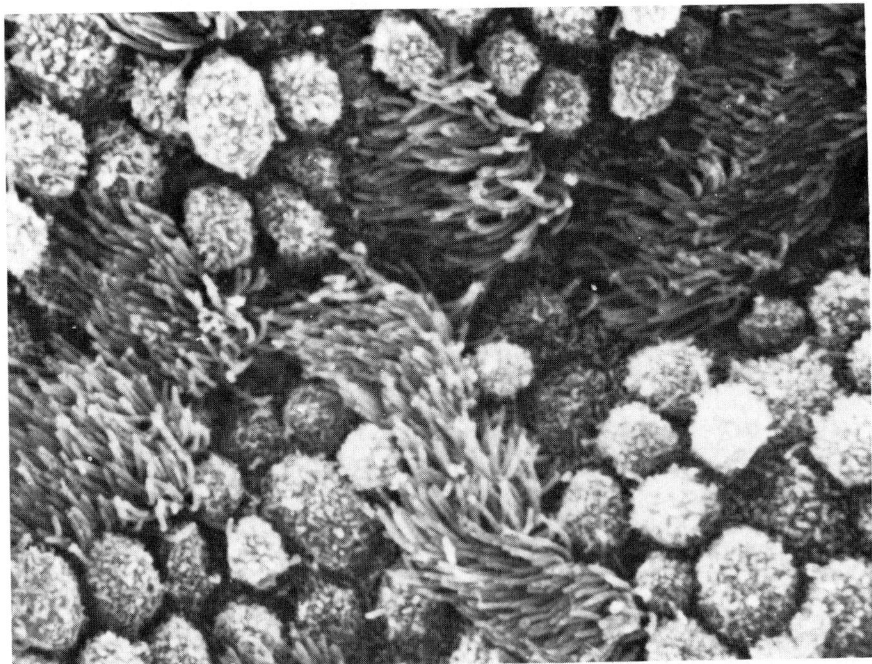

Fig. 8. Scanning electron micrograph of fimbrium of *Pan.* Note the abundance of ciliated cells. Picture width. 30 μm.

comes thinner in older animals. Internal to this layer is a distinct circular muscle layer. There is a variably distinct, age-dependent innermost layer of mixed circular longitudinal and oblique muscle fibers. This basic pattern is seen in *Pan, Pongo, Gorilla,* and *Homo.* It is most obviously developed in all layers in *Gorilla,* and least obviously in *Pongo.*

The endometrium is vascular and composed of stromal and glandular cells. Graham and Bradley (1972) observed patches of stratified squamous epithelium in two out of 16 animals. This suggests a greater incidence than reported in *Homo,* as reviewed by Baggish and Woodruff (1967). Documentation of such an anomaly is not available for *Pongo* or *Gorilla.*

Cyclic changes in the endometrium of *Pan* are similar to those in *Homo,* and the criteria of Hertig and Noyes have been found applicable in dating of the endometrium biopsy (Graham, 1973a), with the exception of intraluminal secretion and edema.

The established pattern of proliferative changes being observed in the follicular phase, and secretory changes occurring in the luteal phase is observed in *Pan.* Increased mitotic activity in glandular and stromal cells, together with pseudostratification of the endometrium, are prominent changes during the follicular phase of the cycle, and tortuosity of the glands could be used as a measure of the degree of proliferation attained (Fig. 9). Changes in the secretory phase closely resemble those

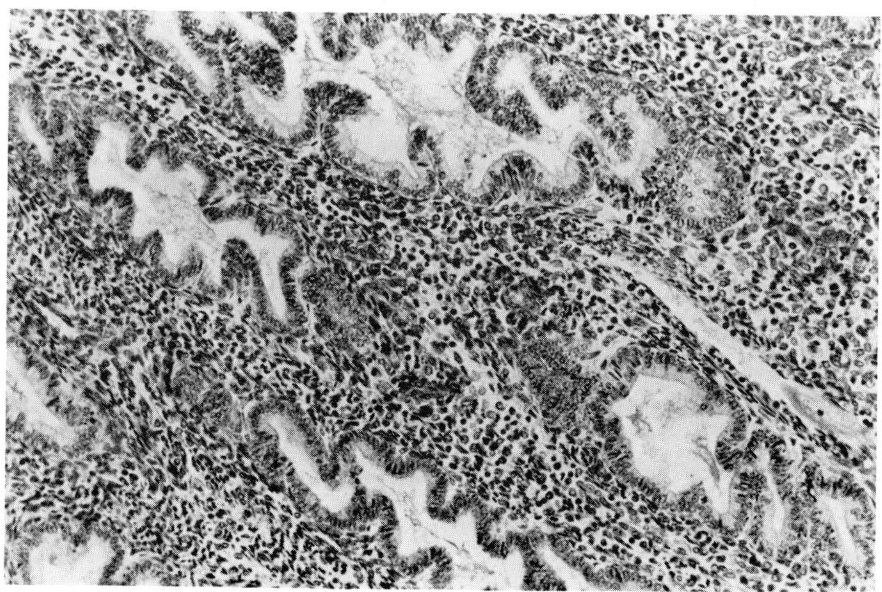

Fig. 9. Endometrium of *Pan,* early luteal phase. Picture width 750 μm. H and E.

in man (Hartman, 1939), with marked development of decidualization and spiral arteries.

Specific dating of the cycle from the endometrial changes is practical only if the changes are related to the time of detumescence (Graham, 1973a). It has been demonstrated that estrogen administration induces proliferative changes, and that combined estrogen and progestin, or progestin alone subsequent to estrogen administration, induces secretory changes. This implies a mechanism of control similar to that in man. *Pan* shares with *Homo* a characteristic endometrial decidualization in every cycle, regardless of coitus or fertilization (Fig. 10).

Studies have been undertaken in the Yerkes laboratory to investigate the role of the luteal estrogen peak and of uterine prostaglandin synthesis in the induction of decidualization in the chimpanzee. Treatment of animals with antiestrogens (Graham *et al.*, 1978) failed to inhibit decidual formation in intact cycling females. Administration of synthetic estrogen (mestranol) in conjunction with chlormadenone acetate, a synthetic progestin, to ovariectomized animals resulted in the formation of typical decidual tissue. Decidualization also could be induced by progestin alone, although the dose required to produce equivalent

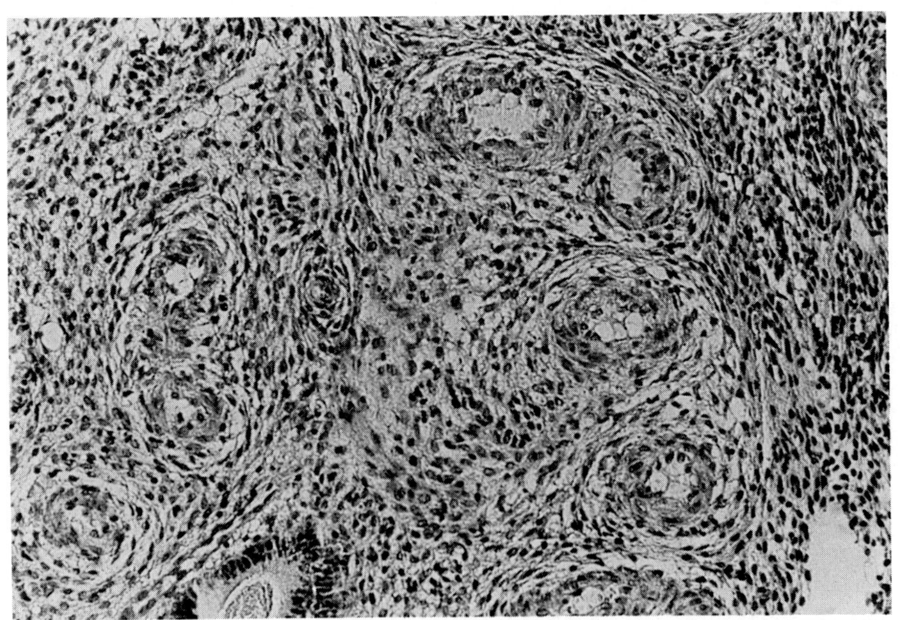

Fig. 10. Endometrial decidualization in *Pan*, Midluteal phase. Picture width 600 μm. H and E.

decidualization was approximately twice that associated with the induction of decidualization in the presence of estrogen (4 mg per os daily *vs.* 2 mg per os daily), and both protocols probably induced a greater circulating level of progestin than normally observed in the early luteal phase. Our interpretation of this data must be that the luteal estrogen peak is not a necessary requisite for the formation of decidual tissue in the chimpanzee. The data do not rule out a role of estrogen that facilitates decidualization.

A study was undertaken to investigate the role of prostaglandin synthesis on spontaneous decidualization by inhibiting prostaglandin activity with indomethacin, either orally or locally administered. Endometrial tissue was obtained by curettage and was subjected to light microscopy and analysis for PGE and PGF levels.

Uterine fluid was obtained by flushing with 2.5 ml saline under standard conditions and was analyzed for PGE and PGF content. We were unable to demonstrate an obligate role for prostaglandins in the luteal phase decidualization. Indomethacin, at 25 or 75 mg *t.i.d.* per os or at 0.5 or 1 mg in oil every other day instilled into the uterus, did not prevent decidual changes as observed with light microscopy of endometrial biopsies. An apparent effect of indomethacin on PG release, as opposed to tissue synthesis, was observed (Table 1), and there is a suggestion that treatment with the oil vehicle alone increased PG release. This effect would be similar to that observed upon uterine stimulation in several nonprimate species. The histological results of the endometrial biopsies were inconclusive, although there was a subjective trend toward reduction of decidualization after indomethacin.

TABLE 1

Uterine prostaglandin levels in the chimpanzee.

	PGE (pg/ml mgww)			PGF (pg/ml mgww)		
	MCT oil	Control	Indomethacin	MCT oil	Control	Indom.
ADA	12,800	10,600	1600	6777	6120	833
LULU	18,824	16,842	7272	8703	9569	1666
MARIA	23,000	8000	2000	6000	6250	1600
ADA[a]	63	56	61	43	48	54
LULU[a]	47	15	9	31	30	27
MARIA[a]	70	9	42	30	6	24
Uterine Cast		10			10	
		40			39	

[a]Tissue levels.

This was not confirmed by statistical analysis of this restricted series. There was a shortened luteal phase under indomethacin treatment when the phase length was compared to that of the preceding control cycle. This alteration was not statistically significant, however, when compared to the length of the luteal phase recorded for the same animal over the previous 24–36 cycles. These data have not yet been analyzed with regard to possible seasonal variations in cycle length and age-related effects. Further experiments need to be conducted to identify the effect, if any, of the reduction in luminal prostaglandins on implantation per se. At this time, our conclusion is that if prostaglandins play a role in luteal decidualization, it is controlled by tissue levels of prostaglandin, for alteration in the amount of PGs released is without effect.

D. Cervix and Vagina

The cervical canal is straight, with a columnar-mucified epithelium in the endocervical region that demonstrates a gradual transition through an area of cuboidal epithelium typical of the ectocervix. Graham and Bradley (1972), from a review of their material, and Graham (1973b) on comparative grounds, argue that this transition is a normal component of the cervix and is not indicative of neoplasia.

The vaginal epithelium is stratified squamous, with the cells overlying a definite basement membrane, which is outside a dense connective tissue layer (Fig. 11).

IV. SECRETIONS OF THE FEMALE TRACT

At this time more is known of the secretions associated with the human female tract than of those in the apes. Polyacrylamide gel electrophoresis of follicular fluid from pre-ovulatory follicles yields a similar pattern in *Homo* and *Pan* (Gould, unpublished data). Other species have not been evaluated. Oviduct fluid composition has been studied in *Macaca* and *Homo*, but not in *Pan, Pongo,* or *Gorilla* (summary in Aitken, 1979). Oviductal fluid provides a nutrient medium for maintenance of the early developing embryo and for spermatozoa at the site of fertilization.

Uterine fluid has not been characterized in any of the apes. No specific studies of uterine fluid changes in the apes during the menstrual cycle or early pregnancy have been reported. The role of prostaglandins,

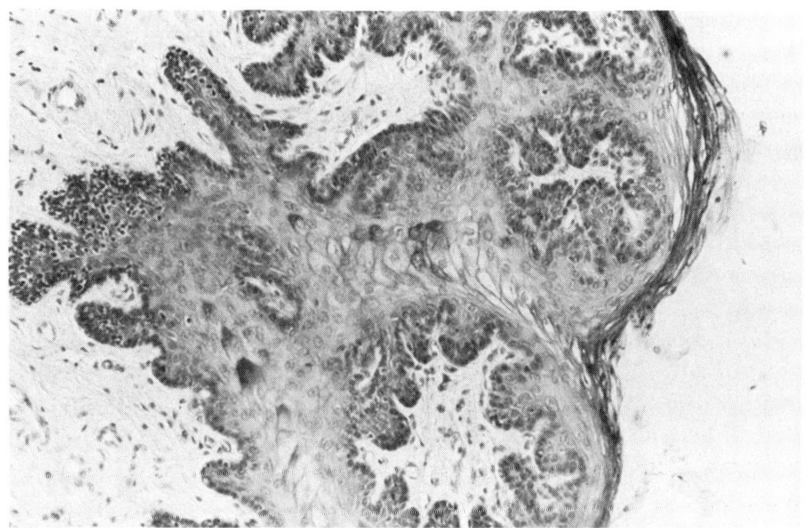

Fig. 11. Vaginal epithelium in *Pan*. Picture width 800 μm. H and E.

which have been detected in the uterine fluid, is not certain. It has been suggested that they influence uterine contractility, and an influence on development of the decidua vera of pregnancy cannot be excluded on the basis of our unpublished experimental data (Singh *et al.*, 1975; Pickles *et al.*, 1965; Aitken, 1979).

With regard to the cervical secretion, more is known, at least with regard to *Pan*. The basic composition of cervical mucus is similar in *Homo* and *Pan* (Table 2). Water is a major component of cervical mucus and varies between 85 and 98% of the total weight. Sperm penetration can be altered and severely reduced if the percentage of water drops below 95% (Bergman, 1953). This water serves as a vehicle for electrolytes, proteins, and the other organic compounds that comprise the cervical plasma.

TABLE 2

Basic components of cervical mucus.

Mucin	Glycoprotein with up to 70% CHO in sidechains
Proteins	Soluble: albumin, IgA, IgG, IgM, lactoferrin enzymes, and inhibitors
Organics	Glucose, maltose, glycogen lipids, and amino acids
Electrolytes	Na, K, Ca, Cl, and trace elements
Cells	Leucocytes, lymphocytes, and so on
Water	

The protein component of cervical mucus is conveniently divided into insoluble and soluble proteins. The insoluble proteins are largely comprised of high molecular weight glycoproteins that contain up to 70% carbohydrate in short chains linked to the peptide backbone. These high molecular weight glycoproteins are rich in serine, threonine, and proline and are reputedly low in sulfur-containing amino acids. Much of the sulfur is present as sulfate in hexosamine adjacent to the peptide chain (André, 1972).

Alterations in electrolyte composition that are associated with changes in the physical properties of cervical mucus have been demonstrated (Fig. 12; Gould and Graham, 1978). The total volume of cervical mucus recovered from *Pan* is lower than in *Homo*, but this cannot be related to specific differences in the histology or gross morphology of the cervix. Unlike the human, in which species there is a single rise in electrolyte

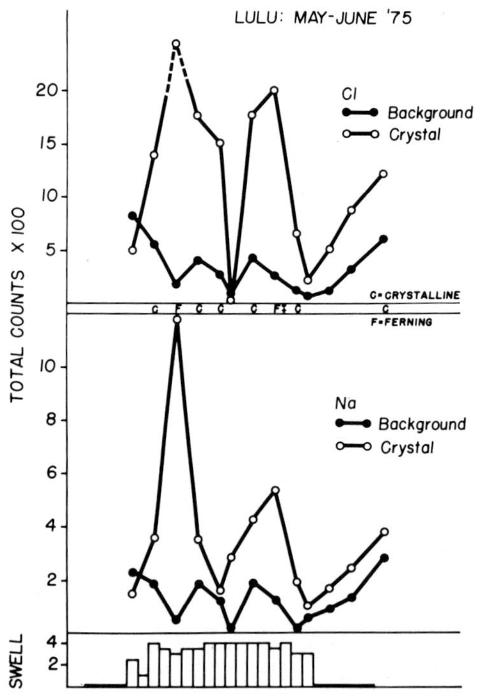

Fig. 12. Relative concentration of Na⁺ and Cl⁻ estimated by energy dispersive x-ray analysis in dried cervical mucus in relation to size of sexual swelling during the menstrual cycle. The concentration of the element measured is approximately proportional to the number of x-rays counted at an energy specific for that element. The graph uses this x-ray count on the Y-axis.

concentration around midcycle, the chimpanzee shows two separate peaks in electrolyte concentration, the latter coinciding with the approximate time of the pre-ovulatory LH peak. The incidence of ferning in this species appears to reflect coincidences of high Cl and Na levels, as lack of coincidence results only in crystallization of cervical mucus, not ferning. The appearance of ferning is probably not required for penetration of the cervix by spermatozoa, as successful artificial insemination has occurred in the presence of nonferning cervical mucus (Martin *et al.*, 1978). Correlation of changes in circulating levels of steroid hormones with changes in electrolyte counts suggests that a difference exists in electrolyte concentration between cervical mucus secreted in response to low and high levels of estrogen stimulation (Gould *et al.*, 1976).

Vaginal fluids and washings have been studied in some primate species with regard to identification of pheromonelike substances. Their existence in the great apes is uncertain. Preliminary investigations into their nature have been conducted, and analysis of vaginal washings has shown the presence of a profile of fatty acids and their derivatives that appears distinct from that of *Homo* (Fox, 1979).

V. FUNCTIONAL ANATOMY

Physical complications affecting the female tract of *Pan* tend to parallel those in *Homo*. In the last few years, it has become evident that the phenomenon of endometrial casting occurs in *Pan* (Solleveld and van Zwieten, 1978). This syndrome carries some similarities to "membranous dysmenorrhea," although it is difficult to demonstrate the pain associated with the human condition. In this situation, subsequent to an apparently normal luteal phase, the entire endometrium is shed in one piece—the "cast" (Fig. 13). Histologically, this material is indistinguishable from an endometrial biopsy recovered just prior to menstruation. The cause of this phenomenon is unknown, but would appear to involve a restriction of circulation to the uterine wall. Subsequent cycles are normal, and pregnancy can occur subsequently in animals that exhibit this phenomenon.

Correlation of the fertile period with alterations in the physical characteristics of cervical mucus are not as precise as those observed in *Homo*. In *Pan*, the cervical secretion is less in volume, and "ferning" is more restricted, with pregnancy occurring in the presence of what would be considered in *Homo* inadequate cervical mucus (Martin *et al.*, 1978).

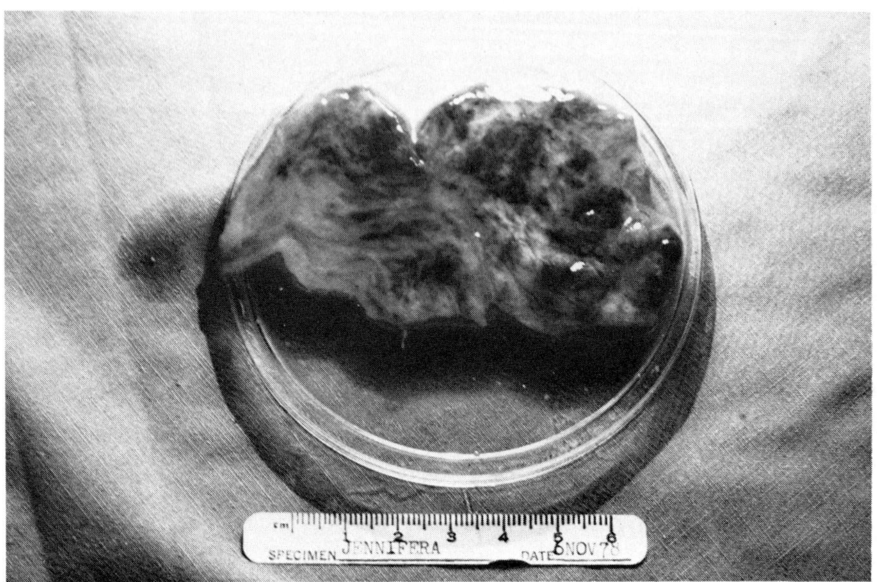

Fig. 13. Endometrial cast from *Pan*. This specimen represents virtually the entire endometrium shed in a single piece.

Functional analysis of the female reproductive tract in the apes reveals further similarities to man. Twinning is relatively rare, although the incidence may be higher than in *Homo* (Martin, Chapter 14, this volume), and appears to be genetically related in the sense that the tendency for twinning can reside in either female or male, and mating of a pair in which both male and female tend to be associated with twinning can result in a very high probability of twinning. Multiple births at the Yerkes Center, for example, have involved the same male in over 60% and the same female in 50% of cases. Frequently, however, twins do not survive. Twin gorillas have been reported on only two occasions.

Ovulation can occur from either ovary, but the presence of a uterus simplex does not permit identification of the ovulating side by fetal position.

As far as can be determined, the mechanism of ovum pickup and transport in the oviduct is similar to that in the human. In any event, the anatomical relationship between fimbria and ovary is similar, and the overall dimensions of the organs approximate those of the human, although there are differences in the degree of development of the ovarian fimbria.

Very little is known of early placentation in the apes (Elder *et al.*, 1938; Hartman, 1939). The gap in this knowledge should be filled if the apes are to be further developed as research models in the area of early development.

VI.   EXPERIMENTAL USE

The reproductive anatomy of the female chimpanzee is so sufficiently similar to that of humans that clinical methodology used in women is largely transferable when *Pan* is the experimental subject. At this time, *Gorilla* and *Pongo* have been less frequently utilized. Modification of human equipment is required for use in *Pan*—vaginal speculae, for example, have to be 15–20 cm long because of the increase in effective vaginal depth associated with the perineal swelling at midcycle. Laparoscopy is a technique of value in the apes, and is subject to the restrictions in the literature pertaining to use in *Homo*. As discussed in Section II above, there are differences in the placement of the ovaries relative to the uterine corpus that make visualization more difficult in *Pan* than in *Homo*. The other apes have not been investigated by us in this regard, but laparoscopy has been used successfully for visualization of the internal genitalia in association with investigation of fertility in the gorilla (Seager, 1979, personal communication).

VII.   CONCLUSIONS

Although there are marked similarities in anatomy and, we suspect, in secretions between the great apes and man, there are also many differences; for example, in overall dimensions, presence and degree of perineal swelling, and oviduct structure. None of the observed differences are sufficiently great as to preclude application of physiological information derived from the apes to man, and vice versa. In no area can we claim to possess a complete understanding of the reproductive anatomy and physiology of the apes, but in some, such as placentation and early development, our data is very meager, and much more is known about the human than the apes. There is opportunity, therefore, for both human and nonhuman science to benefit from continued study in these basic areas.

ACKNOWLEDGMENTS

This work was supported by USPHS Grant RR-00165 from the National Institutes of Health and by Ford Foundation Grant #690-0645B. We are indebted to Dr. Judy Rankin for the prostaglandin assays and wish to acknowledge the typing assistance of Ms. Stephanie Linz.

REFERENCES

Afzelius, B., Eliasson, R., Johnsen, O., Lindholmer, C. (1975). *J. Cell Biol.* **66,** 225–228.
Aitken, R. J. (1979). *J. Reprod. Fertil.* **55,** 247–254.
André, F. (1972). *Carbohyd. Res.* **25,** 395–398.
Baggish, M. S., and Woodruff, J. D. (1967). *Obstet. Gynecol. Surv.* **22,** 69–115.
Bergman, P. (1953). *Fertil. Steril.* **4,** 183.
von Bischoff, T. L. W. (1880). *Akad. Wiss. München Abh.* **13,** 209–267.
Bolk, L. (1907). *Z. Morphol. Anthropol.* **10,** 250–316.
Dukelow, W. R., and Riegle, G. D. (1974). *In* "The Oviduct and Its Functions" (A. D. Johnson and C. W. Foley, eds.), pp. 194–220. Academic Press, New York.
Eckstein, P. (1958). *Primatologia* **3,** 542–629.
Ehlers, E. (1881). *Akad. Wiss. Göttingen Abh.* **28,** 3–77.
Elder, J. H., Hartman, C. G., and Heuser, C. H. (1938). *J. Am. Med. Assoc.* **111,** 1156–1159.
Fox, G. J. (1979). *Am. J. Phys. Anthropol.* **50,** 438.
Gould, K. G. (1974). *In* "The Oviduct and Its Functions" (A. D. Johnson and C. W. Foley, eds.), pp. 271–300. Academic Press, New York.
Gould, K. G., and Graham, C. E. (1978). *Scanning Electron Microsc./1978* **2,** 603–614.
Gould, K. G., Martin, D. E., and Graham, C. E. (1976). *Scanning Electron Microsc./1976* 335–342.
Graham, C. E. (1973a). *Folia Primatol.* **19,** 458–468.
Graham, C. E. (1973b). *In* "Handbook of Physiology" (R. O. Greep and E. G. Astwood, eds.), Vol II, Part 2, pp. 1–24. Waverly Press, Baltimore, Maryland.
Graham, C. E., and Bradley, C. F. (1972). *In* "The Chimpanzee" (G. H. Bourne, ed.), Vol. V, pp. 77–126. Karger, Basel and University Park Press, Baltimore, Maryland.
Graham, C. E., Gould, K. G., Wright, K., and Collins, D. C. (1978). *In* "Recent Advances in Primatology" (D. J. Chivers and E. H. R. Ford, eds.), Vol. IV, pp. 209–211. Academic Press, New York.
Gratiolet, L. T., and Alix, P. H. E. (1865). *Arch. Museum Histoire Natur.* **2,** 244–256.
Hartman, C. G. (1939). *Am. Natur.* **73,** 139–155.
Hill, W. C. O. (1958). *Primatologia* **3,** 630–704.
Horiuchi, K. (1893). *Naturforsch. Gesellsch. Freiburg Breisganu* **7.**
Martin, D. E., Graham, C. E., and Gould, K. G. (1978). *Symp. Zool. Soc. London* No. 43, 249–260.
Mossman, H. W., Koering, M. J., and Ferry, D. (1964). *Am. J. Anat.* **115,** 235–255.
Nadler, R. D. (1975). *Anat. Rec.* **181,** 791–798.
Okamura, H., Harada, O., Morikawa, H., Oshima, M., and Nishimura, T. (1977a). *Acta Obstet. Gynaecol. Jpn. (Engl. Ed.)* **29,** 811–816.
Okamura, H., Morikawa, H., Oshima, M., Man-I, M., and Nishimura, T. (1977b). *Obstet. Gynecol.* **49,** 197–201.

Okamura, H., Morikawa, H., Oshima, M., Man-I, M., and Nishimura, T. (1977c). *Int. J. Fertil.* **22,** 179–183.

Pauerstein, C. J., and Eddy, C. A. (1979). *J. Reprod. Fertil.* **55,** 223–229.

Pickles, V. R., Hall, W. J., Best, F. A., and Smith, G. N. (1965). *J. Obstet. Gynaecol. Br. Commonw.* **72,** 185–192.

Saglic, S. (1938). *Contrib. Embryol. Carnegie Inst.* **27,** 179–189.

Singh, E. J., Baccarini, I. M., and Zuspan, F. P. (1975). *Am. J. Obstet. Gynecol.* **121,** 1003–1006.

Solleveld, H. A., and van Zwieten, M. J. (1978). *J. Med. Primatol.* **7,** 19–25.

Wislocki, G. B. (1932). *Contrib. Embryol. Carnegie Inst.* **23,** 163–204.

Young, W. C. (1944). *Anat. Rec.* **89,** 475–493.

Chapter 6

THE MALE APE GENITAL TRACT AND ITS SECRETIONS

David E. Martin
Kenneth G. Gould

I. INTRODUCTION

There are many basic similarities between man and the great apes concerning both the anatomy and physiology of the male reproductive system. Descended testes with adjoining epididymides allow for sperm production, maturation, and storage. A vas deferens connects these

127

Copyright © 1981 by Academic Press, Inc.
All rights of reproduction in any form reserved.
ISBN 0-12-295020-8

structures to the urinary tract and allows for additional sperm storage. Accessory glands, notably the single prostate gland and paired seminal vesicles, provide additional ingredients to the fluid medium in which sperm are transported. These secretions mix together in the proximal portion of the urethra during ejaculation. A small amount of secretion from the bulbo-urethral glands, released into the distal urethra in advance of the bulk of the ejaculate, may help neutralize the acid environment of the urethra. External genitalia of quite differing size allow for variable vaginal penetration during copulation in the several species.

Variations from this basic plan appear minor, but the information base is very limited. Only the chimpanzee has been studied with some thoroughness in terms of its gross anatomy (Bolk, 1907; Eckstein, 1958; Hill, 1946; Mijsberg, 1923; Wislocki, 1936). Much more restricted descriptions of gorilla and orangutan anatomy are provided in the abovementioned studies, primarily due to unavailability of good specimens for dissection. Microscopic studies of great ape reproductive tract tissues are very few in number. A report by Graham and Bradley (1972) remains the classic paper on the topic, both in terms of literature review and in presentation of original work, but it is restricted to chimpanzees and to gross and light microscopic observations. This chapter will add some recent observations made with scanning electron microscopy, but again it is restricted to the chimpanzee. Genital tract secretions, including gonadal hormones and the ejaculate, have been studied more thoroughly in all of the ape species, allowing a more comparative picture to be presented.

II.   ANATOMY OF THE MALE REPRODUCTIVE TRACT

A.   Testis

A gross view of the chimpanzee testis, with its associated epididymis and pampiniform plexus, is provided in the midsagittal (Fig. 1) and parasagittal (Fig. 2) fresh-frozen sections of an adult male pelvis. The size of the testis varies among the great apes, with no evident relationship to body size. The chimpanzee testis is roughly twice the weight of a human testis, while those of the gorilla and orangutan are roughly 30% less.

As is seen in the human testis, a thick connective tissue capsule called the tunica albuginea covers the testis of each of the great ape species. It has an inner vascular portion termed the tunica vasculosa.

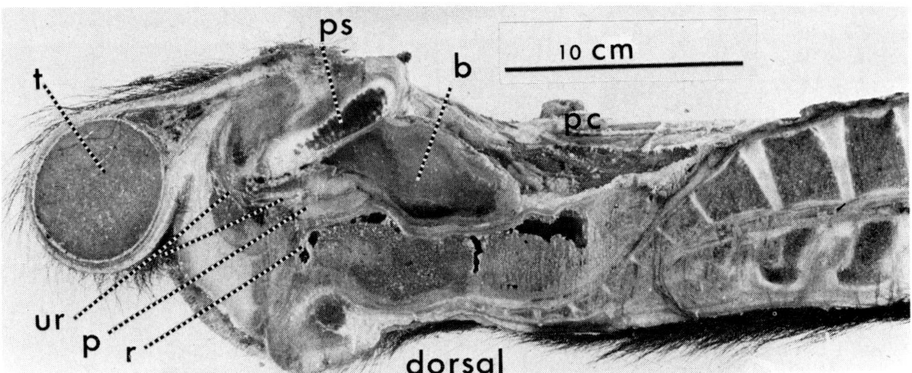

Fig. 1. Gross fresh-frozen midsaggital section through adult male chimpanzee. Abdominal viscera are removed from the peritoneal cavity (**pc**). The urinary bladder (**b**) lies between it and the pubic symphysis (**ps**). The left testis (**t**) was positioned in the midline at the time of freezing. Portions of the rectum (**r**) are clearly visible, as well as the midline prostate gland (**p**), which surrounds the urethra (**ur**).

The entire testis, with tunica albuginea as well as the epididymis, is enclosed by a loose-fitting tunica vaginalis, surrounded in turn by the scrotal wall.

Histologic examination reveals coiled seminiferous tubules with interstitial cells of Leydig and other connective tissue elements in between. The chimpanzee has most of its testicular mass comprised of

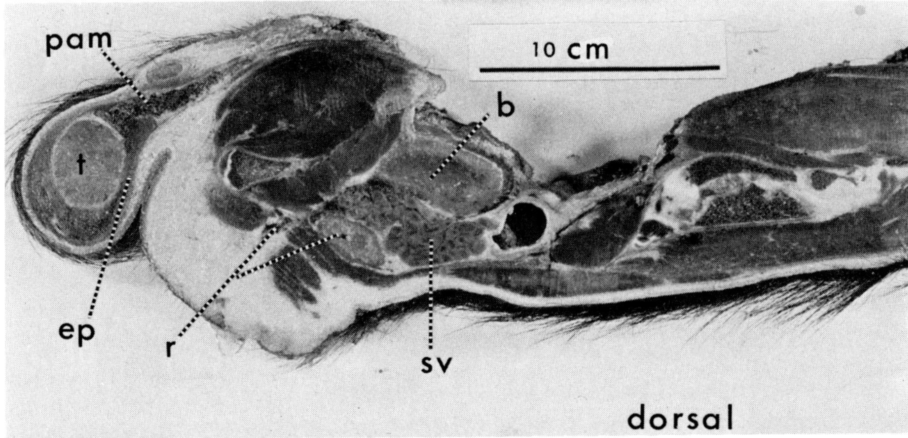

Fig. 2. Gross fresh-frozen parasaggital section through adult male chimpanzee, 2.5 cm away from midline. The pampiniform plexus (**pam**) and epididymis (**ep**) of the left testis (**t**) are visible, as is the large lobulated left seminal vesicle (**sv**) situated between the urinary bladder (**b**) and the rectum (**r**).

seminiferous tubules, with the gorilla testis occupying the opposite end of the spectrum. Not surprisingly, the sperm concentration in a typical chimpanzee ejaculate is many times greater than that seen in the gorilla (Table 1). Graham and Bradley (1972) indicate that the adult stage of testicular development occurs by the age of approximately 10.5 years, as determined by observation of the germinal epithelium. This finding is corroborated by endocrine data in the form of plasma testosterone levels (Martin *et al.*, 1977).

The seminiferous tubules contain the various developmental stages of the spermatogenic epithelium as well as Sertoli cells (Fig. 3). In the chimpanzee the organization of each developmental stage is irregular, which is similar to the human testis (Chowdhury, 1979). That is, instead of any one tubular cross section having only one stage of spermatogenesis visible, several stages may be seen. In contrast, most mammals have a regular pattern of spermatogenesis. The pattern occurring in gorilla and orangutan seminiferous tubules has not been identified. These tubules begin blindly in the outer regions of the testis and course toward a hilum on the medial side. Tubules from each lobe eventually merge into a series of straight tubules (tubuli recti), and these empty into an irregularly channeled network called the rete testis. From here, also in the testicular hilum, a dozen or so coiled tubules, called the efferent ductules, carry the sperm and testicular fluid into a single duct, forming the epididymis.

B. Epididymis

The epididymis is a single tube, very long and convoluted in all of the hominoid species. Estimates of length in the human range between 4 m (Turner and Howards, 1977) and 7 m (Fawcett, 1976, 1979). It forms an organ in its own right, closely apposed to the posterior portion of the testis, since far more happens to sperm here than merely a journey from rete testis to vas deferens.

Not only does maturation of sperm occur, but also a myriad of physiological changes take place that affect their functional ability to fertilize an egg. Structural morphology, cellular metabolism, surface membrane change, motility, and antigenicity are either acquired or developed as sperm pass through the epididymis. White (1973) suggests that while development of normal morphology and motility can occur in the initial regions, termed the caput and corpus, a sojourn in the cauda region is necessary for full fertilizing ability.

The fluid in the epididymal lumen comprises the environment for sperm maturation and storage. Thus, it is certainly reasonable that the

TABLE 1

Comparison of hormone and semen production parameters in the great apes and man.

	Plasma Testosterone (ng/100 ml)			Total Semen Volume (ml)			Sperm Concentration (ml × 10⁶)		
	Mean	*Range*	*N*	*Mean*	*Range*	*N*	*Mean*	*Range*	*N*
Pan troglodytes									
Infant (1–3 yrs)[a]	12	6–20	6	—	—	—	—	—	—
Juvenile (3–7 yrs)[a]	13	3.5–59	20	—	—	—	—	—	—
Adolescent (7–11 yrs)[a]	154	15–323	19	—	—	—	—	—	—
Adult (11 yrs +)[ab]	386	96–680	29	1.1	0.1–2.5	52	548	54–2750	34
Pongo pygmaeus, adult	1003	628–1421	10	1.2	0.2–3.6	23	76	10–165	23
Gorilla gorilla, adult	732	421–1068	5	0.4	0.2–0.8	9	162	29–375	9
Homo sapiens									
Infant (3–8 yrs)[c]		20–23	27	—	—	—	—	—	—
Juvenile (8–12 yrs)[c]		43–163	23	—	—	—	—	—	—
Adolescent (12–15 yrs)[c]		163–300		—	—	—	—	—	—
Adult (15 yrs +)[d]	471	216–726	261	2.5	0.1–12.0	2000	70	0.1–600	2000

N = number of specimens examined; estimates of sperm concentration are determined from the fluid portion of coagulated semen allowed to liquefy for 30 min at 30°C and expressed as numbers per ml fluid.

[a] Martin *et al.* (1977)
[b] Warner *et al.* (1974)
[c] Horst *et al.* (1977)
[d] Smith and Steinberger (1977)

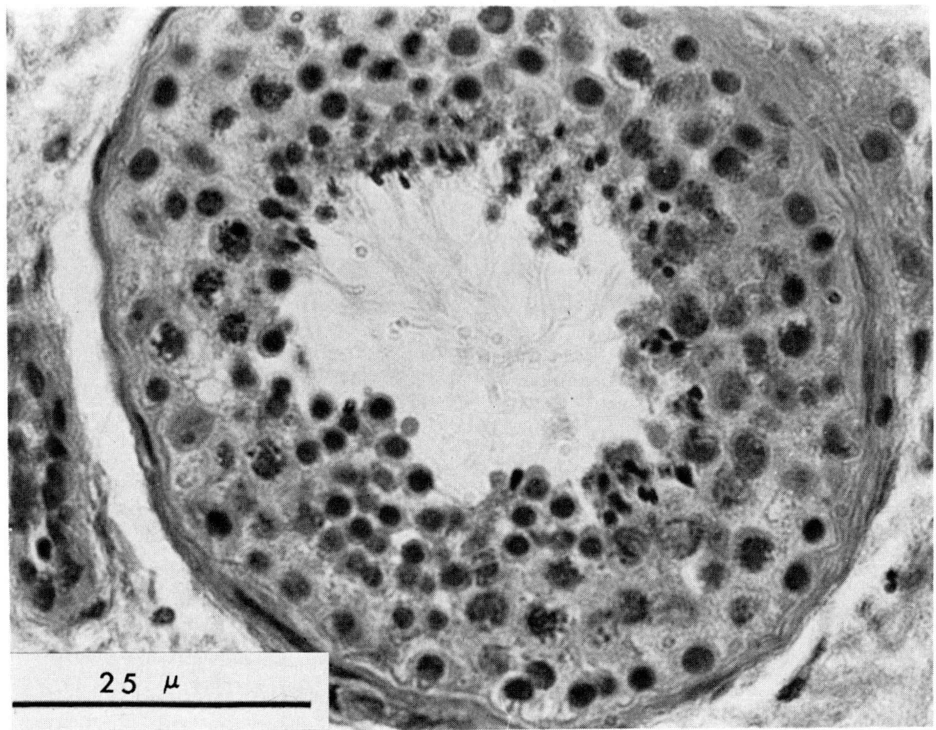

Fig. 3. Light micrograph (LM) of seminiferous tubule from adult chimpanzee. The pattern of spermatogenesis is irregular, with several intermediate cell types visible as well as sperm.

composition of this fluid may play a role in sperm maturation. Much of the fluid produced by the testis is absorbed in the epididymis. Thus, the volume of testicular effluent is greater than the epididymal effluent. Surely there must be chemical components added to or withdrawn from this fluid as it moves through. Even the sperm themselves may contribute to its composition. Hence, analysis of epididymal fluid for its composition will be a frustrating endeavor, and may not shed much light upon the dynamics of epididymal cells themselves. It has long been realized that the epididymis has environmental conditions that are especially conducive to the survival of sperm, giving it a storage as well as a developmental role. The sperm are probably immotile (or nearly so), their metabolism is quiescent, and living conditions are probably aerobic. How this status quo is maintained is just as much a mystery for the great apes as it is for other primates.

There is considerable effort currently being expended to improve and develop contraceptive methods available to the human male. In view of this, it is not surprising to find studies of structure and function of the epididymis proceeding with considerable intensity around the world. A better understanding of basic structure and function will have great benefit in its application to clinical problems in fertility for both physicians and veterinarians.

When viewed microscopically, the chimpanzee caput epididymis is somewhat variable in diameter, with a lining comprised of both ciliated and nonciliated cells. The cilia are in fact stereocilia, which do not have the capacity to initiate directional movement of the luminal contents. They are quite large (Figs. 4, 5, and 6), occasionally giving an impression that they could actually impede sperm movement by entanglement (Fig. 6). The epithelium is very tall, the nonciliated cells have microvilli, and the general appearance of this region is similar to the region termed by Glover and Nicander (1971) as the "initial segment" of the epididymis. Spheres are visible in the lumen (Figs. 5 and 6), which might represent the products of epididymal secretion or the concentrated residue of absorption.

In the corpus epididymis (middle segment), slightly smaller in diameter in the chimpanzee than the caput, there are fewer stereocilia, and the epithelium is not as tall (Fig. 7).

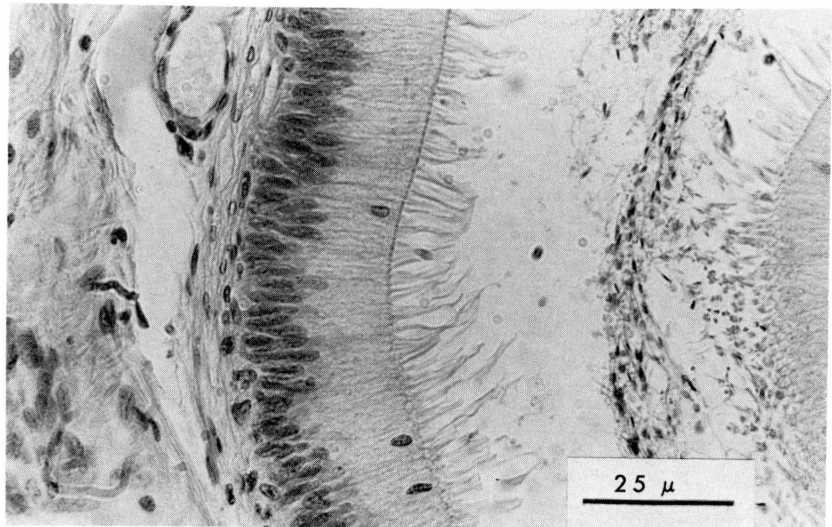

Fig. 4. LM section through caput epididymis from adult chimpanzee. Note the small concentration of sperm in the lumen, the tall columnar epithelium, and large stereocilia.

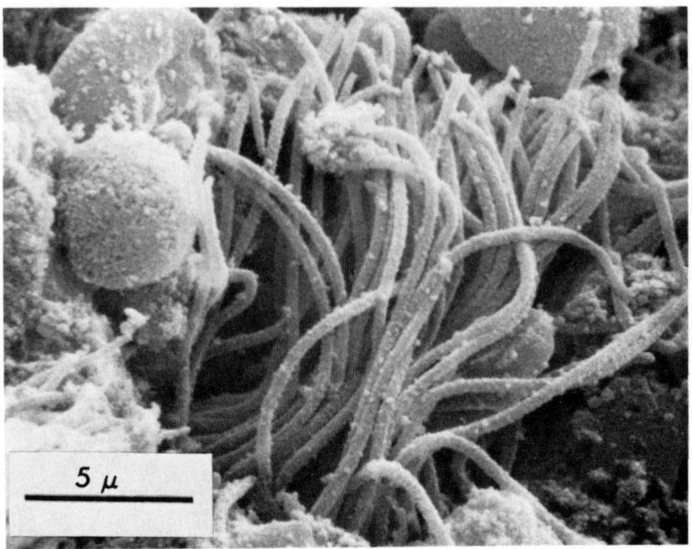

Fig. 5. Scanning electron micrograph (SEM) showing stereocila in interior of caput epididymis from adult chimpanzee. Spheres are probably either epididymal secretion or a concentrated residue from absorption of epididymal fluid.

The net effect of a sizable loss in luminal fluid is seen easily in the cauda epididymis. Here, the tubule lumen is filled with large numbers of sperm, indicating the storage function of this terminal segment (Fig. 9). The duct is larger in diameter than that in the other two regions. Stereocilia are relatively uncommon, and the cilia on other lining cells are shorter than those found elsewhere (Fig. 10).

The entire epididymal duct epithelium is circumscribed by a basement membrane (Figs. 4, 7, and 9). A layer of smooth muscle surrounds this, becoming gradually thicker as one proceeds from the caput to the cauda region.

Chimpanzee epididymal sperm (Figs. 6 and 8) do not appear to have a visible cytoplasmic droplet. This droplet represents redundant cyto-plasm remaining from latter phases of spermiogenesis and is shed following posterior migration along the sperm tail during sperm transit through the epididymis. It is not yet certain whether chimpanzee sperm have a faster rate of maturation within the testis and thus may have shed such redundant cytoplasm prior to entry into the epididymis, or whether the cytoplasmic droplet instead is represented by folds of redundant membrane and cytoplasm that embrace the entire length of the midpiece. Cytoplasmic droplets have not been observed in the

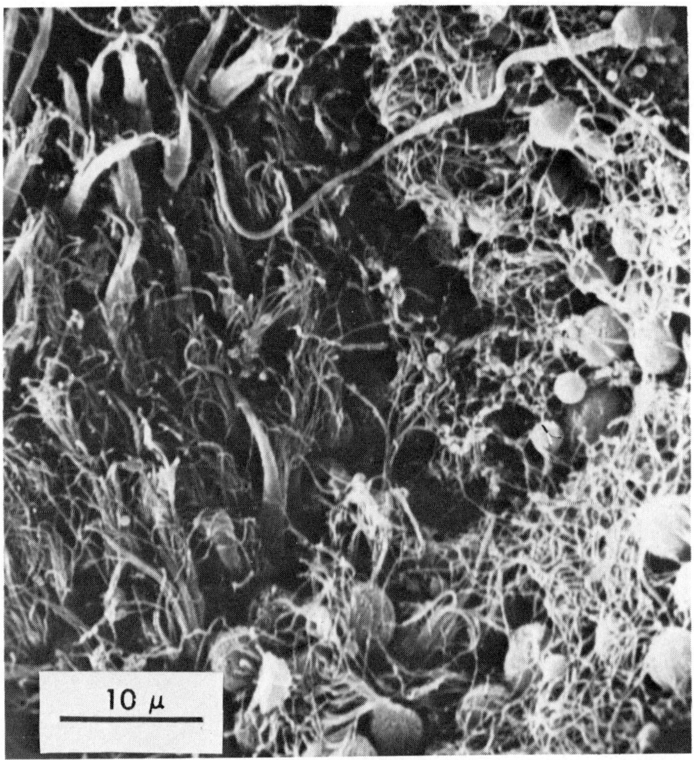

Fig. 6. SEM of lumen of adult chimpanzee caput epididymis. A spermatozoon can be seen among the stereocilia and spheres of tubular fluid material. Other tenacious material, found in this region in addition to spheres, has caused some stereocila to stick together.

orangutan and gorilla sperm collected at ejaculation. In these species, it is not known whether such a droplet exists among immature sperm.

C. Vas Deferens

At the caudal pole of the testis, the epididymal tube straightens, leaves the scrotum, and continues upward as the vas deferens. Coursing into the pelvic cavity, it then extends retroperitoneally, eventually connecting to the urethra in the region of the prostate gland. This junction is very close to the urinary bladder. The tube is quite firm due to an abundance of smooth muscle (one layer circularly arranged placed between two layers longitudinally disposed). The thickness of this muscle is greater, in relation to the luminal diameter, than for any other tubular

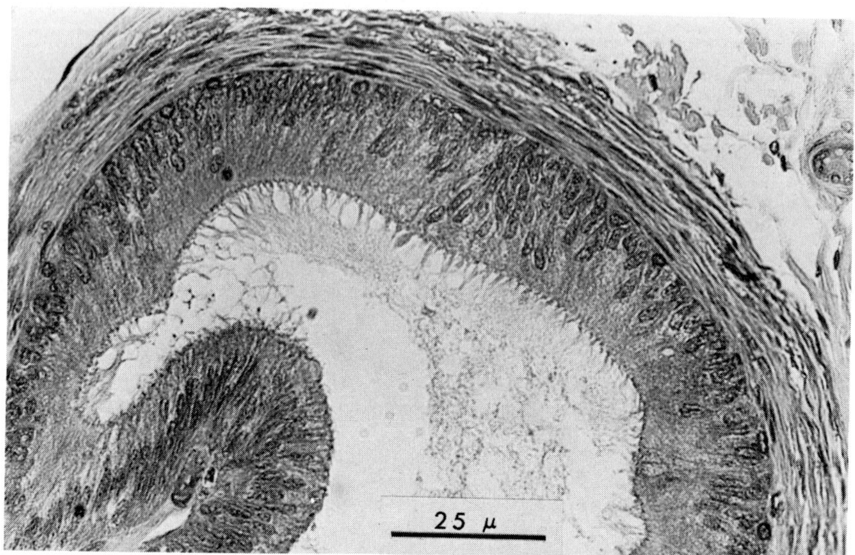

Fig. 7. LM of section through corpus epididymis of adult chimpanzee. The epithelium is not as tall, and the stereocilia are not as large as in the caput region.

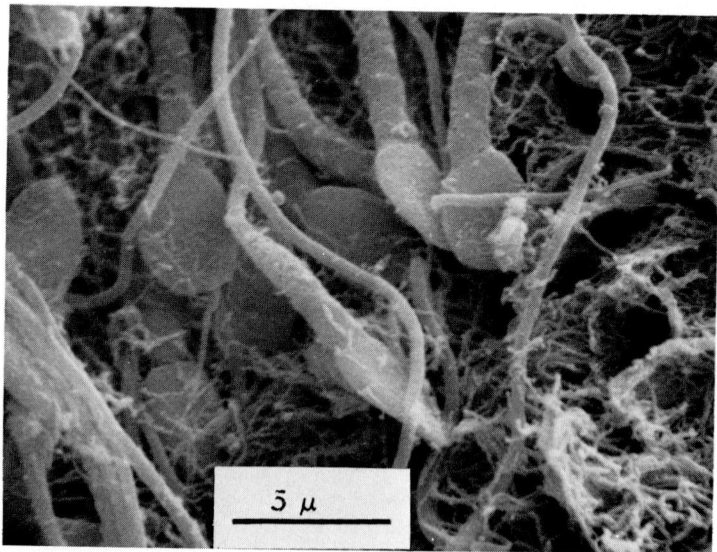

Fig. 8. SEM of lumen of adult chimpanzee corpus epididymis, showing sperm with no cytoplasmic droplets in among stereocilia.

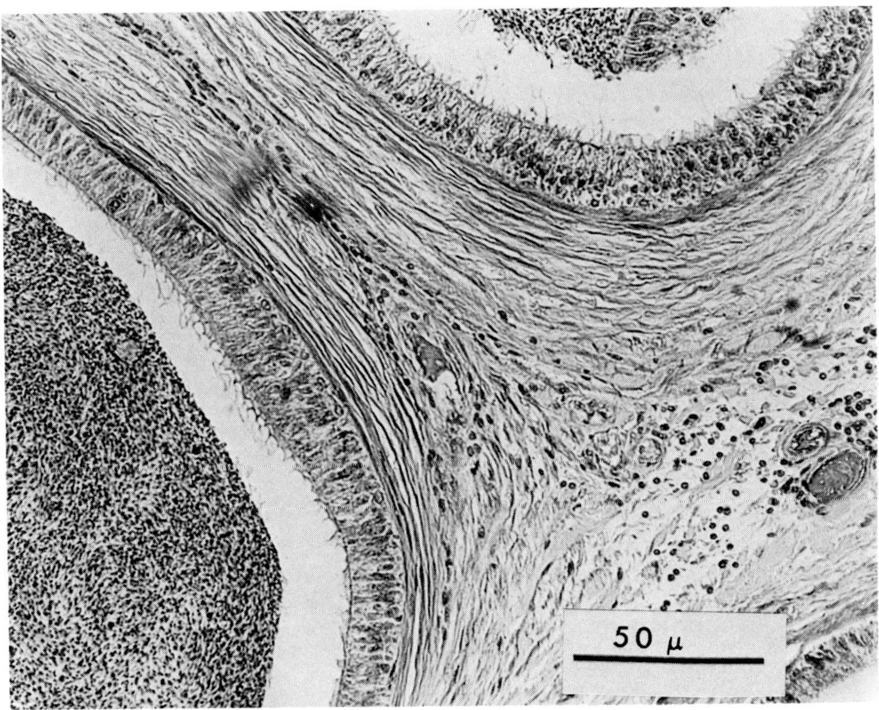

Fig. 9. LM of section through adult chimpanzee cauda epididymis. Note the large concentration of luminal sperm and the continuing convoluted nature of the duct, evidenced by the two cross sections. A basement membrane beneath the ciliated epithelium is clearly visible.

structure in man (Turner and Howards, 1977). The functional value of this muscular endowment is probably related to sperm transport during ejaculation. Lining the tube is a layer of pseudostratified, ciliated columnar cells.

Ultrastructural studies of the human vas deferens indicate an extensive autonomic innervation (predominantly adrenergic) of the muscular layers, and a secretory role for many of the columnar epithelial cells (Popovic *et al.*, 1973). This, in turn, suggests a more physiologically active role for the vas deferens than merely a tube to allow passive movement of sperm from epididymis to urethra.

In man the vas deferens is between 3 and 4 mm in diameter and about 38 cm long. In the apes, comparable dimensions prevail, due to similar anatomical size and structure in this region. At low magnification (Fig. 12) the considerable folding of its lining is seen. This infolding is sufficient to create difficulties with washing all the sperm

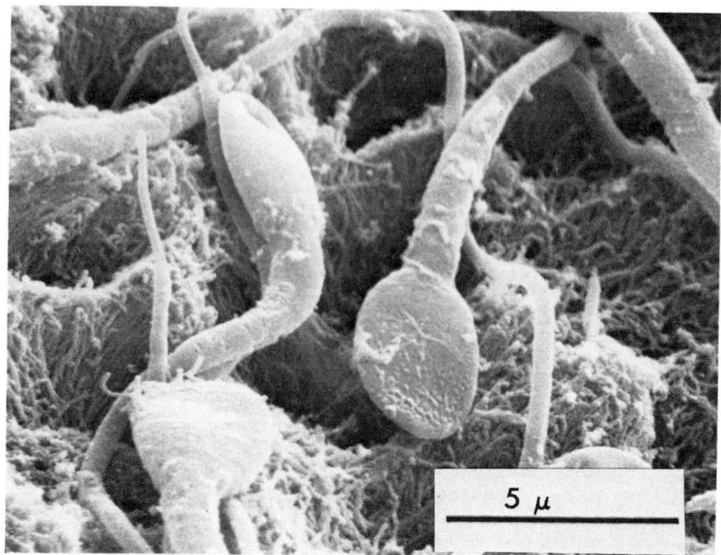

Fig. 10. SEM of lumen of adult chimpanzee cauda epididymis, depicting sperm with no cytoplasmic droplets; ciliated epithelial cells are beneath the sperm.

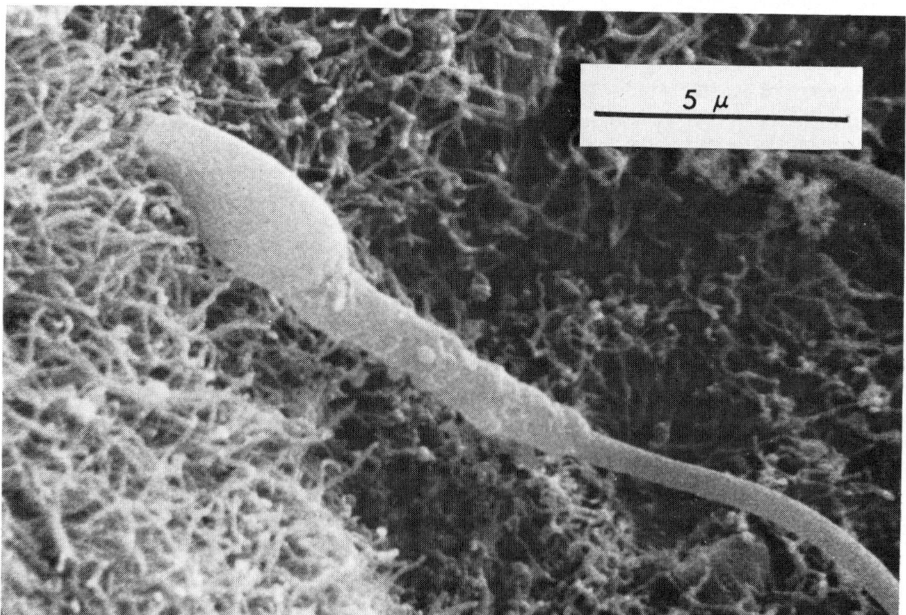

Fig. 11. SEM of lumen of adult chimpanzee vas deferens with its relatively uniform ciliated epithelial lumen. The head, midpiece, and principal piece of the depicted sperm are clearly distinguishable.

Fig. 12. Low-power SEM view of lumen of adult chimpanzee vas deferens, showing the ciliated epithelial lining cells (left) and large numbers of spermatozoa (right).

out of the lumen at the time of tissue fixation for study. In the chimpanzee the individual sperm appear morphologically identical to those in the caput and corpus regions of the epididymis. With the exception of individual protruding cells, the villus lining of the chimpanzee vas deferens is uniform (Figs. 11 and 12). The cilia lining the vas deferens of the chimpanzee are shorter than those observed in man and appear less organized (Fig. 11).

D. Seminal Vesicles

The first of the accessory glandular structures to empty its secretion into the sperm environment are the paired seminal vesicles. When they open into the vas deferens, an ejaculatory duct is formed, which shortly thereafter connects to the urethra. They vary considerably in size among the apes (Short, 1977): slightly smaller in the gorilla than in man, slightly larger in the orangutan than either man or gorilla, and dramatically larger in the chimpanzee.

The glands are generally elongated and lobulated, and filled with small tubules supported by a connective tissue framework. In the chimpanzee the epithelium appears cuboidal (Fig. 13), and it is secretory.

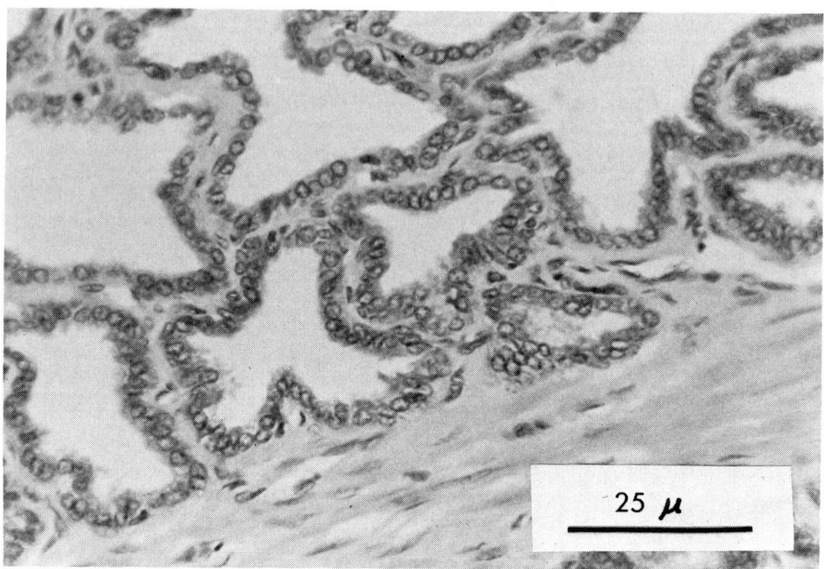

Fig. 13. LM of adult chimpanzee seminal vesicle showing detail of secretory cuboidal epithelium and adjacent acini.

The large quantities of secretory material are difficult to wash from the tubules prior to preservation for viewing with scanning electron microscopy, thereby partially blocking the view of luminal cell surfaces (Fig. 14).

The characteristic secretion products of these glands include fructose and prostaglandins (Eliasson, 1977). This secretion appears to be the last to enter the urethra during ejaculation, at least in man, as demonstrated by measurement of its chemical components in the various portions of split ejaculates (Eliasson, 1963).

Another unidentified secreted substance, which promotes coagulation of semen, is apparently produced by the seminal vesicles. This is evidenced by the finding of Amelar and Hotchkiss (1965) that humans with congenital absence of their vasa deferentia and seminal vesicles produce an ejaculate with no evidence of coagulation. Among the apes, suggestive evidence exists for a similar situation. Chimpanzee semen, upon ejaculation, is almost entirely coagulated, and often will remain so. This ape species has the largest seminal vesicles of all the hominoid species. Gorilla semen is usually liquid, and the seminal vesicles of *Gorilla* are smaller than those of man.

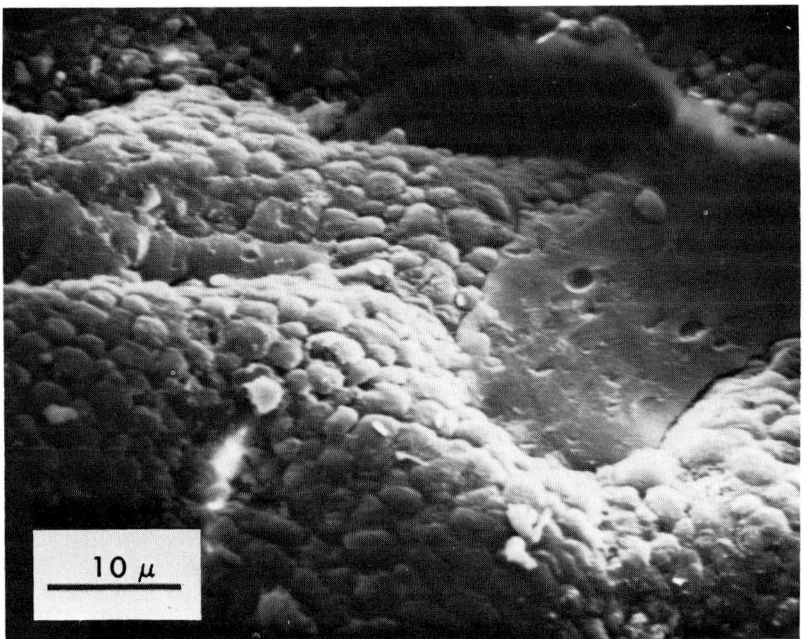

Fig. 14. Low-power SEM view of lumen of seminal vesicle from adult chimpanzee, depicting epithelial lining cells and interspersed tenacious secretory material.

E. Prostate Gland

This is a midline structure, surrounding completely the cephalic portion of the urethra. In man the ducts of the seminal vesicles meet the vasa deferentia prior to penetrating the prostatic tissue, while in the chimpanzee the four ducts course separately through much of the prostate before eventually uniting to form the ejaculatory duct. Graham and Bradley (1972) have described chimpanzee prostatic histology, but no similar studies have been done for the other ape species. The chimpanzee resembles man in the anatomy of the prostate; there are no well-defined cranial and caudal lobes. This is quite different from the situation in the rhesus monkey, which has an active cranial lobe that apparently produces the seminal fluid-coagulating substance.

At least two types of secretory tissue are evident in the chimpanzee prostate gland. One type (Fig. 15) has a low cuboidal epithelium, with large acini often so close to adjacent glands that only a thin connective tissue layer separates them. A second type (Fig. 16) has a tall cuboidal

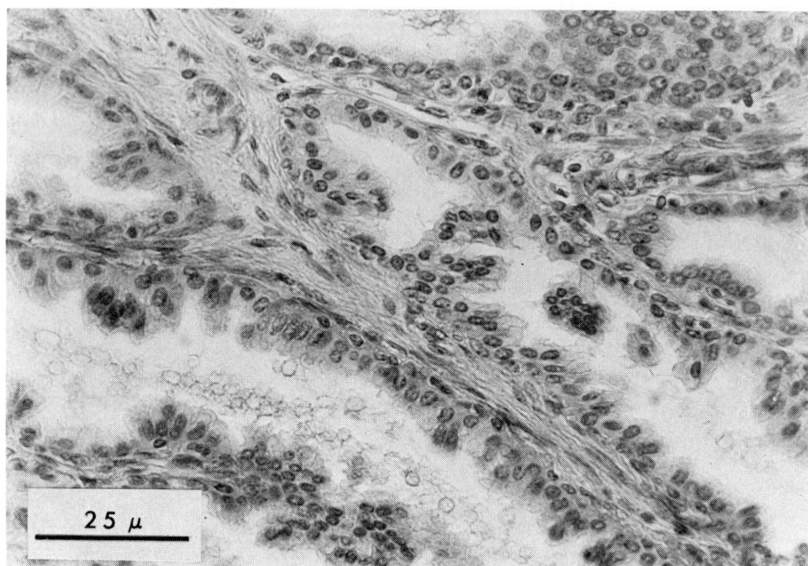

Fig. 15. LM of section through adult chimpanzee prostate gland. Note the low cuboidal epithelium enclosing a large acinus, with only sparse connective tissue separating adjacent acini.

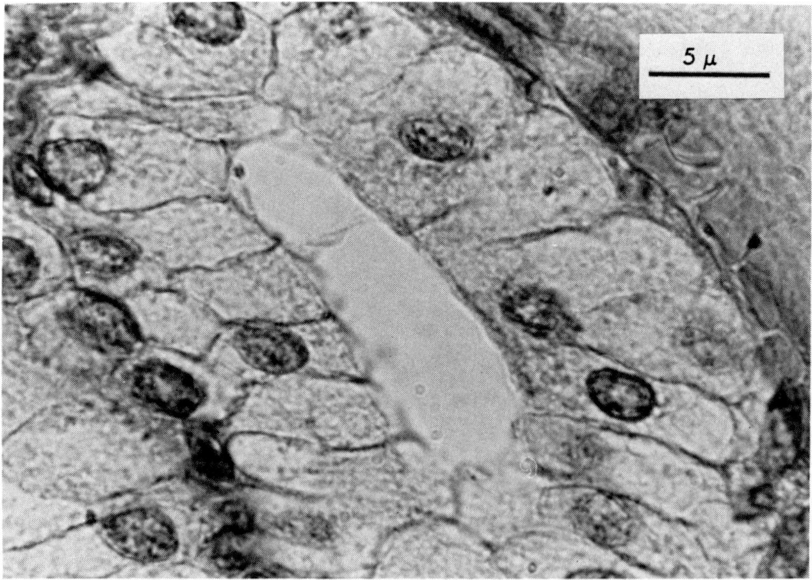

Fig. 16. LM showing detail of prostate gland from adult chimpanzee. Here, a tall secretory cuboidal epithelium surrounds an obvious central collecting duct.

epithelium arranged around an obvious central collecting duct. There are at least four important secretory products of the human prostate gland: magnesium, zinc, acid phosphatase, and uric acid. The presence and importance of these substances in the seminal fluid of the great apes is unknown.

F. Bulbo-urethral Glands

It is known definitely that these glands occur in the chimpanzee (Graham and Bradley, 1972). They have not been observed in the other apes, though the reason for this is probably because of the paucity of available material for dissection and not due to their actual absence. The glands are not large, and the secretion has not been analyzed. The lining epithelium is similar to that of the urethra.

G. Penis

A good review of the literature comparing penile appearance, dimensions, and general gross anatomy is provided by Short (1977). All ape species possess a bony os penis, in contrast to man, who has no such element. The long body hair of the gorilla and orangutan, coupled with small penile size and penile retraction into a prepuce having its orifice flush with the body wall, makes the organ essentially invisible unless searched for when the animal is anesthetized. During erection, penile engorgement in the orangutan is visible, but that of the gorilla remains inconspicuous.

The chimpanzee penis is much larger and lighter in color; hence, it can be seen easily, especially against the background of dark body hair. Short (1977) notes that the large penile dimensions of the chimpanzee are probably related to the considerably increased distance from the external vaginal orifice to the cervical os resulting from enlarged sexual swelling during the ovulatory phase.

The chimpanzee is unique among man and the anthropoid apes, however, in having neither gross nor histological evidence of a glans penis (Graham and Bradley, 1972). There is erectile tissue in the penis of all the apes, however. In the chimpanzee the corpus cavernosum begins essentially where the os penis ends (about 6 mm behind the tip) and continues dorsally the entire length of the penis. Below it is the corpus spongiosum. Detailed studies of penile anatomy in the orangutan and gorilla are lacking.

III. GENITAL TRACT SECRETIONS

A. Hormone Production by the Testis

Testicular androgen production (testosterone, androstenedione, and dehydroepiandrosterone) has been more clearly elucidated in the rhesus monkey and in the human than in the great ape species, where only fragmentary data exist. In the human, it has been clearly demonstrated that fetal testosterone and androstenedione levels are higher than those in the prepubertal male, presumably due to stimulation by human chorionic gonadotropin (Maguelone and Cathiard, 1975). This elevation continues for a short period postpartum, apparently as a result of decreased inhibitory feedback on gonadotropin secretion subsequent to reduction of estrogen levels at birth. Studies are underway in our laboratory to identify similar patterns in the great apes.

Male circulatory testosterone levels are at their lowest values during the prepubertal years (Table 1). During puberty, characterized most precisely for the apes in the chimpanzee as occurring between seven and 10 years of age (Martin *et al.*, 1977), there is a rapid rise in testosterone levels.

During the adult years, blood levels of testosterone in man fluctuate widely due to individual variation, to episodic release into the circulation (Nankin and Troen, 1971), and perhaps also to diurnal and seasonal fluctuations. This wide individual variation is seen in the apes as well and is depicted in Table 1. Diurnal variation has been studied little in the great apes, although it should be a topic of considerable interest in view of the fact that man (Southren and Gordon, 1975) and the rhesus monkey (Perachio *et al.*, 1977) both exhibit such a rhythm, but with the zenith and nadir reversed in the two species. Highest daily values in man appear in the morning and the lowest values in the evening. Evidence for such fluctuation has not been reported for the other testicular androgens.

Seasonal variation in testosterone has been difficult to demonstrate for the apes, in contrast to a definite observed annual rhythm in the rhesus monkey matching the seasonal breeding pattern (Plant *et al.*, 1974). McCormack (1971) provided some evidence for seasonality in the chimpanzee, which was, however, not statistically significant. Our unpublished observations on monthly blood withdrawal from six mature chimpanzee males, collected at the same hour of the day to minimize diurnal fluctuation, indicated a definite trend toward decreased plasma testosterone levels during January, February, and March. The interanimal variability was too large, however, to allow a conclusion that this monthly variation was statistically significant. Seasonal variation in plasma testosterone levels, if any, appears not to result in a seasonality in fertility. During the 13-year history of the Yerkes Primate Center in Atlanta, Georgia, there has been no period or periods during the year when statistically significant differences have occurred in conception rates.

While evidence has appeared to indicate an episodic secretory pattern of testosterone in the human (Naftolin *et al.*, 1973), data from the great apes are inadequate for comparison.

Data are not sufficient to allow correlations to man (August *et al.*, 1972) among plasma testosterone, testicular size, and body size. Table 2 summarizes preliminary and hitherto unpublished data obtained from seven adult male chimpanzees at the Yerkes Primate Research Center. No significant correlations among body size, testis size, and plasma testosterone are evident in this small group of individuals. However, there is a trend for the left testis to occupy more volume than the right.

When mean adult levels of the primary testicular androgen (i.e., testosterone) are compared among the great ape species and with similar values for man (Table 1), it is seen that while the range of values for each species is quite large, the chimpanzee most closely approximates man. Orangutan levels are much higher, and gorilla values are marginally higher than those of man.

B. Ejaculate Volume and Composition

1. *Collection of seminal fluid*

In order for useful evaluation of semen to be accomplished, physiologically normal specimens must be collected using repeatable meth-

TABLE 2

Relationship of age, body weight, testis volume, and plasma testosterone in adult chimpanzees.

Animal	Age at Examination (yrs)	Plasma Testosterone (ng/100 ml)	Body Weight at Examination (kg)	Calculated Rt. Testis Volume (cc)	Calculated Left Testis Volume (cc)
Hal	41	482	59	28.5	38.2
Iyk	14	352	49	29.2	35.5
Jimoh	12	372	46	20.7	24.2
Harv	13	673	46	38.6	38.2
Frank	12	360	43	24.8	31.4
Dobbs	13	451	66	30.9	37.2
John	13	460	57	26.8	20.3
Mean ± S.D.		450 ± 21.2	53 ± 7.2	28.4 ± 5.3	32.1 ± 5.7

odology. The four generally employed techniques for semen collection in primates have been the artificial vagina, automasturbation, electro-ejaculation, and removal of the ejaculate from an impregnated female immediately following copulation. The latter method has been the least successful, but problems remain with the other three methods as well.

With chimpanzees, one group (Fussell *et al.*, 1973) employed an artificial vagina constructed from a plastic tube with inner latex rubber sleeve surrounded by warm water. Another group (Hardin *et al.*, 1975) used a smaller artificial vagina, prepared from automobile radiator hose and latex tubing, which fitted around the penis, but left the head free for insertion into a collection tube. Because of the large size of chimpanzee genitalia, in direct contrast to that of the gorilla and orangutan, the logistics of semen collection by enticing the male to thrust his genitalia through his cage fencing make this a feasible method for collecting physiologically natural specimens. Unfortunately, the risk of working with these animals in such close-up conditions, as well as the difficulties with training technicians and animals to work together productively, add problems to the routine use of this technique. To our knowledge, this technique has not been utilized with success in the gorilla or orangutan, probably due as much to their small genitalia as to their small numbers in appropriate training environments.

In addition to the use of the artificial vagina, which our group at the Yerkes Primate Center has used successfully, we have trained some males to exchange automasturbated semen specimens for a food reward. Since masturbation is a common practice among male chimpanzees, the training has essentially been one of developing technician-animal interaction for safe and successful exchange of specimen for reward.

Electro-ejaculation was first employed for the collection of semen from primates in 1963; in that year Mastroianni and Manson successfully used penile stimulation to electro-ejaculate rhesus, pigtail, and stumptail monkeys. Two years later, Weisbroth and Young (1965) utilized rectal probe electro-ejaculation in the rhesus and pigtail macaques and in Celebes apes. One laboratory reported successful electro-ejaculation of the chimpanzee (Fussell *et al.*, 1967; Roussel and Austin, 1968). In no report was there sufficient discussion of the important electrical parameters involved, such as the amplitude *and* density of electric current, *and* the impedance of the electrode/tissue interface. Without such information it is difficult, if not impossible, for investigators in other laboratories to duplicate the stimulation regimen successfully and to be certain of animal safety.

Because of a great concern for the protection of our animals and a desire for more knowledge about the electrical parameters required for electro-ejaculation of primates, we embarked on a serious development of functional hardware for electrostimulation. Our first report (Warner *et al.*, 1974) demonstrated conclusively that sperm-rich ejaculates could be obtained safely from the great apes.

We constructed a bipolar rectal probe assembly and accurately monitored both stimulus current and voltage. Electrode/tissue interface resistance could then be calculated. This technique ensures a controlled increase in stimulus strength with complete knowledge of the electrical parameters involved in producing erection and emission of the seminal fluid. Penile erection occurs at lower current values, with release of sperm-rich seminal fluid following predictably at somewhat higher values.

The standard 60-Hz electrical main frequency was employed initially, but recently we evaluated the efficacy of frequencies between 10 and 100 Hz. Our results indicate that a frequency between 20 and 25 Hz will elicit erection and seminal emissions at lower current densities than other frequencies in the 10 to 100 Hz range. Also, more of the seminal emissions are forcibly expelled, suggesting a closer approach to the initiation of normal physiologic reflex mechanisms.

Another means for achieving emission without increasing the stimulation current involves the energization of adjacent electrode pairs (each 90° apart), instead of opposite electrode pairs (180° apart). This configuration imparts greater directionality to the stimulation current. We have now collected semen from 27 species of primates, including all of the great apes (Gould *et al.*, 1978).

2. Ejaculate Analysis

Short (1977) advanced the hypothesis that a distinct relationship exists among the great apes concerning ejaculate volume, sperm production, gonadal size, and copulatory frequency. For example, the gorilla, with the lowest copulatory frequency of any of the apes, has a smaller gonadal size and smaller ejaculate volume than does the orangutan. The chimpanzee, however, has the largest gonadal size, highest copulatory frequency, and highest ejaculate sperm concentration of any of the great apes. Short (1977) also provided a small number of individual testis weights for the great ape species to support his contention. Data from our laboratory concerning gonadal size (Table 2), as well as ejaculate volume and sperm concentration (Table 1), as determined through collection by rectal probe electro-ejaculation (Warner *et al.*, 1974; Gould *et al.*, 1978) also contribute favorable evidence for this hypothesis.

A problem of considerable importance in evaluation of primate semen is dissolution of the coagulum, in which varying numbers of sperm are embedded. This coagulation renders difficult not only determination of sperm concentration, but also analysis of the various biochemical constituents of the accessory glands. In human semen, liquefaction generally occurs within 20 minutes following ejaculation. Gorilla semen liquefies more rapidly than semen from the orangutan and chimpanzee; however, it is common to find specimens from all three species that liquefy very little even following an hour of incubation at 37°C.

It is likely that both the prostate and seminal vesicle secretions contribute to semen coagulation. This is evidenced by Amelar's finding (1962) in humans with congenital absence of the vasa deferentia and seminal vesicles that the ejaculate is entirely fluid. Also, in the rhesus monkey, removal of the prostate gland, especially the caudal portion, results in production of liquid semen (Greer *et al.*, 1968).

Many chemical agents have been utilized in an attempt to liquefy primate semen quickly and without diminishing sperm motility or fertilizing ability. These agents include pancreatic deoxyibonuclease (Gersh, 1970), α-amylase (Bunge and Sherman, 1954; Wilson and Bunge, 1975), Alevaire, a tertiary octylphenolformaldehyde polymer with mucolytic properties (Amelar, 1962), chymotrypsin (Hoskins and Patterson, 1967), trypsin and chymotrypsin (Roussel and Austin, 1967a, b), and trypsin and lipase (Roussel and Austin, 1968).

Of these studies, only the last two involved coagulum from the great apes. In the former study, a 1% solution of either α-chymotrypsin or trypsin liquefied 75% of chimpanzee coagulum within 30 minutes at room temperature, and 100% of the coagulum within one hour. Temporary storage at 5°C (24 hours) indicated no loss in motility with the trypsinized specimens when compared to spontaneously liquefied specimens, but chymotrypsinized specimens showed reduced motility. The fertilizing ability of the trypsinized specimens was not tested by their use in insemination. Our laboratory has been unsuccessful at discovering either a solution of these enzymes or other enzymes that will allow routine coagulum dissolution with maintenance of normal sperm motility.

Little is known about the actual mechanism of semen coagulation, its autolysis, or its ultrastructure. Scanning electron microscopy (SEM) has been used to characterize some of the features of coagulum undergoing spontaneous liquefaction in human seman (Zaneveld *et al.*, 1974) and in other primates (Martin and Gould, 1975; Plate 1).

Coagulum retrieval from the vagina several hours following copulation indicates that complete liquefaction does not occur here. However, adequate numbers of sperm must be released for migration into the female reproductive tract and subsequent fertilization of the egg. Data are needed on two specific points, namely: (1) the duration of release of motile sperm from the site of deposition and (2) the rate of migration of released sperm in the female reproductive tract. This information will be important in determining whether inseminations should be intravaginal or intrauterine, and in estimating the time required for sperm to reach the site of fertilization. If the suggestion of White *et al.* (1977) is correct (namely, that where a cervical sperm reservoir is very quickly established from a concentrated ejaculate, a copulatory plugging should not be needed) then the presence of such a viscous coagulum in the great ape species appears enigmatic.

The potential use of collected semen for artificial insemination has made it desirable to record objective parameters of semen fertility. Many parameters are routinely measured that involve physical aspects of the specimen (volume, initial concentration and motility of sperm, numbers of morphologically abnormal sperm, percent live and dead sperm using vital staining, resistance to temperature change, and resistance to dilution) and metabolic aspects (oxygen uptake, rate of fructolysis, and enzymatic activity). The unequivocal superiority of any single parameter for determining potential fertility of a given semen specimen has not been determined, even for the large domesticated species where artificial insemination is practiced on a large scale. This is explained by the fact that no one of these parameters directly measures the fertilizing ability of a spermatozoon, but only provides an indication of quantity, mobility, or response to environmental changes. Measurements of physical activity are generally emphasized in preference to biochemical enzymatic parameters, in part because of simplicity of determination, but also because of the lack of conclusive evidence that measurements of specific aspects of metabolism would be of distinctly greater value than physical measurements.

In our evaluation of the ejaculates obtained from the great apes, we have made a series of measurements of physical parameters and have established a set of baseline data to utilize in our work. There is considerable variation from one specimen to another, even after a period of several days' abstinence. This is true of all species of primates that have been examined thus far, including the human (Freund, 1968; Smith and Steinberger, 1977). Masturbatory practices among the chimpanzees, and perhaps among the other apes as well, make it difficult to ensure that specimens collected have been collected after a predictable period of ejaculatory abstinence.

After specimens are collected, they are transferred immediately to

                                    David E. Martin and Kenneth G. Gould

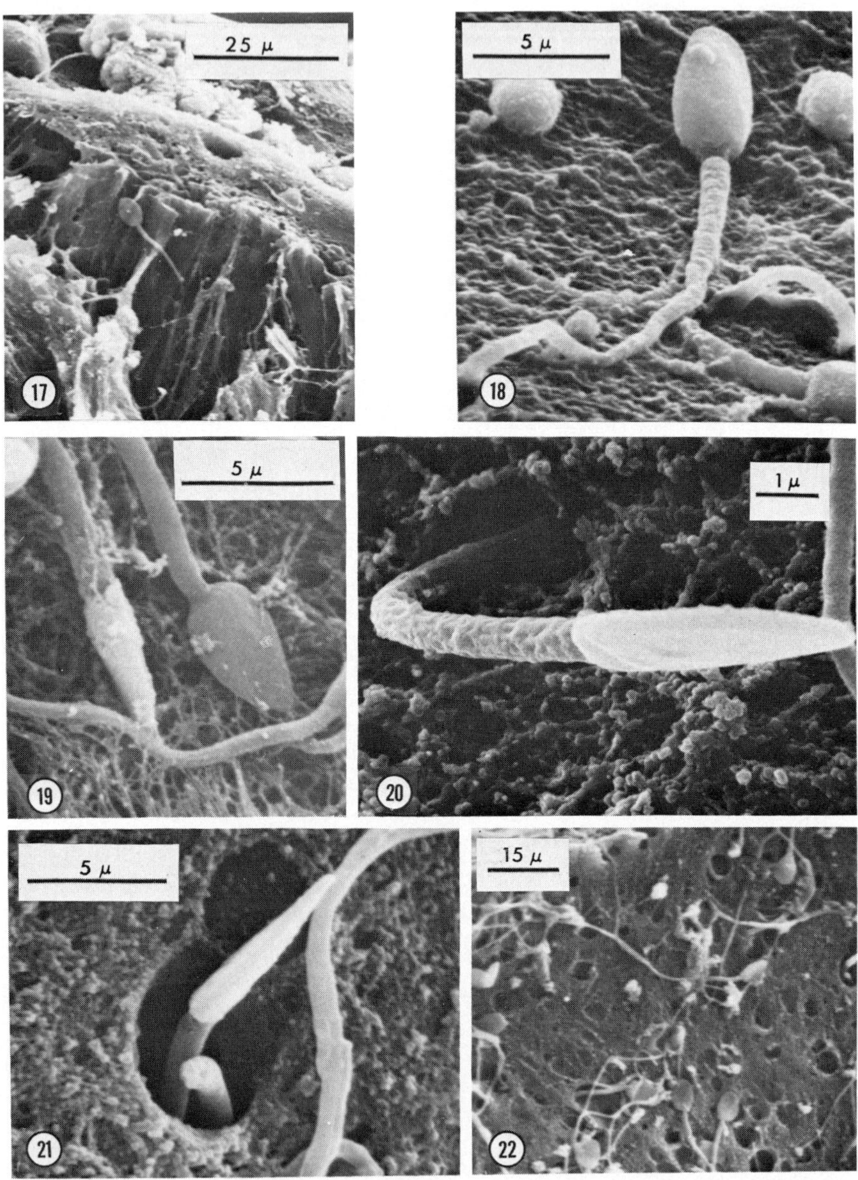

Plate 1 (Figs. 17–22) Ultrastructure of the great ape ejaculate, viewed using scanning electron microscopy (SEM). The proportions of coagulum, fluid, and sperm in the semen at ejaculation is variable among primate species, but in the great apes virtually the entire seminal mass is coagulated as soon as, and perhaps prior to, exit from the urethra. This is also observed among the baboons (*Papio*). Upon ejaculation, the coagulum has a

a 37°C environment and kept in a closed container for 30 minutes; often this is sufficient for considerable coagulum liquefaction to occur. Three types of evaluative procedures are then employed to describe the specimen. First, a general description of physical characteristics is made, including volume, percentages of fluid and coagulum, and pH of the fluid portion. Second, the sperm concentration per ml of fluid is determined using a hemocytometer. Third, an objective estimate is made of the percentages of motile and live sperm in the liquefied portion, using, respectively, visual counting and vital staining. Stains employing eosin B as the cell stain, with either nigrosin, methyl blue, or aniline blue as background stain have proven acceptable for live/dead smears, provided both specimen and stain are at similar temperatures, pH, and osmolarity (Martin and Davidson, 1976). Sperm that are live at the time of application of the stain exclude eosin B from adsorption onto their cell membrane or from penetration into their interior, and hence appear unstained. In addition to indicating the percent of live and dead sperm, these smears are useful for indicating the presence of abnormal sperm, e.g., those with bent midpieces and looped tails, and sperm with cytoplasmic droplets.

Once these evaluative data have been examined, a decision is made concerning the appropriateness of the specimen in question for immediate artificial insemination or for cryopreservations for further study. Recently, it has become possible in our laboratory to perform additional biochemical evaluation of the major accessory gland constituents in seminal fluid collected from the same chimpanzees by automasturbation and by rectal probe electrostimulation. It is thus possible to use an animal as its own control to more accurately discern whether

---

complex structure (Fig. 17, *Papio cynocephalus* and Fig. 18, *Pan paniscus*), appearing as a dense, fibrous meshwork. Sperm are effectively trapped and unable to move unless some liquefaction or dissolution occurs. Incubation of coagulum at 37°C in a covered container for 30 minutes results in increasing numbers of sperm on its surface (Fig. 19, *Pan troglodytes*). The spaces visible in the fibrous meshwork suggest that some liquefaction has begun, but the interstices are not large enough to permit sperm passage. Whether liberation of sperm from the coagulum is accomplished by a combination of their own motility and enzymatic activity, by autolysis of coagulum, or both, is presently unclear. Figures 20 (*Pongo pygmaeus*) and 21 (*Papio cynocephalus*) indicate that both sperm motility and autolysis may be responsible, since the space through which these sperm are moving appears larger than,   but similar in form to, other spaces in the coagulum. Figure 22 (*Papio cynocephalus*) is a lower power view of an area surrounding and including Fig. 21; it depicts several other holes in the coagulum through which sperm could have passed.

(Figures 17, 18, and 19 reprinted from Martin and Gould, 1975; reproduced with permission of the *Journal of Reproductive Medicine*.)

ejaculates produced electrically do in fact differ from those produced naturally.

3. Sperm Structure

We are indebted to Retzius for the first definitive descriptions of great ape sperm structure. Beginning with a report on the orangutan (Retzius, 1910a, b) and continuing with the chimpanzee (Retzius, 1911, 1912, 1914) and gorilla (Retzius, 1913), the meticulous detail and clarity of his sketches have challenged scanning electron microscopists of the 1970s to discover anything new. A basic picture of the surface and three-dimensional structure of great ape sperm has developed from recent scanning electron microscopic investigations (Martin *et al.*, 1975; Gould *et al.*, 1975; Plate 2). Detailed comparative investigation of ape sperm morphology utilizing transmission electron microscopy has not been published. Table 3 summarizes basic dimensional data for great ape sperm, and allows comparison to man and the rhesus monkey.

Sperm from *Gorilla gorilla* and *Homo sapiens* are uniquely characterized among primate sperm thus far examined by the presence of a marked size pleomorphism (Seuanez *et al.*, 1977). A few studies have been directed toward explaining this pleomorphism, but essentially no information is available either for the human or for the great ape species concerning the fertilizing capacity of morphologically abnormal or variant forms. Seuanez *et al.* (1977) measured the total dry mass in the head as an indicator of the DNA content of sperm. Whereas the gorilla and human have an incidence of slightly more than 1% diploid sperm, the chimpanzee and orangutan essentially have none. However, chimpanzee sperm are more similar to gorilla sperm in their total DNA content. Also, the gorilla has the largest amount of haploid DNA, while the human has the smallest amount.

Chromosome banding techniques have been recently employed to study relationships among great ape sperm with a view to identifying phylogenetic similarities among the genera (Miller, 1977), since banding reveals a degree of organization well above the molecular level. Quinacrine staining (*Q*-banding) allows visualization of the *Y* chromosome in sperm as a brilliant fluorescent spot, called the *F*-body. *F*-bodies can also be visualized in the chimpanzee (*Pan paniscus* and *Pan troglodytes*) and gorilla, but not in the orangutan (Seuanez *et al.*, 1976). The fact that the orangutan has no brilliant fluorescence on any chromosomes suggests that they diverged from the ancestral primate stock prior to the emergence of that branch of the evolutionary tree that led to man and the great apes.

Whereas in man the *F*-body is always indicative of the *Y* chromosome,

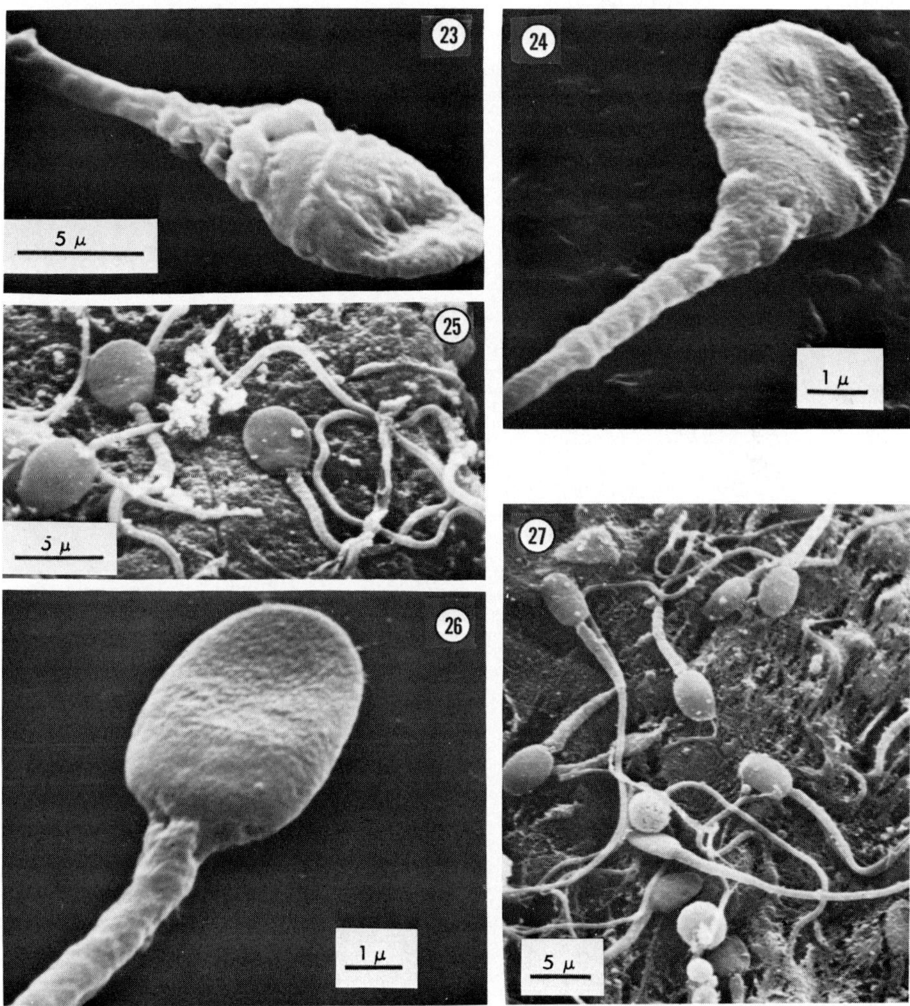

Plate 2 (Figs. 23–27) General surface appearance of ape spermatozoa as viewed using scanning electron microscopy. Sperm from both *Gorilla gorilla* (Fig. 23) and *Homo sapiens* (Fig. 24) exhibit marked size pleomorphism. They are similar in total length, but the midpiece of *Gorilla* is generally much longer, and the head is larger in both width and length. Sperm from *Pongo pygmaeus* (Fig. 25) are remarkably uniform in shape, as are those from *Pan paniscus* (Fig. 26) and *Pan troglodytes* (Fig. 27). Sperm from the two species of chimpanzee, however, are indistinguishable from each other (Figs. 26 and 27). The mitochondria of all the apes are irregularly arranged, and critical point drying, employed here, leaves the covering plasma membrane intact but loosely apposed to the gyres instead of drawn more firmly against them as with air drying.

(Figures 23, 24, 26, and 27 reprinted from Martin *et al.*, 1975; reproduced with permission of the *Journal of Human Evolution.* Copyright © by Academic Press, Inc. (London) Ltd. Figure 25 reprinted from Martin and Gould, 1975; reproduced with permission of the *Journal of Reproductive Medicine.*)

TABLE 3[a]

Dimensions of great ape spermatozoa (Adopted from Martin et al., 1975).

Species	N	Head Length	Head Width	Midpiece Length	Principal Piece Length	Total Length
Homo sapiens[b]	99	6.1 ± 0.2	3.5 ± 0.1	4.7 ± 0.1	47.1 ± 0.8	58.4 ± 0.7
Gorilla gorilla	94	9.8 ± 0.3	7.1 ± 0.1	11.7 ± 0.3	41.0 ± 0.8	61.2 ± 1.1
Pan troglodytes	106	4.7 ± 0.01	2.9 ± 0.01	6.3 ± 0.01	46.4 ± 0.1	57.4 ± 0.2
Pongo pygmaeus	102	5.3 ± 0.1	3.8 ± 0.01	9.0 ± 0.03	52.3 ± 0.1	66.6 ± 0.12
Macaca mulatta[b]	108	5.3 ± 0.01	3.2 ± 0.01	11.1 ± 0.02	57.7 ± 0.2	73.8 ± 0.2

[a] All dimensions are expressed as mean ± standard error of the mean.
[b] Data for man (*Homo*) and rhesus monkey (*Macaca*) provided for comparison.

in the chimpanzee the Y chromosome has no such brilliant fluorescence, and thus the F-bodies represent brilliant regions in the autosomes. The gorilla is intermediate between the two, since its F-bodies correspond to both the Y chromosome and to some brilliant autosomal regions.

Still another approach to investigating genetic and phylogenetic similarities among the great apes and man has involved the study of satellite DNA, or highly repetitive DNA sequences, isolatable from the main band DNA. Their function is not exactly known, but they appear to be involved in speciation. Highly repeated DNA sequences can be transcribed *in vitro*, and a complementary RNA can be prepared with the labeled nucleotides. The complementary RNA is then incubated with chromosomes, which contain DNA, and RNA–DNA hybrids will be formed at sites where there are base matchings. This matching can be shown by autoradiographic techniques.

In a recent study (Mitchell *et al.*, 1977), we transcribed four human satellite DNAs and incubated RNAs with chromosomes of chimpanzee, gorilla, and orangutan. The results showed that man and the great apes have homologous, highly repeated DNA sequences. Hybridization was detected at many sites in the ape chromosomes, but only the gorilla showed all four human satellite DNA. The chimpanzee had three, and so did the orangutan, except at its Y chromosome where it had all four. These results do not yet indicate which great ape is phylogenetically closest to man, but they do suggest that these sequences are present in a single copy or in a few copies in the common ancestor, amplified later independently after speciation.

C. Cryopreservation of the Ejaculate

Spermatozoa from several species of primates have been successfully frozen and thawed, with recovery of living cells (Human—Behrman and Ackerman, 1969; Matheson *et al.*, 1969; Sherman, 1973; Barwin, 1974. Rhesus monkey—Anonymous, 1972; Leverage *et al.*, 1972. Baboon—Kraemer and Vera Cruz, 1969. Five species of primates, including the chimpanzee—Roussel and Austin, 1967a, b. Chimpanzee—Sadleir, 1966). The methodology employed by these workers was essentially adapted from techniques already in use for freezing domestic farm-animal semen, modified to fit the idiosyncracies of primate sperm. The most important details that seem to vary among the species involve (1) the recipe for the extender, (2) the final dilution of sperm per ml of extender, and (3) the optimal rate of freezing and thawing.

Sadleir's study must be considered preliminary. He reported some success with mixtures of (1) egg yolk, citrate, and 10% glycerol; (2) egg yolk, milk, and 10% glycerol; and (3) egg yolk, glucose, and 7% glycerol. However, specific data as to postthaw motility after one to three days freezing were not presented.

Roussel and Austin (1967a) used an extender composed of 20% avian egg yolk, 64% of a 3% aqueous sodium glutamate solution, and 14% glycerol. Liquid nitrogen was used to freeze the sperm with the semen extended tenfold and sealed in glass ampules. Survival rate varied between 37% and 75%, averaging 51%. These frozen sperm were not utilized subsequently for artificial insemination, hence their fertilizing ability was not tested.

In at least two species, notably the rhesus monkey (Anonymous, 1972) and human (Behrman and Sawada, 1966; Sherman, 1973; Barwin, 1974), artificial insemination of cryopreserved sperm has initiated successful pregnancies. Because of our established interest in developing an artificial breeding program at the Yerkes Primate Center (Martin *et al.*, 1978), we began introductory studies to develop an adequate freezing regimen for chimpanzee and orangutan semen. We developed an extending medium consisting of a solution of 7% glycerol, 20% egg yolk, and 73% 0.2 M tris buffer containing citric acid and glucose, with osmolarity adjusted to 290–310 mOsm/liter and pH adjusted to 7.2–7.4. After the initial dilution with extender to a concentration of 50–80 million live sperm per ml, the diluted semen is removed from its 37°C environment (where it resides from the time of collection, optimally about 30 minutes), is transferred to a refrigerator for slow cooling to 10–12°C, and then is frozen in pellets on a dry-ice block (Nagase and Niwa, 1964; Nagase *et al.*, 1964). These pellets are subsequently immersed in liquid nitrogen.

We have recovered an encouragingly high percentage of motile and progressively motile cells after thawing frozen chimpanzee and orangutan semen specimens (as little as 5% loss in progressive motility after three weeks in liquid nitrogen), and thus believe our methodology may be applicable in artificial insemination. However, a major problem resides in rendering the semen specimens appropriate for freezing. The presence of undissolved coagulum, other proteinaceous debris, and dead cells in ejaculates with otherwise sizable numbers of progressively motile sperm often make the overall specimen undesirable for preservation. Further studies are underway to develop techniques for washing the semen specimen and concentrating motile sperm in an appropriate medium for cryopreservation.

In other studies, as well as in ours, it appears certain that cryopreservation does result in a measureable amount of cell death. It does not appear that this cell death is restricted simply to the weaker members of the sperm population, since there is also a reduction in fertility. A crucial problem to be solved if semen banking is to provide specimens for artificial insemination that are equivalent in fertilizing ability to fresh sperm is to identify changes that occur in the healthy members of the cryopreserved population and that are responsible for this loss. By identifying these changes, one can more intelligently consider which parameters of the specimen preparation regimen should be altered in order to reduce the fertility loss.

These changes are very likely to be structural, since cryopreservation even under the most optimal conditions presents a great stress to living cells. We have developed techniques for studying sperm ultrastructural morphology with a view toward elevating potential fertility of fresh sperm, and plan to apply these techniques to cryopreserved sperm in order to further develop an artificial breeding capability.

We initially applied the technique of SEM to the study of cryopreserved sperm to confirm observations made in an earlier study (Leverage *et al.*, 1972) using transmission electron microscopy (TEM), which indicated deleterious changes in surface structure (Plate 3). It is presumed that the observed morphological changes may have physiological significance. Loss of the acrosome membrane, for example, would reduce or remove the fertilizing ability of cryopreserved sperm. There may be internal changes as well that could be observable only by TEM. There appear to be no visible alterations in surface morphology seen in cryopreserved sperm that are unique, however. Rather, structural defects seen in nonpreserved sperm appear only to be exaggerated.

It remains very desirable to have some means for determining whether a sperm being viewed by SEM was alive or dead prior to preparation and observation. The problem has been solved in light microscopy by use of staining techniques. Certain stains, e.g., eosin blue, are taken up by dead cells, but are excluded from living cells. We have applied this principle, combined with the technique of energy dispersive X-ray (EDX) analysis to solve this problem (Gould and Martin, 1975).

A conventional SEM beam excites X-radiation from the specimen, and since the energy spectrum of the X rays emitted is characteristic of specific elements, analysis of this spectrum may be used to identify the elements present in the scanned specimen. We exposed fresh sperm to a conventional live/dead stain (eosin yellow) and then examined the preparation by SEM and EDX analysis. Dead sperm are identifiable by high bromine levels (eosin yellow is bromine-rich), whereas sperm that were living at the time of preparation for SEM have low bromine levels

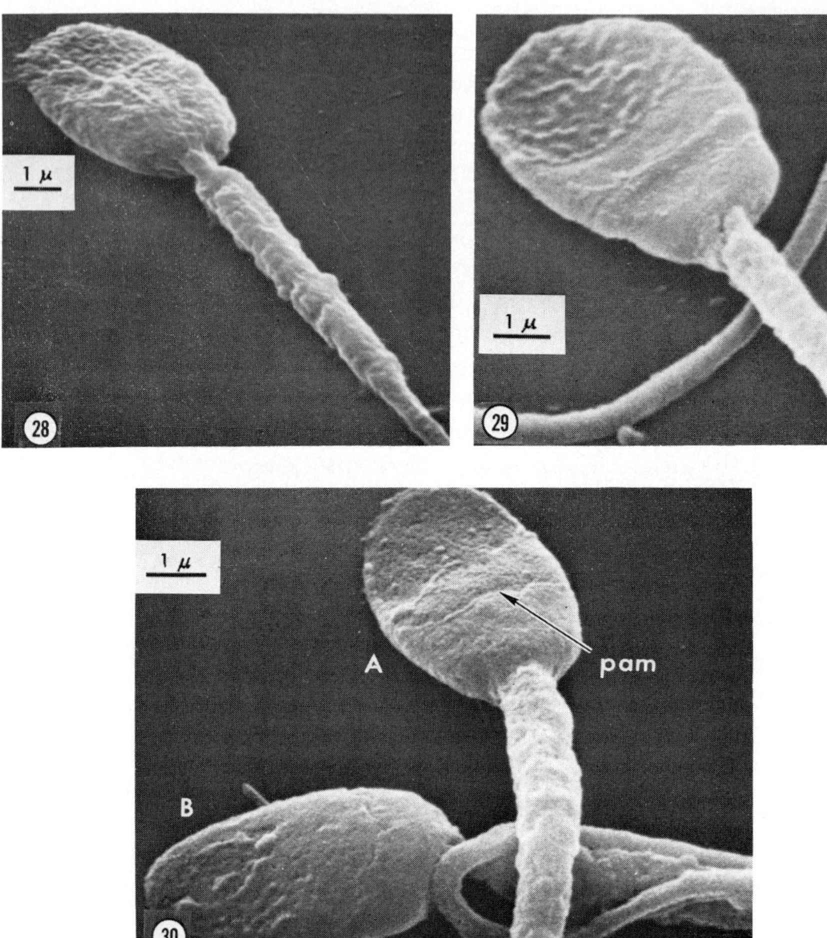

Plate 3 (Figs. 28–30) Cryopreserved (liquid N$_2$) ape spermatozoa, illustrated by scanning electron microscopy. Considerable wrinkling and disorganization of the acrosome membrane is seen in Fig. 28 (*Pan troglodytes*) and Fig. 29 (*Pongo pygmaeus*), but other structural features of the cell surface appear relatively intact. (In Fig. 29, the breakage of the junction piece where the midpiece inserts into the head is probably an artifact of fixation.) The visible alteration of acrosomal integrity is obvious when these sperm are compared with unfrozen sperm of the same species (cf. Fig. 19 for *Pan troglodytes;* cf. Fig. 25 for *Pongo pygmaeus*). The postacrosomal region of the head in these depicted specimens is relatively smooth and intact. Figure 30 portrays a more drastic change seen in some cryopreserved sperm, i.e., a loss of the entire acrosome membrane. This is seen most clearly in Fig. 30A by the apparent depression of the anterior head surface at a line coincident with the postacrosomal margin (**pam**).

(Reprinted from Martin and Gould, 1975; reproduced with permission of the *Journal of Reproductive Medicine.*)

(Gould and Martin, 1974). Thus, the use of SEM to evaluate cell damage, and the use of EDX analysis to distinguish between live or dead cells at the time of fixation, provide a basis for objectively evaluating the extent to which various extending media serve to protect sperm from the stress of cryopreservation. Studies are presently underway to evaluate the effects of cryopreservation on great ape sperm using this new technique.

Results of these studies might answer several specific questions. First, what is the correlation between the percentage of sperm that upon light microscopic observation have been stained with differential vital dyes and considered dead, and the percentage of spermatozoa that upon SEM/EDX analysis appear to have large amounts of the dye as part of their elemental composition? Second, are there acrosomal membrane abnormalities or other visible surface structural changes that appear to have occurred on all cells or that are restricted to live cells that could be correlated with a decreased ability to fertilize the ovum? Third, by using EDX analysis to systematically map the elemental composition of sperm, are there regional changes in membrane composition not detectable solely by secondary electron emission?

Changes in these parameters are predicted to occur during cryopreservation and aging, and could correlate with changes in the fertilizing capacity of sperm. By appropriate changes in preservation media and rates of freeze/thaw, these changes might be minimized. Once this information is known, semen cryobanking using the chimpanzee and other great apes could proceed on a more sound basis, providing a repository of maximum quality sperm for use in an artificial breeding program. Clearly, much work needs to be done in this area.

Cryopreservation of semen could be a very useful adjunct to an insemination program involving fresh semen. It could provide the potential for continued siring of infants by viable males after their death or after their loss of normal sperm production. Sperm transport to other colonies for use in insemination would also be possible, allowing production of infants even in situations where a breeding pair of animals is nonexistent.

IV. SUMMARY

Much information, both basic and applied, remains to be learned about the gross and microscopic anatomy of the great ape male reproductive tract. It is important that such information be acquired, not only to add to the fund of primate knowledge, but also to enhance our appreciation of the problems of taxonomic relations and of physiologic processes.

Studies of the genital tract secretions are relatively advanced in some areas, notably sperm morphology and hormonal relationships. However, many fundamental problems still need resolution. The biochemical differences among semen specimens collected by various methods, the liquefaction of seminal coagulum with release of sperm trapped within, and the effects of cryopreservation on sperm viability are only three of the most noteworthy. Further knowledge in these areas will be of significance in breeding these ape species and thus preserving their integrity.

ACKNOWLEDGMENTS

The authors would like to thank in particular those colleagues who have collaborated in the acquisition of some of the original data described in this chapter. Chimpanzee plasma testosterone assays were performed in the laboratory of Dr. Delwood C. Collins, Departments of Medicine and Biochemistry, Emory University School of Medicine. Biomedical engineering problems in the design and implementation of rectal probe electro-ejaculation equipment were solved by Harold Warner, Chief of the Biomedical Engineering Unit, Yerkes Primate Research Center of Emory University.

REFERENCES

Amelar, R. D. (1962). *J. Urol.* **87,** 187–190.

Amelar, R. D., and Hotchkiss, R. S. (1965). *Fertil. Steril.* **16,** 46–60.

Anonymous (1972). *Primate Record* 3(2), 7–9.

August, G. P., Grumbach, M. M., and Kaplan, S. L. (1972). *J. Clin. Endocrinol.* **34,** 319–326.

Barwin, B. N. (1974). *J. Reprod. Fertil.* **36,** 101–106.

Behrman, S. J., and Ackerman, D. R. (1969). *Am. J. Obstet. Gynecol.* **103,** 654–661.

Behrman, S. J., and Sawada, Y. (1966). *Fertil. Steril.* **17,** 457–466.

Bolk, L. (1907). *Z. Morphol. Anthropol.* **10,** 250–316.

Bunge, R. G., and Sherman, J. K. (1954). *Fertil. Steril.* **5,** 353–356.

Chowdhury, A. K. (1979). Personal communication.

Eckstein, P. (1958). *In* "Primatologia" (H. Hofer, A. H. Schultz, and D. Starck, eds.), Vol. 3, pp. 542–629. Karger, Basel.

Eliasson, R. (1963). *Biochem. Pharmacol.* **12,** 405.

Eliasson, R. (1977). *In* "Male Fertility" (A. T. K. Cockett and R. L. Urry, eds.), pp. 189–204. Grune and Stratton, New York.

Fawcett, D. W. (1976). *In* "Reproduction and Human Welfare: A Challenge to Research" (R. O. Greep, M. A. Koblinsky, and F. S. Jaffe, eds.), pp. 165–227. Ford Foundation, New York.

Fawcett, D. W. (1979). *Perspect. Biol. Med.* **22,** S56–S74.

Freund, M. (1968). *In* "Progress in Infertility" (S. J. Behrman and R. W. Kistner, eds.), pp. 593–627. Little, Brown, Boston, Massachusetts.

Fussell, E. N., Roussel, J. D., and Austin, C. R. (1967). *Lab. Anim. Care* **17**, 528–530.
Fussell, E. N., Franklin, L. E., and Frantz, R. C. (1973). *Lab. Anim. Sci.* **23**, 252–255.
Gersh, I. (1970). *Fertil. Steril.* **21**, 147–150.
Glover, T. D. and Nicander, L. (1971). *J. Reprod. Fertil. Suppl.* **13**, 39–50.
Gould, K. G., and Martin, D. E. (1974). *IRCS Med. Sci.* **2**, 1445.
Gould, K. G., and Martin, D. E. (1975). *J. Reprod. Med.* **14**, 197–200.
Gould, K. G., Martin, D. E., and Hafez, E. S. E. (1975). *In* "SEM Atlas of Mammalian Reproduction" (E.S.E. Hafez, ed.), pp. 42–57. Igaku Shoin, Tokyo.
Gould, K. G., Warner, H., and Martin, D. E. (1978). *J. Med. Primatol.* **7**, 213–222.
Graham, C. E., and Bradley, C. F. 1972. *In* "The Chimpanzee" (G. H. Bourne, ed.), Vol. V, pp. 77–126. Karger, Basel.
Greer, W. E., Roussel, J. D., and Austin, C. R. (1968). *J. Reprod. Fertil.* **15**, 153–155.
Hardin, C. J., Liebherr, G., and Fairchild, O. (1975). *Int. Zoo Yearb.* **15**, 132–134.
Hill, W. C. O. (1946). *Proc. Zool. Soc. London* **116**, 129–132.
Horst, H.-J., Bartsch, W., and Dirksen–Thedeus, I. (1977). *J. Clin. Endocrinol. Metab.* **45**, 522–527.
Hoskins, D. D., and Patterson, D. L. (1967). *J. Reprod. Fertil.* **13**, 337–340.
Kraemer, D. C., and Vera Cruz, N. C. (1969). *J. Reprod. Fertil.* **20**, 345–348.
Leverage, W. E., Valerio, D. A., Schultz, A. P., Kingsbury, E., and Dorey, C. (1972). *Lab. Anim. Sci.* **22**, 882–889.
Maguelone, G. F., and Cathiard, A. M. (1975). *J. Clin. Endocrinol. Metab.* **41**, 977–980.
Martin, D. E., and Davidson, M. W. (1976). *Proc. Int. Congr. Anim. Reprod. Artif. Insemin., 8th, Krakow* **4**, 919–922.
Martin, D. E., and Gould, K. G. (1975). *J. Reprod. Med.* **14**, 204–209.
Martin, D. E., Gould, K. G., and Warner, H. (1975). *J. Hum. Evol.* **4**, 287–292.
Martin, D. E., Swenson, R. B., and Collins, D. C. (1977). *Steroids* **29**, 471–481.
Martin, D. E., Graham, C. E., and Gould, K. G. (1978). *Symp. Zool. Soc. London* **43**, 249–260.
Mastroianni, L., Jr., and Manson, W. A., Jr. (1963). *Proc. Soc. Exp. Biol. Med.* **112**, 1025–1027.
Matheson, G. W., Carlborg, L., and Gemzell, C. (1969). *Am. J. Obstet. Gynecol.* **104**, 495–501.
McCormack, S. A. (1971). *Endocrinology* **89**, 1171–1177.
Mijsberg, W. A. (1923). *Verh. K. Akad. Ned.* **23**, 1–92.
Miller, D. A. (1977). *Science* **198**, 1116–1124.
Mitchell, A. R., Seuanez, H. N., Lawrie, S. S., Martin, D. E., and Gosden, J. R. (1977). *Chromosoma* **61**, 345–358.
Naftolin, F. S., Judd, H. L., and Yen, S. C. (1973). *J. Clin. Endocrinol. Metab.* **36**, 285–288.
Nagase, H., and Niwa, T. (1964). *Proc. Int. Congr. Anim. Reprod. Artif. Insemin., 5th* **4**, 410–415.
Nagase, H., Niwa, T., Yamashita, S., and Irie, S. (1964). *Proc. Int. Congr. Anim. Reprod. Artif. Insemin., 5th* **3**, 498–502, 503–506.
Nankin, H. R., and Troen, P. (1971). *J. Clin. Endocrinol. Metab.* **33**, 558–560.
Perachio, A. A., Alexander, M., Marr, L. D., and Collins, D. C. (1977). *Steroids* **29**, 21–33.
Plant, T. M., Zumpe, D., Sauls, M., and Michael, R. P. (1974). *J. Endocrinol.* **62**, 403–404.
Popovic, N. A., McLeod, D. G., and Borski, A. A. (1973). *Invest. Urol.* **10**, 266–277.
Retzius, G. (1910a). *Ark. Zool.* **6**(8), 1–6.
Retzius, G. (1910b). *Biol. Untersuch.* **15**, 83–86.
Retzius, G. (1911). *Biol. Untersuch.* **16**, 79–81.
Retzius, G. (1912). *Biol. Untersuch.* **17**, 100–108.
Retzius, G. (1913). *Anat. Anz.* **43**, 577–582.

Retzius, G. (1914). *Biol. Untersuch.* **18**, 91–94.

Roussel, J. D., and Austin, C. R. (1967a). *J. Reprod. Fertil.* **13**, 333–335.

Roussel, J. D., and Austin, C. R. (1967b). *Int. J. Fertil.* **12**, 288–290.

Roussel, J. D., and Austin, C. R. (1968). *J. Inst. Anim. Tech.* **19**, 22–32.

Sadleir, R. M. F. S. (1966). *Lab. Pract.* **15**, 413.

Seuanez, H., Robinson, J., Martin, D. E., and Short, R. V. (1976). *Cytogenet. Cell Genet.* **17**, 317–326.

Seuanez, H. N., Carothers, A. D., Martin, D. E., and Short, R. V. (1977). *Nature (London)* **194**, 345–347.

Sherman, J. K. (1973). *Fertil. Steril.* **24**, 397–412.

Short, R. V. (1977). *In* "Reproduction and Evolution" (J. H. Calaby and C. H. Tyndale-Biscoe, eds.), pp. 3–19. Australian Academy of Science, Canberra.

Smith, K. D., and Steinberger, E. (1977). *In* "The Testis in Normal and Infertile Men" (P. Troen and H. R. Nankin, eds.), pp. 489–503. Raven Press, New York.

Southren, A. L., and Gordon, G. G. (1975). *J. Steroid Biochem.* **6**, 809–813.

Turner, T. T., and Howards, S. S. (1977). *In* "Male Infertility" (A. T. K. Cockett and R. L. Urry, eds.), pp. 29–57. Grune and Stratton, New York.

Warner, H., Martin, D. E., and Keeling, M. E. (1974). *Ann. Biomed. Eng.* **2**, 419–432.

Weisbroth, S., and Young, F. A. (1965). *Fertil. Steril.* **16**, 229–235.

White, I. G. (1973). *J. Reprod. Fertil. Suppl.* **18**, 225–235.

White, I. G., Rodger, J. C., Morris, S. R., and Marley, P. B. (1977). *In* "Reproduction and Evolution" (J. H. Calaby and C. H. Tyndale-Biscoe, eds.), pp. 183–192. Australian Academy of Science, Canberra.

Wilson, V. B., and Bunge, R. G. (1975). *J. Urol.* **113**, 509–510.

Wislocki, G. B. (1936). *Hum. Biol.* **8**, 309–347.

Zaneveld, L. J. D., Tauber, P. F., Port, C., Propping, D., and Schumacher, G. F. B. (1974). *J. Reprod. Fertil.* **40**, 223–225.

Chapter 7

COMPARATIVE ASPECTS OF APE STEROID HORMONE METABOLISM

Kristina Wright
Delwood C. Collins
Paul I. Musey
John R. K. Preedy

I. INTRODUCTION

Steroid hormone metabolism in great apes has received little attention, particularly when compared with the information available on this

163

Copyright © 1981 by Academic Press, Inc.
All rights of reproduction in any form reserved.
ISBN 0-12-295020-8

subject in Old World monkeys. Various aspects of steroid metabolism in primates have been reviewed (Dorfman, 1970; Ryan and Hopper, 1974; Hobson *et al.*, 1976; Schubert and Schade, 1975). The following discussion will focus on steroid hormone metabolism in great apes (chimpanzees, gorillas, and orangutans), as well as the taxonomic differences observed among these apes, humans, and monkeys. The endocrine patterns occurring in different physiologic states are discussed in greater detail in other chapters.

Most information on steroid hormone metabolism in great apes has been obtained from three experimental procedures. (1) The concentrations of endogenous hormones in plasma or, more frequently, in urine have been measured by various chemical methods. These findings have been used to deduce pathways of steroid metabolism by extrapolation from our knowledge of metabolic pathways in other species. (2) Exogenous steroids, usually radioactive, have been administered *in vivo*, and urinary metabolites isolated and identified. (3) Tissue homogenates and cell fractions have been incubated *in vitro* with radioactive precursor, and the products isolated and identified. The *in vivo* methods are advantageous in that the metabolic sequences observed are more closely related to those occurring physiologically. However, the collection of specimens, particularly urinary specimens, introduces several sources of error that must be considered in evaluating the data thus obtained.

Steroid hormones may occur as the free steroid, or they may be chemically linked to a conjugating group, such as glucosiduronic acid or sulfuric acid. In addition to the biologically active hormone, precursors and metabolites also appear in plasma and urine as the free or conjugated steroid. This schema is depicted in Fig. 1. It previously has been assumed that the conjugation or further metabolism of the active hormone was the initiating, irreversible step in a chain of metabolic events ultimately leading to elimination of the hormone from the body. However, our current knowledge suggests that this view may be incorrect. Steroid conjugates such as dehydroisoandrosterone sulfate (Siiteri and MacDonald, 1963), estrone sulfate (Stern and Givner, 1978), and estrone glucosiduronate (Collins *et al.*, 1975a) have been shown to serve as storage forms for physiologically active hormones. Thus, the occurrence of a hormone in the conjugated form does not preclude its further endocrine activity. No information is available on plasma concentrations of steroid conjugates in apes. The steroid forms that have been most studied in apes are unconjugated plasma steroids and total, or conjugated and unconjugated, urinary steroids.

Aspects of steroid metabolism in great apes will be compared, where

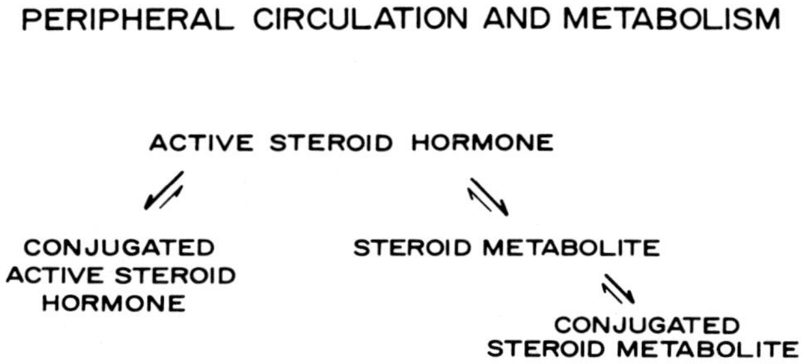

PERIPHERAL CIRCULATION AND METABOLISM

ACTIVE STEROID HORMONE

CONJUGATED
ACTIVE STEROID
HORMONE

STEROID METABOLITE

CONJUGATED
STEROID METABOLITE

Fig. 1. General patterns of steroid hormone metabolism. Active steroid hormones in both the unconjugated and the conjugated forms circulate in plasma. In addition, unconjugated and conjugated metabolites of the steroid nucleus may occur in plasma. All of these forms are excreted in the urine, although the amount of active hormone in the urine is generally much smaller than the amount of metabolized hormone. In humans little unconjugated steroid is excreted in the urine. However, in apes a significant fraction of urinary steroid may be in the unconjugated form.

possible, with data obtained in humans and monkeys using similar techniques. This should eliminate apparent variations that actually result from different methodologies employed. Many studies of steroid metabolism in great apes have utilized analytical procedures that have subsequently become somewhat outmoded due both to a general increase in our knowledge of steroid metabolism and to the development of more sensitive, specific analytical techniques. However, as will be discussed in this chapter, studies of steroid hormone metabolism in great apes using both older and newer techniques indicate that these primates have many similarities to humans and thus may prove to be very useful models for humans.

II.   THE TESTES

A.   Testicular Androgens

1.   Endogenous Hormone Levels

Plasma testosterone (Fig. 2) has been measured in male chimpanzees (*Pan troglodytes*) by McCormack (1971) and Martin *et al.* (1977) and correlated with age. These and comparable data from a study of humans (August *et al.*, 1972) are presented in Table 1. In chimpanzees, a significant increase in testosterone to adolescent levels occurred at seven to nine years; the rise to adult levels occurred at 11 years. Although maturity occurs at an earlier age in chimpanzees, plasma levels of

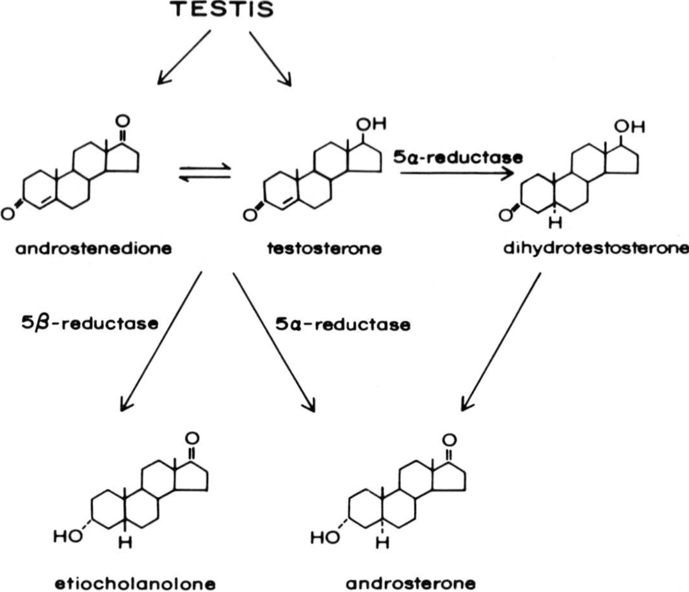

Fig. 2. Metabolism and excretion of androgens. Androgens are secreted principally by the testes, although secretion occurs from the ovaries and adrenals as well. Testosterone and androstenedione, the androgens secreted by the testes, serve as precursors for the active androgen 5α-dihydrotestosterone. The primary urinary androgens that have been identified in humans, apes, and monkeys are the 5α- and 5β-reduced steroids androsterone and etiocholanolone.

testosterone are similar to those of humans at comparable stages of maturity. Testosterone levels in adult rhesus monkeys (*Macaca mulatta*) are also in the same range (Plant *et al.*, 1974).

No annual variation has been shown in plasma testosterone levels in adult humans. However, an annual rhythm in plasma testosterone levels in adult males has been found in the seasonal-breeding rhesus monkey, with a nadir in April during the birth season and peaks in September and December during the mating season (Plant *et al.*, 1974; Gordon *et al.*, 1976). In the nonseasonal-breeding pigtail monkey (*Macaca nemestrina*), no annual rhythm of plasma testosterone in adult males was found (Bernstein *et al.*, 1978). Plasma testosterone in adult male chimpanzees varied from month to month, but the data were not adequate to detect seasonal trends (McCormack, 1971). Similar results were obtained from a second series of 10 adult male chimpanzees (D. E. Martin, R. B. Swenson, and D. C. Collins, unpublished observations). A nadir of plasma testosterone concentration was seen in January, but it was not significantly different from the peak levels.

TABLE 1

The correlation between plasma testosterone and age in male chimpanzees and humans.

	Chimpanzees			Humans	
Age (years)	Testosterone (ng/100 ml) mean ± SE (n)	Age (years)	Testosterone (ng/100 ml) mean ± SE (n)	Age (years ± SE)	Testosterone (ng/100 ml) mean ± SE (n)
		1–3	11.7 ± 2.3(6)	7.7 ± 0.3	18 ± 4.4(20)
		3–5	11.1 ± 2.3(9)	12 ± 0.2	71 ± 19(7)
0.25–2	4.7 ± 6.2(11)	5–7	15.1 ± 4.7(11)	13.7 ± 0.2	248 ± 46(14)
3–6	14.7 ± 11.8(25)	7–9	162 ± 27(13)	15.7 ± 0.3	482 ± 27(10)
		9–11	197 ± 29(6)		
6–9	263 ± 144(15)	11–13	409 ± 45(12)	25–35	625 ± 28(12)
		13–15	404 ± 46(8)		
>11	495 ± 143(27)	>15	374 ± 54(9)		
McCormack (1971)		Martin *et al.* (1977)		August *et al.* (1972)	

Circulating levels of dihydrotestosterone (Fig. 2) have not been reported for any of the apes. In humans dihydrotestosterone is mostly formed in the target organ, and circulating concentrations are generally low and may not be reflective of hormonal status. Literature on this androgen has been recently reviewed by Bardin (1978).

In humans the level of sex steroid binding globulin plays a significant role in regulating biologically available dihydrotestosterone, testosterone, and 17β-estradiol. As reported for humans (Vermeulen *et al.*, 1969), at maturity male chimpanzees demonstrate not only a decrease in the levels of sex steroid binding globulin, but also an increase in the percentage of unbound testosterone in plasma (McCormack, 1971).

The presence of androgenic substances in the urine of male and female chimpanzees was first demonstrated by Fish *et al.* (1941). Using a bioassay standardized to androsterone, these investigators found that adult male chimpanzees excreted twice as much androgens as females. However, the urinary androgens in male chimpanzees represented only 10% of the androgens found in human urine using the same bioassay. These low values may reflect differences in the biological activity of the urinary androgen metabolites in chimpanzees when compared with humans.

2. *Androgen Metabolism*

Fish and Dorfman (1944) administered 300 mg testosterone propionate daily for five days to two adult male chimpanzees. The animals were housed in metabolism cages, and urine was collected during treatment and for 4 days thereafter. The urine was pooled and analyzed for ketosteroids by using a chemical method and for androgen activity by bioassay. Three compounds, representing a total of 18.3% of the recovered ketosteroids, were identified: etiocholanolone (2%), androsterone (5%), and 17-oxoandrostenone (11%) (Fig. 2). If 17-oxoandrostenone were a dehydration product of androsterone formed during the hydrolysis and extraction procedures, as suggested by Fish and Dorfman, then the ratio of androsterone to etiocholanolone in these adult male chimpanzees would have been 8:1. In contrast, a similar study in normal men showed androsterone and etiocholanolone to be excreted in a ratio of approximately 1:1 (Callow, 1939) (Fig. 2).

A comprehensive study of the metabolism of testosterone in the immature male chimpanzee was reported by Layne *et al.* (1963). This experimental design will be described in some detail, because it is typical of the protocols routinely used in studies of radiolabeled steroid metabolism. Four ml of a solution of ^{14}C-testosterone in saline was injected intravenously into each of four chimpanzees. The animals were placed in metabolism cages, and urine was collected. The urine samples were analyzed for 17-ketosteroids to measure total androgen excretion, for total radioactivity to assess urinary recovery of the injected dose, and for creatinine to correct for incomplete urine collections. The sample was then subjected to sequential hydrolysis: steroids soluble in organic solvent before hydrolysis are designated as unconjugated; steroids soluble after treatment with β-glucuronidase represent the glucosiduronate fraction; those soluble after acid hydrolysis form the sulfate fraction. The identities of the radiolabeled steroids in each fraction were established by chromatography, followed by recrystallization to a constant specific activity with added steroid of confirmed structure.

One major drawback of this experimental design is the collection of the urine samples, because the possibility of contamination with fecal material exists. Of more critical concern, however, is the further metabolism or destruction of urinary steroids by chemicals, by enzymes present in the urine, or by contaminating bacteria. Hydrolysis of steroid conjugates can occur in significant amounts under these conditions. In addition, because it is difficult to insure a complete urine collection, creatinine is frequently used as an excretory index, although creatinine excretion is itself subject to a number of variables. The possibility of artifacts introduced by these experimental errors should be considered in the interpretation of the data thus obtained.

When ^{14}C-testosterone was administered to four immature male chimpanzees, Layne *et al.* (1963) found the major urinary metabolite to be etiocholanolone (Fig. 2). The ratio of etiocholanolone to androsterone ranged from 2:1 to 4.2:1. This ratio contrasts to the data obtained earlier by Fish and Dorfman (1944), in which the excretion of androsterone was eight times that of etiocholanolone. The reason for this discrepancy is unclear. It should be noted that the data obtained by Layne *et al.* (1963) were from immature animals, whereas the animals used by Fish and Dorfman (1944) were adults.

When the conjugation pattern of urinary metabolites of ^{14}C-testosterone was examined, surprisingly high proportions of the metabolites were recovered as unconjugated steroids (Layne *et al.*, 1963). In general, the proportion of unconjugated steroids was approximately equal to that of the glucosiduronates, whereas the sulfated fraction was 10 times less. Men excrete very low levels of unconjugated testosterone relative to the amount of conjugated testosterone excreted (Sandberg and Slaunwhite, 1956). The relatively high excretion of unconjugated steroid appears to be a recurring aspect of great ape steroid metabolism (Sections III,A,1; III,B,1; and IV,A). This is strikingly different from the pattern of excretion occurring in humans and suggests differences among primates in the renal clearance of steroids. However, hydrolysis of steroid conjugates during urine collection could result in artificially high excretion of unconjugated steroid and cannot be completely excluded in these experiments.

B. Testicular Estrogens

The testes of most species have been shown to secrete both androgens and estrogens. The occurrence and metabolism of estrogens in male great apes have received little attention. Fish *et al.* (1941) identified estrogens in the urine of male chimpanzees, although the amount of estrogens excreted by males was lower than that excreted by female chimpanzees. Immature male chimpanzees excrete approximately equal amounts of estrogen glucosiduronates and sulfates. The excretion of estrone sulfate exceeded that of estrone glucosiduronate, whereas excretion of estriol glucosiduronates exceeded that of estriol sulfates (Elmadjian and Forchielli, 1965). In addition, the proportion of estrogen excreted as estriol was only slightly less than that excreted as estrone (2.07 *versus* 2.16 µg/0.5 g creatinine in one animal; 1.34 *versus* 1.61 µg/ 0.5 g creatinine in a second).

Estrogens in the plasma have not been studied in the male of any of the apes. The *in vivo* and *in vitro* metabolism of estrogens will be considered in Section III,A,2.

III.   THE OVARIES

A.   Ovarian Estrogens

1.   *Endogenous Hormone Concentrations*

Estrogenic substances in the chimpanzee were first identified by Allen *et al.* (1935,1936) in pregnancy urine and in placental tissue by using the ovariectomized mouse bioassay. During the menstrual cycle, the highest amounts of estrogens were excreted at midcycle and the lowest amounts during menstruation (Allen *et al.*, 1936). In addition to the midcycle peak, luteal estrogen peaks similar to those occurring in women were described in the urine of two female chimpanzees.

Reyes *et al.* (1975) characterized the pattern of serum 17β-estradiol (Fig. 3) throughout the menstrual cycle in four chimpanzees. Serum 17β-estradiol was low during menses (10–54 pg/ml) and increased to a peak of as much as 352 pg/ml at midcycle. The midcycle peak was followed by a rapid decrease to follicular phase levels, before a luteal phase peak of as much as 181 pg/ml was reached. Similarly, during the menstrual cycle in three gorillas (*Gorilla gorilla*), a midcycle peak of plasma 17β-estradiol (180–500 pg/ml) as well as a luteal phase peak occurred (Nadler *et al.*, 1979). The presence of both a midcycle and a luteal phase peak and the absolute levels observed in chimpanzees and gorillas are similar to plasma 17β-estradiol patterns observed in women (Abraham *et al.*, 1972). The monkey differs markedly from apes and women in that, although follicular phase and midcycle plasma 17β-estradiol levels are comparable, no obvious luteal phase peak has been demonstrated in rhesus monkeys (Hotchkiss *et al.*, 1971; Hodgen *et al.*, 1972). Because the luteal phase peak of progestin and estrogen is attributable to steroid secretion by the corpus luteum, these findings suggest that the corpus luteum of Old World monkeys lacks the ability to produce estrogens in significant quantities.

Both the growing follicle and the corpus luteum of women secrete estrone (Fig. 3) as well as 17β-estradiol. Estrone and 17β-estradiol are rapidly metabolized to estrone sulfate, and these three steroids are the major plasma estrogens in men and women. Neither estrone nor estrone

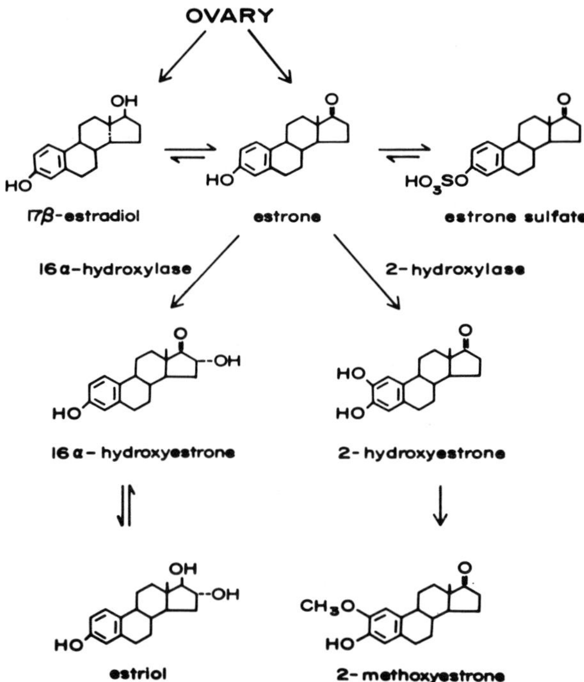

Fig. 3. Metabolism and excretion of estrogens. Estrogens are secreted principally by the ovaries, although secretion occurs from the testes and adrenals as well. In humans, secreted estrone and 17β-estradiol are rapidly converted to estrone sulfate, quantitatively the most important plasma estrogen. Estrone sulfate has not been studied in either apes or monkeys. The principal urinary estrogens that have been identified in primates are estrone (monkeys), estrone and estriol (apes and humans), and the catechol estrogens (humans and possibly apes). The 16α-hydroxylase pathway is of little importance in monkeys but is quite significant in humans and some apes. Although the 2-hydroxylase pathway has not been studied in monkeys, it is a major pathway of estrogen metabolism in humans and may be equally important in apes.

sulfate has yet been studied in apes. Plasma concentrations of these estrogens are of particular interest, since they appear to be important intermediates in the metabolism of 17β-estradiol, as well as storage forms for active estrogens.

Total estrogen excreted as estrone, 17β-estradiol, and estriol was studied throughout the menstrual cycle in three chimpanzees (Graham *et al.*, 1972) and two orangutans (*Pongo pygmaeus;* Collins *et al.*, 1975b). The animals were housed in metabolism cages, and daily urine samples were obtained. A portion of the urine sample was subjected to acid hydrolysis. Hydrolyzed estrone, 17β-estradiol, and estriol were then extracted, purified, and quantitated by chemical assay. The remainder of the specimens was combined, and the endogenous estrogens and estrogen conjugates present were identified by sequential hydrolysis, chromatography, and recrystallization to constant specific activity after addition of radioactive estrogen. The sources of error inherent in these procedures are described in Section II,A,2 and will be discussed further in Section III,A,2.

In the three chimpanzees studied by Graham *et al.* (1972), total estrone excretion in the follicular phase was 2 μg/24 hr; in midcycle, 12 to 20 μg/24 hr; and in the luteal phase, 8 to 15 μg/24 hr. The excretion of 17β-estradiol was much lower, ranging from 0.5 to 5 μg/24 hr in the follicular phase and from 1 to 9 μg/24 hr in the luteal phase. Estriol was excreted in the lowest amounts, varying from 0.1 to 2.5 μg/24 hr during the cycle. Although midcycle and luteal peaks in estrone excretion were clearly defined, peaks for 17β-estradiol and estriol were not as apparent (Graham *et al.*, 1972).

The pattern of estrogen excretion was similar during the menstrual cycles of two orangutans (Collins *et al.*, 1975b). Follicular phase excretion was 1 to 8 μg/24 hr for estrone, 0.5 to 6 μg/24 hr for 17β-estradiol, and 1 to 8 μg/24 hr for estriol. Excretion was higher in the luteal phase, ranging from 3 to 21 μg/24 hr for estrone, 2 to 10 μg/24 hr for 17β-estradiol, and 1 to 9 μg/24 hr for estriol. In both cycles, peaks of estrone excretion occurring at midcycle and in the luteal phase were clearly defined. Midcycle and luteal peaks of 17β-estradiol and estriol were not apparent.

The biphasic pattern of estrogen excretion during the menstrual cycle in chimpanzees and orangutans resembles the pattern occurring during the menstrual cycle in women (Brown *et al.*, 1958; Goebelsmann *et al.*, 1969) but differs strikingly from that of rhesus monkeys, in which there is no apparent peak of estrogen excretion in the luteal phase (Hopper and Tullner, 1970). This is not unexpected, as there is also no luteal phase peak of plasma estrogens in rhesus monkeys (Hotchkiss *et al.*, 1971; Hodgen *et al.*, 1972). Furthermore, although the excretion of total estrone greatly exceeds that of total estriol in rhesus monkeys (Hopper and Tullner, 1970), estriol is the major urinary estrogen in menstruating women (Brown *et al.*, 1958; Goebelsmann *et al.*, 1969). The proportion of urinary estrone is high compared to estriol in chimpanzees, while estrone is only slightly higher than estriol in orangutans. The source of these taxonomic differences is unknown at the present time, but may prove to be due to differences in secretion, metabolism, or both.

The relative size of the midcycle and luteal peaks of urinary estrogen in chimpanzees, orangutans, and women is noteworthy. In women the midcycle peak is usually much larger than the luteal peak (Brown *et al.*, 1958; Goebelsmann *et al.*, 1969), but the reverse is true in orangutans (Collins *et al.*, 1975b). In chimpanzees the two peaks are about equal (Graham *et al.*, 1972). Again, the taxonomic differences remain unexplained. Considerable variation has been shown to exist among women, however, and these apparent differences in excretion patterns among apes may be attributable to the small sample sizes in studies of apes.

With respect to total amount of urinary estrogens, chimpanzees and orangutans are comparable to women (Graham *et al.*, 1972; Collins *et al.*, 1975b; Brown *et al.*, 1958; Goebelsmann *et al.*, 1969). Total urinary estrogens excreted by rhesus monkeys, however, are approximately an order of magnitude lower (Hopper and Tullner, 1970).

Graham *et al.* (1972) identified the following estrogens in a pool of urine from an adult female chimpanzee: estrone (15% of total urinary estrogens), estriol (4%), estrone sulfate (7%), estrone glucosiduronate (55%), and 17β-estradiol glucosiduronate (25%). Unconjugated 17β-estradiol, 17β-estradiol sulfate, and estriol sulfate and glucosiduronate were not detected. Similarly, in an adult female orangutan, estrone (4% of total urinary estrogens), 17β-estradiol (11%), estriol (<1%), estrone sulfate (23%), 17β-estradiol sulfate (11%), estrone glucosiduronate (27%), 17β-estradiol glucosiduronate (18%), and estriol glucosiduronate (6%) were identified (Collins *et al.*, 1975b). The most striking finding in both apes again is the high proportion of unconjugated steroid that is excreted when compared with women (Crowell *et al.*, 1967). Orangutans excreted only slightly more estrogen as the glucosiduronate than as the sulfate, whereas chimpanzees excreted significantly more estrogen glucosiduronates than estrogen sulfates. In this respect the pattern seen in chimpanzees more closely resembled that seen in women, in whom the glucosiduronate is the major urinary estrogen conjugate (Crowell *et al.*, 1967; Wright *et al.*, 1978).

2. Estrogen Metabolism

The metabolism of intravenously administered ^{14}C-estrone was studied in a pregnant chimpanzee (Jirku and Layne, 1965). Seventy-two percent of the injected isotope was recovered in the urine within 72 hr, of which 45% was present as unconjugated steroid, 20% as glucosiduronates, 22% as sulfates, and 17% as conjugates not hydrolyzable by conventional techniques. In a rhesus monkey at midcycle, only 44% of an injected dose of ^{3}H-estrone was recovered in the urine (Breckwoldt *et al.*, 1972). After the administration of ^{14}C-estrone to two normal women and one normal man, 63%, 72%, and 59% of the isotope were recovered in the urine (Migeon *et al.*, 1959). Only 1–2% appeared as unconjugated steroid, with 5% in the sulfate fraction and 33–53% in the glucosiduronate fraction. Again, chimpanzees appear to excrete a large amount of free steroid when compared with humans.

Metabolites of ^{14}C-estrone identified in chimpanzee urine included estrone, 17β-estradiol, estriol, and 2-methoxyestrone (Jirku and Layne, 1965) (Fig. 3). In addition, a steroid conjugate more polar than estriol, which appeared to be an estrone glycoside, was isolated from the chimpanzee urine (Jirku and Layne, 1965). Later *in vitro* studies suggest that this metabolite is estrone glucoside (Labow *et al.*, 1975). With the exception of the high proportion of steroid excreted in the unconjugated form, the metabolism of estrone in chimpanzees is quite similar to that in humans.

The metabolism of ^{3}H-estradiol-17β in a pygmy chimpanzee (*Pan paniscus*) presents some interesting aspects (Musey *et al.*, 1979). The pygmy chimpanzee, unlike the chimpanzee studied by Jirku and Layne (1965), excreted negligible amounts of labeled material as unconjugated steroid, 68% as glucosiduronates, and 26% as sulfates. This ape metabolizes 17β-estradiol primarily via the 2-hydroxy pathway, with 2-hydroxyestrone being quantitatively the most important urinary metabolite (Fig. 3). As much as 35% of the total urinary glucosiduronates was accounted for by 2-hydroxyestrone glucosiduronates, and 62% of the sulfate pool was identified as 2-hydroxyestrone sulfates. Estrone glucosiduronate and estrone sulfate accounted for 25% each of their respective fractions. Other metabolites isolated from both fractions included 2-methoxyestrone, 17β-estradiol, estriol, and 16β,17β-epiestriol.

The importance of this study lies not only in the identification of the catechol estrogens, 2-hydroxyestrone and 2-methoxyestrone, in the pygmy chimpanzee, but also in the use of the antioxidant ascorbic acid throughout the analysis. Catechol estrogens are extremely labile and are subject to oxidative breakdown during the routine procedures described here unless such precautions are taken (Gelbke *et al.*, 1977). Although some 2-methoxyestrone was identified in chimpanzee urine by Jirku and Layne (1965), significant loss of catechol estrogens could have occurred. All studies of endogenous urinary estrogens in apes and in women described in Section III,B,1 would probably have failed to detect catechol estrogens due to procedural losses. The 2-hydroxy pathway (Fig. 3) is a major route of estrogen metabolism in humans (reviewed by Gelbke *et al.*, 1977) and may be of equal importance in other primates. In addition, catechol estrogens have been shown to be biologically active in some species (Gelbke *et al.*, 1977). The formation and metabolism of catechol estrogens in nonhuman primates also merits attention because of the competitive nature of the C-2 and C-16 hydroxylations. If catechol estrogens are found to be major urinary estrogens in nonhuman primates, this finding could explain the rather low estriol levels reported for these species.

The tentative identification of an estrogen glycoside in chimpanzee urine (Jirku and Layne, 1965) prompted an investigation into the ability of microsomes prepared from chimpanzee liver to transfer glucose to estrogens. This preparation transferred glucose from uridine-diphosphoglucose to the 17α-hydroxyl group of 17α-estradiol and of 17α-estradiol-3-glucosiduronate, and to the phenolic hydroxyl group of estrone, but not to estriol (Labow *et al.*, 1975). The microsomal glucose transferase of human liver utilized estriol, but not 17α-estradiol-3-glucosiduronate, as a substrate (Williamson *et al.*, 1972). Thus, the specificity of the glucosyl transferase differs in chimpanzees and humans.

B. Ovarian Progestins

1. Endogenous Hormone Levels

The mammalian antral follicle secretes small but significant amounts of progesterone and 17α-hydroxyprogesterone. However, the active corpus luteum is the major source of progestins during the menstrual cycle, and elevated plasma progesterone is generally accepted as evidence of ovulation and corpus luteum formation.

The measurement of plasma progesterone throughout the menstrual cycles of three chimpanzees was reported by Graham *et al.* (1974). Plasma concentration was less than 0.5 ng/ml during the follicular phase; luteal phase peaks occurred of 16.0 ng/ml, 11.6 ng/ml, and 7.5 ng/ml. Similar values were reported by Reyes *et al.* (1975) for chimpanzees. The maximal concentrations of plasma progesterone during the luteal phase of four menstrual cycles in gorillas were 10.8 to 12.2 ng/ml (Nadler *et al.*, 1979). These levels are comparable to those reported for the luteal phase of human menstrual cycles (Yoshimi and Lipsett, 1968; Abraham *et al.*, 1972). Similar, although somewhat lower, values have been reported for rhesus monkeys (3 to 5 ng/ml; Neill *et al.*, 1967; Monroe *et al.*, 1970), cynomolgus monkeys (*Macaca fascicularis*; 7.8 ng/ml; Stabenfeldt and Hendrickx, 1973), and bonnet monkeys (*Macaca radiata*; 3.5 ng/ml; Stabenfeldt and Hendrickx, 1972). The pattern of plasma progesterone, characterized by low follicular levels and a broad luteal phase peak, appears to be quite consistent among primates.

Plasma concentrations of 17α-hydroxyprogesterone have not been reported for any of the apes. It might be expected from studies in women (Abraham *et al.*, 1972) and in yellow baboons (*Papio cynocephalus;* Kling and Westfahl, 1978) that the plasma levels of 17α-hydroxyprogesterone would be parallel to, but lower than, the levels of progesterone.

Pregnanediol (Fig. 4) has been identified as the major urinary metabolite of progesterone in humans (Klopper and Michie, 1956; Section III,B,2). Fish *et al.* (1942) first isolated pregnanediol from the urine of a pregnant chimpanzee. The excretion of pregnanediol during the menstrual cycles of three chimpanzees was reported by Graham *et al.* (1972). Urinary levels were low during the follicular phase, ranging from less than 20 μg/24 hr to 700 μg/24 hr. A luteal phase peak of as much as 1860 μg/24 hr occurred. Similar amounts of pregnanediol were excreted during the menstrual cycles of two orangutans (Collins *et al.*, 1975b). Excretion during the follicular phase ranged from 20 to 206 μg/24 hr;

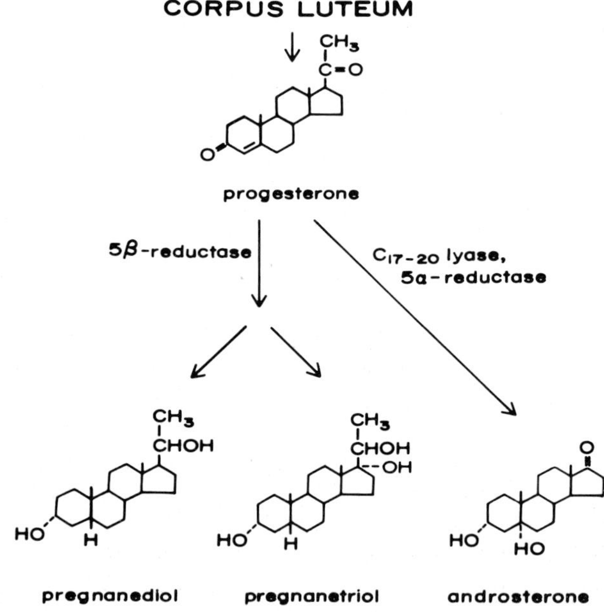

Fig. 4. Metabolism and excretion of progesterone. Although progestins are secreted by the ovaries, the testes, and the adrenals, the principal source of plasma progesterone is the corpus luteum. Unmetabolized progesterone has not been detected in the urine of any primate. The major urinary metabolites of progesterone in humans are pregnanediol and pregnanetriol, both 5β-reduced C_{21} steroids. In monkeys, 5α-reduction and cleavage to the C_{19} steroid androsterone is a major metabolic pathway for progesterone. In apes, pregnanediol and pregnanetriol, but not androsterone, have been identified as urinary progesterone metabolites.

maximal excretion during the luteal phase peak was 800 μg/24 hr. Although in this respect chimpanzees resemble women, the absolute amounts of pregnanediol excreted by both apes are markedly lower than that excreted by women, in whom maximal excretion ranges from 3000 to 6000 μg/24 hr (Brown *et al.*, 1958; Metcalf, 1968; Goebelsmann *et al.*, 1969).

Unconjugated pregnanediol, pregnanediol sulfate, and pregnanediol glucosiduronate were isolated and identified from orangutan urine and accounted for 14%, 39%, and 47%, respectively, of the total pregnanediol (Collins *et al.*, 1975b). Again, when compared with humans (Harkness and Fotherby, 1963), an unusually high proportion of unconjugated steroid was excreted by the orangutan. A significant proportion of pregnanediol was excreted as the sulfate conjugate by the orangutan, whereas little pregnanediol sulfate was excreted by the chimpanzee (Graham *et al.*, 1972).

Although plasma progesterone concentrations are similar among women, chimpanzees, and gorillas, the amount of pregnanediol excreted in the urine by chimpanzees and orangutans is significantly lower than that excreted by women. If pregnanediol is the major urinary metabolite of progesterone, then a lower production rate of progesterone could account for the relatively low urinary excretion of pregnanediol. The production rate of progesterone has not been measured in any adult female ape. However, the progesterone production rate in immature chimpanzees of both sexes, in which progesterone is largely of adrenal origin, is lower than that in either young or elderly men (Romanoff *et al.*, 1963a,b).

The presence of other important urinary metabolites of progesterone in addition to pregnanediol could also account for low urinary pregnanediol. Androsterone (Fig. 4) has been shown to be a significant urinary metabolite of progesterone in rhesus monkeys (Plant *et al.*, 1971). Accordingly, androsterone was measured throughout the menstrual cycles of two orangutans (Collins *et al.*, 1975b). Excretion in the follicular phase ranged from 120 to 522 μg/24 hr and in the luteal phase from 90 to 1158 μg/24 hr. Although an apparent luteal phase peak did occur, significant amounts of androsterone were excreted in the follicular phase when the excretion of pregnanediol was low. The urinary androsterone observed in these cycles may represent a progesterone metabolite (Section III,B,2), but it may also arise from androgens secreted by the ovary.

The elimination of progesterone metabolites by routes other than urinary excretion would also lead to comparatively low urinary pregnanediol with comparatively normal plasma progesterone concentrations. The fecal excretion of progesterone and its metabolites has not been studied in any of the apes (see also Section III,B,2).

2. *Progesterone Metabolism*

The *in vivo* metabolism of ^{14}C-progesterone has been studied in immature chimpanzees of both sexes (Romanoff *et al.*, 1963b), in the adult female chimpanzee (YoungLai *et al.*, 1975), and in the adult female gorilla (YoungLai *et al.*, 1977). In none of these studies was the injected dose recovered quantitatively in the urine, typical recoveries being 30% to 50%. This low urinary recovery of exogenous progesterone seems to be a feature typical of progesterone metabolism in all primates studied, including humans. An obvious alternative mode of excretion is via the feces. In only one report has fecal excretion been examined; in five adult female rhesus monkeys 41% to 57% of the injected dose was recovered in the feces (Plant *et al.*, 1971). Thus, the measurement and characterization of only the urinary metabolites may give an incomplete picture of progesterone metabolism in primates. However, some interesting comparative aspects do emerge from these urinary studies, which are summarized in Table 2.

In six immature chimpanzees 6.2% to 47.4% (mean of 26%) of the radioactivity recovered in urine after intravenous injection of ^{14}C-pro-

TABLE 2

Urinary metabolites of radioactive progesterone in primates.

Species	% of Recovered Dose		Reference
	Pregnanediol	Androsterone	
Human	26	—	Romanoff *et al.* (1963a)
Chimpanzee	26	—	Romanoff *et al.* (1963b)
Chimpanzee	45	NS[a]	YoungLai *et al.* (1975)
Gorilla	41	NI[b]	YoungLai *et al.* (1977)
Baboon	41	36	Goldzieher and Axelrod (1969)
Rhesus monkey	3	18	Plant *et al.* (1971)
Pigtail monkey	NI	42	Jeffery (1966)

[a] *NS* = not a significant metabolite.
[b] *NI* = not identifiable as a metabolite.

gesterone was identified as pregnanediol (Romanoff *et al.*, 1963b). Pregnanediol was identified as the principal metabolite in two adult female chimpanzees, representing 39.8% and 49.5% of the recovered radioactivity (YoungLai *et al.*, 1975). Pregnanolone (0.6% and 2.1% of the recovered dose) and pregnanetriol (0.1% and 1.5% of the recovered dose) were also identified, but neither androsterone nor etiocholanolone could be identified as a significant urinary metabolite of progesterone in the chimpanzee. Pregnanediol was identified as the major urinary progesterone metabolite in adult female gorillas also (43.5% and 37.8% of the recovered dose; YoungLai *et al.*, 1977). Androsterone was not identified as a urinary metabolite of progesterone in this primate either. Pregnanolone accounted for less than 1% of the recovered radioactivity, and pregnanetriol was not identified.

Although pregnanediol was the only major urinary progesterone metabolite identified in humans, it accounted for only 26% of the recovered dose (Romanoff *et al.*, 1963a). Pregnanediols (both 5α and 5β) were the major urinary metabolites (41% of the recovered dose) in baboons (*Papio sp.*), but androsterone (36% of the recovered dose) was also a major metabolite (Goldzieher and Axelrod, 1969). Androsterone (1% to 12% of the injected dose) was the predominant urinary metabolite in rhesus monkeys, whereas pregnanediol accounted for only 0.2% to 1.2% of the injected dose (Plant *et al.*, 1971). In pigtail monkeys androsterone was the major metabolite, and pregnanediol could not be identified (Jeffery, 1966). Thus, progesterone metabolism in humans appears to be more similar to that in apes than to that in monkeys, in which there is an increasing production of C_{19} metabolites of progesterone.

The principal urinary metabolite of injected ^{3}H-hydroxyprogesterone-17α in immature chimpanzees was pregnanetriol (Romanoff *et al.*, 1963c). The average urinary recovery of this steroid was also low, ranging from 20% to 55%. If progesterone were metabolized to C_{19} steroids (e.g., androsterone), 17α-hydroxyprogesterone would be an intermediate in this pathway. The failure to identify androsterone as a major urinary metabolite of either progesterone or 17α-hydroxyprogesterone in chimpanzees suggests that the urinary androsterone measured in adult female orangutans (Collins *et al.*, 1975b) arose from a precursor other than progesterone.

The data obtained from these *in vivo* radioisotope studies fail to identify the cause of the relatively low urinary pregnanediol in apes when compared with women. Measurement of production rates of progesterone as well as investigation of other routes of excretion in adult apes would help clarify this aspect of progesterone metabolism.

C. Ovarian Androgens

In addition to the well-known secretion of estrogens and progestins, the ovaries of most mammals secrete significant amounts of androgens, principally androstenedione and testosterone. The metabolism of androgens has not been studied in the female of any species of ape. Nadler *et al.* (1979) measured plasma testosterone concentrations throughout the menstrual cycle in three gorillas and reported a midcycle peak of 350–500 pg/ml compared to follicular and luteal phase levels (150–300 pg/ml). These values are comparable to those seen in women (Ito and Horton, 1970).

IV. THE ADRENAL

A. Endogenous Adrenal Hormones

The adrenal gland of most mammals secretes a variety of steroids, including progestins, glucocorticoids, mineralocorticoids, and androgens. Much of the preceding discussion of the metabolism of androgens and progestins can be applied to the adrenal. For example, the production rate for progesterone in immature chimpanzees of 0.8 mg/24 hr presumably represents adrenal secretion.

Endogenous adrenal hormones have not been much studied in apes. Cutler *et al.* (1978) measured the plasma concentrations of dehydroisoandrosterone, dehydroisoandrosterone sulfate, androstenedione, and cortisol in a series of chimpanzees ranging in age from <1 to >11 years old. The concentration of each androgen increased progressively from 0–3, 3–7, and 7–11 yr, indicating that chimpanzees may undergo adrenarche, a period of increased adrenal androgen secretion associated with sexual maturation. Thus far, adrenarche has been demonstrated only in humans (Parker *et al.*, 1978) and chimpanzees (Cutler *et al.*, 1978). Plasma levels of androstenedione and dehydroisoandrosterone sulfate in chimpanzees are similar to levels in humans at comparable stages of maturity (Parker *et al.*, 1978). However, dehydroisoandrosterone and cortisol were significantly higher in chimpanzees than in humans (Cutler *et al.*, 1978; Parker *et al.*, 1978). Plasma cortisol concentrations in chimpanzees did not show any age-related changes, remaining fairly constant at 200 to 300 ng/ml.

Urinary adrenal steroids in apes generally have been measured as 17-ketosteroids and 17-hydroxycorticosteroids by using fairly nonspecific colorimetric methods. Urinary 17-ketosteroids are considered to be androgen metabolites, and 17-hydroxycorticosteroids represent glucocorticoid metabolites. These quantitative methods do not provide adequate information on urinary adrenal steroids in apes, and specific compounds should be measured.

B. Adrenal Steroid Metabolism

The metabolism of the adrenal androgen 11β-hydroxyandrostenedione has been studied in immature chimpanzees (Layne *et al.*, 1963) and in humans (Bradlow and Gallagher, 1957). Following administration of ^{3}H-hydroxyandrostenedione-11β to four immature chimpanzees, 54% to 100% of the dose was recovered in the urine within 84 hr. The urinary excretion of unconjugated steroid was again high in chimpanzees (7% to 37% of the recovered isotope) when compared with humans (less than 5%). In two animals, 11β-hydroxyetiocholanolone was the major metabolite (11β-hydroxyetiocholanolone: 11β-hydroxyandrosterone = 5:1 and 4:1), while in two other animals, 11β-hydroxyandrosterone was the primary metabolite (2:1 in both animals). The primary metabolite identified in humans, 11β-hydroxyandrosterone, represented almost 60% of the recovered steroid. A secondary metabolite in humans (approximately 10%) was 11-ketoetiocholanolone. The metabolism of 11β-hydroxyandrostenedione appears to be somewhat different in chimpanzees when compared to humans, although the large variation among the chimpanzees studied makes a definite conclusion difficult.

Following the injection of ^{14}C-cortisol into six immature chimpanzees, 5% or less of the isotope was recovered in the urine as unconjugated steroid, while a mean of 53% was recovered as the glucosiduronate conjugate (Layne *et al.*, 1964). In this instance, metabolism in chimpanzees is similar to that in humans (Migeon *et al.*, 1956; Flood *et al.*, 1961). The major metabolites of cortisol in both chimpanzees and humans were tetrahydrocortisol and tetrahydrocortisone. Calculation of the cortisol secretion rate in chimpanzees based on each of these compounds yielded 3.82 mg/24 hr and 3.87 mg/24 hr, respectively (Layne *et al.*, 1964). The cortisol secretion rate is much higher in humans (16 to 18 mg/24 hr; Flood *et al.*, 1961). However, when the secretion rates are expressed on a weight basis, values for chimpanzees and humans are comparable.

Kirdani *et al.* (1963) studied the metabolism of ^{3}H-aldosterone in immature chimpanzees and found that a mean of 67% of the dose was recovered in the urine after 24 hr. Urinary recovery of ^{3}H-aldosterone administered to humans was 91% (Flood *et al.*, 1961), a finding that suggests chimpanzees may excrete significant amounts of aldosterone by routes other than the urine. Ten percent of the radioactive dose was recovered as unconjugated steroid, 28% as steroid sulfates, and 37% as glucosiduronates. The radioactivity consisted of a small amount of unmetabolized aldosterone (<10%) and significant amounts of two other steroids, one of which was tentatively identified as 3α,5β-tetrahydroaldosterone. In humans this isomer is the major urinary glucosiduronate metabolite (Ulick, 1961). The secretion rate of aldosterone in chimpanzees calculated from urinary aldosterone was 26 to 63 μg/24 hr with a mean of 40 μg/24 hr. The absolute secretion rate of aldosterone in chimpanzees is lower than that in humans, although values for the two species are again comparable when expressed on a weight basis.

V. THE FETOPLACENTAL UNIT

During gestation in primates, the fetoplacental unit assumes part or all of the endocrine activities of the corpus luteum and the pituitary. Unlike the case in most other mammals, the fetuses of women and monkeys will survive to term after removal of both the ovaries and the pituitary of the mother, even if the surgery is performed during the first half of gestation. The fetoplacental unit of primates produces both steroid and protein hormones and also performs nonendocrine functions that permit the fetus to develop. This section will focus only on the steroid hormones produced by the fetoplacental unit. Other aspects of fetoplacental functions are discussed by Faiman *et al.* (Chapter 2, this volume).

A. Fetoplacental Progestins

1. *Endogenous Hormone Levels*

Progesterone circulating in maternal plasma is synthesized principally by the placenta (Fig. 5). Plasma progesterone concentrations during gestation have been reported for chimpanzees (Reyes *et al.*, 1975) but not for gorillas or orangutans. Progesterone reached a plateau of 10 to 50 ng/ml by Day 40 of gestation and increased only slightly throughout

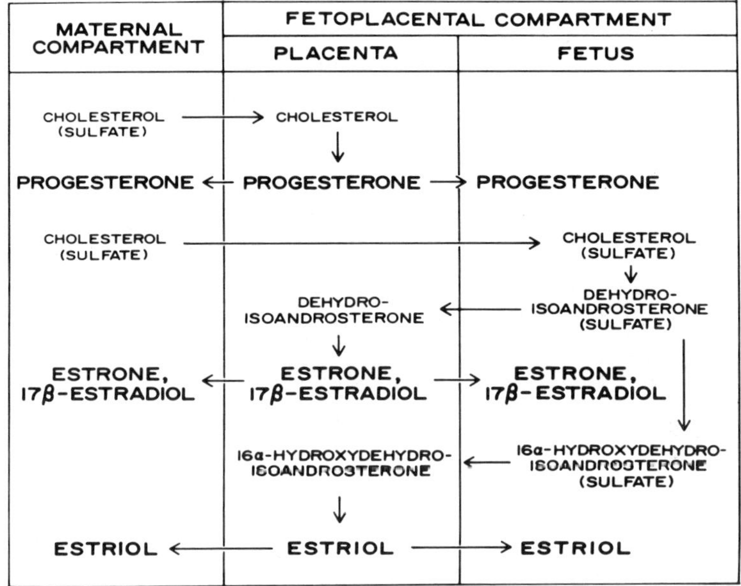

Fig. 5. Steroidogenic activity of the human fetoplacental unit. Progesterone is synthesized primarily by the placenta from maternal precursors. Estrogen synthesis requires participation by both the placenta (for the production of C_{21} steroids and the aromatization of C_{19} steroids) and the fetus (for the cleavage of C_{21} steroids to form C_{19} steroids and for 16α-hydroxylation). Both the placenta and the fetal adrenal gland have the capacity to synthesize cholesterol *de novo*, but most of the progestins and estrogens produced by the fetoplacental unit appear to be derived from maternal cholesterol or cholesterol sulfate. Although little information is available on fetoplacental steroidogenesis in apes, the pathways appear to be quite similar to those in humans.

the remainder of gestation. In two animals plasma progesterone levels at term were 49 and 120 ng/ml (Reyes *et al.*, 1975). These values are slightly lower than those generally seen in women at term (100–150 ng/ml; Johansson, 1969) but are much higher than in rhesus monkeys (2 to 6 ng/ml; Neill *et al.*, 1969).

Urinary progestins during gestation in apes have received little attention. Fish *et al.* (1942) isolated about 2 mg/day of pregnanediol from the urine of two chimpanzees late in pregnancy. As in nonpregnant chimpanzees, pregnanediol excretion in the urine was surprisingly low in view of the concentration of plasma progesterone (Section III,B,1). It is not known whether this apparent discrepancy is due to low production rates, to the occurrence of other metabolites, or to other modes of excretion.

2. Progesterone Metabolism

The lower plasma progesterone levels during gestation in monkeys when compared with women and apes (reviewed by Ryan and Hopper, 1974) had been thought to be of only quantitative importance, but it now appears that substantial qualitative differences may also exist among monkeys, apes, and women. Following incubation *in vitro* of placental tissue from a chimpanzee with pregnenolone, the major radiolabeled compounds identified were pregnenolone and progesterone (Shinada and Ryan, 1973). A similar incubation with progesterone yielded mainly unmetabolized starting material (Shinada and Ryan, 1973). Little radioactivity was present as 5α- or 5β-reduced metabolites. Similar results were obtained using placental tissue from an orangutan, except that, following incubation with pregnenolone, small but significant amounts of 5α-pregnanedione and 5α-pregnane-3β-ol-20-one were identified (Ainsworth and Ryan, 1969).

A Δ^4-5α-reductase has not been identified in human placenta. On the other hand, a wide variety of metabolites, including Δ^4-5α-reduced steroids, are produced when placental tissue from rhesus monkeys is incubated with pregnenolone or progesterone (Ainsworth *et al.*, 1968; Ryan and Hopper, 1974). Thus, progesterone may not be the major progestin secreted by the rhesus monkey placenta, and the lower plasma progesterone concentrations observed in rhesus monkeys when compared with apes and women probably reflect taxonomic differences in steroidogenic pathways. The lower plasma progesterone concentration in apes when compared with women probably reflects quantitative differences only, while the biosynthesis of progesterone is similar among apes and women.

B. Fetoplacental Estrogens

Definitive studies in humans have shown that progesterone is produced by the placenta without any fetal contribution. Thus, progesterone may remain elevated in maternal plasma in the face of intrauterine fetal distress or death. However, the production of estrogens requires the joint effort of both the fetus and the placenta (Fig. 5), and maternal estrogen levels may decrease precipitously in the event of intrauterine fetal distress or death. The placenta produces pregnenolone sulfate, which is transported to the fetus. The fetal adrenal metabolizes pregnenolone sulfate to dehydroisoandrosterone sulfate, which may be metabolized to 16α-hydroxydehydroisoandrosterone sulfate by the fetal liver. The placenta then aromatizes these androgens to estrone and

17β-estradiol, in the case of dehydroisoandrosterone sulfate, and estriol, in the case of 16α-hydroxydehydroisoandrosterone sulfate (Fig. 5).

1. Endogenous Hormone Levels

After a transient decline between 30 and 60 days of gestation, plasma estrogens in chimpanzees increased to term (Reyes *et al.*, 1975), when plasma estrone ranged from 1.8–3.0 ng/ml, 17β-estradiol ranged from 5–8 ng/ml, and plasma estriol ranged from 4–10 ng/ml. In women at term, the mean plasma estrone was 12 ng/ml; mean 17β-estradiol, 26 ng/ml; and mean estriol, 6 ng/ml (Loriaux *et al.*, 1972). Plasma estrone and 17β-estradiol are three to four times higher in women at term than in chimpanzees, although estriol levels are comparable.

The excretion of estrogens, and particularly of estriol, has been shown to increase steadily throughout gestation in women (Cohen *et al.*, 1935). In women at term, estriol represents at least 90% of the urinary estrogens and is excreted at the rate of 20 mg/24 hr (Cohen *et al.*, 1935; Brown, 1956). Estriol excretion in a pregnant gorilla increased rapidly from 20 μg/g creatinine at nine weeks of gestation to 2890 μg/g creatinine at term (Hopper *et al.*, 1968). The excretion of estrone also increased significantly from 50 μg/g creatinine to 1200 μg/g creatinine at term; that of 17β-estradiol remained low with very little change (20 to 100 μg/g creatinine). Estrogen excretion fell markedly by three days postpartum, although the excretion of estriol was four times that of estrone. No estriol was detected in a sample collected five months postpartum (Hopper *et al.*, 1968). Using a nonspecific estrogen radioimmunoassay, Seaton (1978) also showed an increase in urinary estrogens with gestation in a gorilla. In rhesus monkeys, the excretion of estrone, 17β-estradiol, and estriol increased with gestation (Hopper and Tullner, 1967). However, the amounts of each estrogen excreted were much lower, and estrone was the major urinary estrogen (estrone, 3.96–45.47 μg/24 hr at term; 17β-estradiol, 0.19–3.84 μg/24 hr; estriol, 0.08–2.18 μg/24 hr; Hopper and Tullner, 1967). Merkatz and Beling (1969) have reported similar findings for yellow baboons (estrone, 80–167 μg/24 hr at term; 17β-estradiol, 8–19 μg/24 hr; estriol, 8–19 μg/24 hr).

The amount of estrogens excreted increases progressively from monkeys to apes to women. Although pregnant monkeys excreted negligible amounts of estriol, a pregnant gorilla excreted twice as much estriol as estrone, and estriol represented 90% of urinary estrogens in women at term. Both qualitative and quantitative differences exist between monkeys and humans in estrogen excretion during pregnancy, whereas the differences between apes and humans are largely quantitative.

Despite the apparent similarity of fetoplacental unit function in apes and in women, little information is available on estrogens during pregnancy in apes. Of particular interest would be studies of the plasma concentrations of estrogen conjugates. Estriol-3-sulfate and estriol-3-sulfate,16α-glucosiduronate have been shown to be quantitatively the most important plasma estrogens in women during the third trimester of pregnancy (Young *et al.*, 1976). Although the concentrations of both conjugates generally exceeded that of the unconjugated estrogens, their biochemical and physiological significance is unknown. Apes may prove to be an excellent model for studying estrogens and estrogen conjugates during pregnancy.

2. Estrogen Metabolism

On the basis of histological studies, the human fetal adrenal gland has been divided into two zones, called the definitive zone and the fetal zone. During fetal and early neonatal life, the fetal zone comprises the bulk of the adrenal cortex. The fetal zone regresses after birth, however, and the adult adrenal cortex is composed of the definitive zone. Recently, Serón-Ferré *et al.* (1978) demonstrated that, on incubation *in vitro* with or without ACTH, the fetal zone of the human fetal adrenal produced mainly dehydroisoandrosterone sulfate, while the definitive zone produced mainly cortisol. Histological evidence indicates that the fetal adrenal of apes may also contain both a fetal and definitive zone (Lanman, 1957). If the fetal zone of the ape adrenal is responsible for the production of dehydroisoandrosterone sulfate, then the similarities in estrogen patterns in apes and women during gestation may be explained.

Shackleton *et al.* (1970), Manson *et al.* (1971), Shackleton and Gustafsson (1971), and Shackleton (1974) have examined the qualitative pattern of urinary and fecal steroid excretion by neonatal primates. The majority of the steroids identified in a human, a chimpanzee, and an orangutan were 3β-hydroxy-5-ene steroids, most of which contained 16α- or 16β-hydroxyl groups. Steroids excreted by a neonatal rhesus monkey and a neonatal cynomolgus monkey were generally 5α- or β-reduced, and few 16α-hydroxylated steroids were detected. These findings support the suggestion that the high estrogen levels during pregnancy in apes and humans may be attributed to the production of aromatizable androgens by the fetal zone of the adrenal via the 3β-hydroxy-5-ene pathway. Thus, the steroidogenic properties of the fetoplacental units of humans and apes appear to be quite similar.

In vitro studies of estrogen metabolism using chimpanzee placental tissue (Shinada and Ryan, 1973) and orangutan placental tissue (Ainsworth and Ryan, 1969) have been reported. The chimpanzee placenta could aromatize androstenedione to estrogens, but could not convert pregnenolone or progesterone to C_{19} steroids or to estrogens (Shinada and Ryan, 1973). The ability of placental enzymes to convert C_{19}, but not C_{21}, steroids to estrogens has been demonstrated in all primates studied, including humans (Ryan and Hopper, 1974). These findings also suggest that the steroidogenic pathways in the fetoplacental compartment (Fig. 5) are similar in apes and humans.

VI. SUMMARY

Steroid hormone metabolism can be compared among humans, apes, and monkeys from several aspects. The patterns and concentrations of steroids circulating in the plasma in apes are similar to those in humans. The patterns of urinary steroids are generally similar in apes and humans, but the total amounts of steroid excreted in the urine by apes are frequently lower than those excreted by humans. In monkeys the levels of both plasma and urinary steroids are generally lower than in humans, and marked differences in patterns of steroids are apparent. As examples, monkeys (1) do not exhibit a luteal plasma or urinary estrogen peak, (2) excrete negligible amounts of estriol, (3) have a low production of progesterone during gestation, and (4) excrete androsterone as a major metabolite of progesterone.

Steroid hormone metabolism in apes more closely resembles that in humans than does steroid metabolism in monkeys. Many of the taxonomic differences in the patterns and amounts of steroid hormones among primates can now be attributed to differences in metabolic pathways. However, many aspects of steroid metabolism in apes remain unclear or unknown. The formation of catechol estrogens in apes has been reported only for the pygmy chimpanzee, and the methodology generally used would have destroyed most catechol estrogens present. The role of the enterohepatic circulation in steroid metabolism has not been studied in apes, although it plays a major role in estrogen metabolism in humans and baboons. The occurrence, nature, and function of steroid conjugates in apes have received little attention. Differential hydrolysis of conjugates has provided some basic information, but more sophisticated methods of steroid conjugate analysis and identification should be employed for both urine and plasma samples from apes. The relatively low excretion of steroids in the urine by apes when plasma

                                                      Kristina Wright et al.

concentrations are comparable to those in humans needs to be explained. The reason for the relatively high urinary excretion of unconjugated steroid should be clarified. Pathways of steroid biosynthesis in apes also have received little attention.

Therefore, while sufficient information is available to indicate that the metabolism of steroids in apes is significantly different from that in monkeys and more similar to that in humans, considerably more information is necessary for a full understanding of steroid biosynthesis and metabolism in apes. This knowledge could give important clues about the evolution of the reproductive process in primates.

ACKNOWLEDGMENTS

The authors gratefully acknowledge the editorial assistance of Dr. Charles E. Graham, Dr. Mark E. Wilson, and Ms. Ava Wilhite, and the secretarial assistance of Ms. Deana Perry and Ms. Sandra Milline.

REFERENCES

Abraham, G. E., Odell, W. D., Swerdloff, R. S., and Hopper, K. (1972). *J. Clin. Endocrinol. Metab.* **34,** 312.

Ainsworth, L. and Ryan, K. J. (1969). *Steroids* **14,** 301.

Ainsworth, L., Daenen, M., and Ryan, K. J. (1968). *Endocrinology* **84,** 1421.

Allen, E., Diddle, A. W., and Elder, J. H. (1935). *Am. J. Physiol.* **110,** 593.

Allen, E., Diddle, A. W., Burford, T. H., and Elder, J. H. (1936). *Endocrinology* **20,** 546.

August, G. P., Grumbach, M. M., and Kaplan, S. L. (1972). *J. Clin. Endocrinol. Metab.* **34,** 319.

Bardin, C. W. (1978). *In* "Reproductive Endocrinology" (S. S. C. Yen and R. B. Jaffe, eds.), pp. 110–125. Saunders, Philadelphia, Pennsylvania.

Bernstein, I. S., Gordon, T. P., Rose, R. M., and Peterson, M. S. (1978). *Behav. Biol.* **24,** 400.

Bradlow, H. L. and Gallagher, T. F. (1957). *J. Biol. Chem.* **229,** 505.

Breckwoldt, M., Flickinger, G. L., Murawec, T., and Touchstone, J. C. (1972). *Endocrinology* **91,** 1.

Brown, J. B. (1956). *Lancet* **270,** 704.

Brown, J. B., Klopper, A., and Loraine, J. A. (1958). *J. Endocrinol.* **17,** 401.

Callow, N. H. (1939). *Biochem. J.* **33,** 559.

Cohen, S. L., Marrian, G. F., and Watson, M. (1935). *Lancet* **i,** 674.

Collins, D. C., Balikian, H. M., and Preedy, J. R. K. (1975a). *Endocrinology* **96,** 1343.

Collins, D. C., Graham, C. E., and Preedy, J. R. K. (1975b). *Endocrinology* **96,** 93.

Crowell, G. C., Turner, M. E., Jr., Schmidt, F. H., Howard, C. M., and Preedy, J. R. K. (1967). *J. Clin. Endocrinol. Metab.* **27,** 807.

Cutler, G. B., Jr., Glenn, M., Bush, M., Hodgen, G. D., Graham, C. E., and Loriaux, D. L. (1978). *Endocrinology* **103,** 2112.

Dorfman, R. I. (1970). *In* "The Chimpanzee" (G. H. Bourne, ed.), Vol. 3, pp. 167–182. Karger, New York.

Elmadjian, F. and Forchielli, E. (1965). *In* "Hormonal Steroids, Biochemistry, Pharmacology and Therapeutics" (*Proc. Int. Congr. Hormonal Steroids, 1st*), Vol. 2, pp. 535–544. Academic Press, New York.

Fish, W. R. and Dorfman, R. I. (1944). *Endocrinology* **35**, 22.

Fish, W. R., Young, W. C., and Dorfman, R. I. (1941). *Endocrinology* **28**, 585.

Fish, W. R., Dorfman, R. I., and Young, W. C. (1942). *J. Biol. Chem.* **143**, 715.

Flood, C., Layne, D. S., Ramcharan, S., Rossipal, E., Tait, J. F., and Tait, S. A. (1961). *Acta Endocrinol. (Kbh.)* **36**, 237.

Gelbke, H. P., Ball, P., and Knuppen, R. (1977). *Adv. Steroid Biochem. Pharmacol.* **6**, 81.

Goebelsmann, U., Midgley, A. R., Jr., and Jaffe, R. B. (1969). *J. Clin. Endocrinol. Metab.* **29**, 1222.

Goldzieher, J. W. and Axelrod, L. K. (1969). *Gen. Comp. Endocrinol.* **13**, 201.

Gordon, T. P., Rose, R. M., and Bernstein, I. S. (1976). *Horm. Behav.* **7**, 229.

Graham, C. E., Collins, D. C., Robinson, H., and Preedy, J. R. K. (1972). *Endocrinology* **91**, 13.

Graham, C. E., Wright, K., Collins, D. C., and Preedy, J. R. K. (1974). *IRCS* **2**, 1697.

Harkness, R. A., and Fotherby, K. (1963). *Biochem. J.* **88**, 308.

Hobson, W., Coulston, F., Faiman, C., Winter, J. S. D., and Reyes, F. (1976). *J. Toxicol. Environ. Health* **1**, 657.

Hodgen, G. D., Dufau, M. L., Catt, K. J., and Tullner, W. W. (1972). *Endocrinology* **91**, 896.

Hopper, B. R., and Tullner, W. W. (1967). *Steroids* **9**, 517.

Hopper, B. R., and Tullner, W. W. (1970). *Endocrinology* **86**, 1225.

Hopper, B. R., Tullner, W. W., and Gray, C. W. (1968). *Proc. Soc. Exp. Biol. Med.* **129**, 213.

Hotchkiss, J., Atkinson, L. E., and Knobil, E. (1971). *Endocrinology* **89**, 177.

Ito, T., and Horton, R. (1970). *J. Clin. Endocrinol. Metab.* **31**, 362.

Jeffery, J. D.'A. (1966). *J. Endocrinol.* **34**, 387.

Jirku, H. and Layne, D. S. (1965). *Steroids* **5**, 37.

Johansson, E. D. B. (1969). *Acta Endocrinol. (Kbh.)* **61**, 607.

Kirdani, R. Y., Gleason, T. L., and Layne, D. S. (1963). *Gen. Comp. Endocrinol.* **3**, 616.

Kling, O. R. and Westfahl, P. K. (1978). *Biol. Reprod.* **18**, 392.

Klopper, A. and Michie, E. A. (1956). *J. Endocrinol.* **13**, 360.

Labow, R. S., Williamson, D. G., Layne, D. S., and Collins, D. C. (1975). *Can. J. Biochem.* **53**, 1028.

Lanman, J. T. (1957). *Endocrinology* **61**, 684.

Layne, D. S. *et al.* (1963). Technical Documentary Rep. No. ARL-TDR-63-25.

Layne, D. S., Kirdani, R. Y., Gleason, T. L., and Pincus, G. (1964). *Gen. Comp. Endocrinol.* **4**, 155.

Loriaux, D. L., Ruder, H. J., Knab, D. R., and Lipsett, M. B. (1972). *J. Clin. Endocrinol. Metab.* **35**, 887.

Manson, M. E., Shackleton, C. H. L., Mitchell, F. L., Gustafsson, J.-Å., and Sjövall, J. (1971). *Steroids* **18**, 51.

Martin, D. E., Swenson, R. B., and Collins, D. C. (1977). *Steroids* **29**, 471.

McCormack, S. A. (1971). *Endocrinology* **89**, 1171.

Merkatz, I. R. and Beling, C. G. (1969). *J. Reprod. Fertil. Suppl.* **6**, 129.

Metcalf, M. G. (1968). *Anal. Biochem.* **25**, 510.

Migeon, C. J., Sandberg, A. A., Decker, H. A., Smith, D. F., Paul, A. C., and Samuels, L. T. (1956). *J. Clin. Endocrinol. Metab.* **16**, 1137.

Migeon, C. J., Wall, P. E., and Bertrand, J. (1959). *J. Clin. Invest.* **38**, 619.

Monroe, S. E., Atkinson, L. E., and Knobil, E. (1970). *Endocrinology* **87**, 453.

Musey, P. I., Gould, K. G., Graham, C. E., and Collins, D. C. (1979). *Biol. Reprod. Suppl. 1* **20**, 142A.

Nadler, R. D., Graham, C. E., Collins, D. C., and Gould, K. G. (1979). *Endocrinology* **105**, 290.

Neill, J. D., Johansson, E. D. B., and Knobil, E. (1967). *Endocrinology* **81**, 1161.

Neill, J. D., Johansson, E. D. B., and Knobil, E. (1969). *Endocrinology* **84**, 45.

Parker, L. N., Sack, J., Fisher, D. A., and Odell, W. D. (1978). *J. Clin. Endocrinol. Metab.* **46**, 396.

Plant, T. M., James, V. H. T., and Michael, R. P. (1971). *J. Endocrinol.* **51**, 751.

Plant, T. M., Zumpe, D., Sauls, M., and Michael, R. P. (1974). *J. Endocrinol.* **62**, 403.

Reyes, F. I., Winter, J. S. D., Faiman, C., and Hobson, W. C. (1975). *Endocrinology* **96**, 1447.

Romanoff, L. P., Morris, C. W., Welch, P., Grace, M. P., and Pincus, G. (1963a). *J. Clin. Endocrinol. Metab.* **23**, 286.

Romanoff, L. P., Grace, M. P., Sugarman, E. M., and Pincus, G. (1963b). *Gen. Comp. Endocrinol.* **3**, 649.

Romanoff, L. P., Grace, M. P., Sugarman, E. M., and Pincus, G. (1963c). *Gen. Comp. Endocrinol.* **3**, 655.

Ryan, K. J. and Hopper, B. R. (1974). *Contrib. Primatol.* **3**, 258.

Sandberg, A. A. and Slaunwhite, W. R., Jr. (1956). *J. Clin. Invest.* **35**, 1331.

Schubert, K. and Schade, K. (1975). *Endokrinologie* **66**, 228.

Seaton, B. (1978). *J. Reprod. Fert.* **53**, 231.

Serón-Ferré, M., Lawrence, C. C., Siiteri, P. K., and Jaffe, R. B. (1978). *J. Clin. Endocrinol. Metab.* **47**, 603.

Shackleton, C. H. L. (1974). *J. Steroid Biochem.* **5**, 113.

Shackleton, C. H. L. and Gustafsson, J.-Å. (1971). *Steroids* **18**, 175.

Shackleton, C. H. L., Mitchell, F. L., Gustafsson, J.-Å., and Sjövall, J. (1970). *FEBS Lett.* **11**, 129.

Shinada, T. and Ryan, K. J. (1973). *Steroids* **21**, 233.

Siiteri, P. K. and MacDonald, P. C. (1963). *Steroids* **2**, 713.

Stabenfeldt, G. H. and Hendrickx, A. G. (1972). *Endocrinology* **91**, 614.

Stabenfeldt, G. H. and Hendrickx, A. G. (1973). *Endocrinology* **92**, 1296.

Stern, M. D. and Givner, M. L. (1978). *Endocrinology* **102**, 223A.

Stevens, V. C., Sparks, S. J., and Powell, J. E. (1970). *Endocrinology* **87**, 658.

Ulick, S. (1961). *J. Biol. Chem.* **236**, 680.

Vermeulen, A., Verdonck, L., Van der Straeten, M., and Orie, N. (1969). *J. Clin. Endocrinol. Metab.* **29**, 1470.

Williamson, D. G., Layne, D. S., and Collins, D. C. (1972). *J. Biol. Chem.* **247**, 3286.

Wright, K., Collins, D. C., Musey, P. I., and Preedy, J. R. K. (1978). *J. Clin. Endocrinol. Metab.* **47**, 1092.

Yoshimi, T., and Lipsett, M. B. (1968). *Steroids* **11**, 527.

Young, B. K., Jirku, H., Kadner, S., and Levitz, M. (1976). *Am. J. Obstet. Gynecol.* **126**, 38.

YoungLai, E. V., Graham, C. E., and Collins, D. C. (1975). *Steroids* **25**, 465.

YoungLai, E. V., Collins, D. C., and Graham, C. E. (1977). *J. Endocrinol.* **75**, 439.

Chapter 8

LABORATORY RESEARCH ON SEXUAL BEHAVIOR OF THE GREAT APES

Ronald D. Nadler

191

Copyright © 1981 by Academic Press, Inc.
All rights of reproduction in any form reserved.
ISBN 0-12-295020-8

I.   INTRODUCTION

Laboratory research on sexual behavior of the great apes was initiated more than 50 years ago in relation to two basic issues: (1) the comparative relationship among the higher primates, monkey, ape, and human, especially with respect to the frequency and distribution of mating during the menstrual cycle, i.e., the issue of estrus or the female's willingness to copulate with a male; and (2) the relative influence of experiential factors, in contrast to innate genetic factors, on the development of species-typical mating patterns (Bingham, 1928; Tinklepaugh, 1933; Yerkes and Elder, 1936). Controversy regarding the biological relatedness of species derived, of course, from the work of Darwin (1859), but was heightened following Huxley's (1893) assertion that the apes more closely resembled humans (anatomically) than they did the monkeys.

Early reports describing the sexual relations of monkeys and apes were divided, however, with respect to the patterns of mating attributed to these animals. Some authors reported that monkeys, as well as apes and humans, mated acyclically, irrespective of female sexual status (Hamilton, 1914; Kempf, 1917; Miller, 1928, for monkeys; Montané, 1915; Sokolowsky, 1923, for chimpanzees). Others reported that female chimpanzees and monkeys exhibited fairly discrete periods of estrus associated with swelling of the perineal area (Keith, 1899; Köhler, 1925; Zuckerman, 1930). It was, in part, as response to this controversy that Bingham (1928) conducted the first longitudinal study of sexual behavior development in chimpanzees.

Prior to initiating his study, Bingham accompanied R. M. Yerkes to the private estate of Madam Rosalía Abreu in Cuba, where he spent several months observing the sexual relations of monkeys and chimpanzees in the owner's personal collection. Yerkes arranged the visit because he was convinced that experimental research on monkeys and apes could make a unique contribution to the study of experiential determinants of behavior, "facilitate the development of human biology, and thus help to provide a factual basis for biological engineering"

(Yerkes, 1925, p. 289). As a result of his conviction, Yerkes conducted, inspired, and/or facilitated much of the early research on chimpanzees, nonsexual as well as sexual, and established one of the first research laboratories devoted to scientific investigation of nonhuman primates (Yerkes, 1932). The purpose of his and Bingham's visit to the Abreu estate was to gather information on the behavior and captive maintenance of nonhuman primates, in preparation for the establishment of a research colony. Madam Abreu was one of the first individuals to establish adequate conditions in captivity for long-term maintenance and breeding of primates (Yerkes, 1925) and successfully reared in her colony one of the first chimpanzees born in captivity (Montané, 1915).

II.   THE ONTOGENY OF SEXUAL BEHAVIOR IN CHIMPANZEES

A.   Early Social Experience

During his observations of the Abreu primates, Bingham noted several behavioral characteristics of comparative significance toward which his subsequent study was directed. He suspected that "the copulatory adjustments of monkeys tend more towards precision and uniformity than those of the apes," and he proposed that "the quicker arrival of monkeys at reproductive maturity implies a more rapid perfection of sexual adjustments. . . ." (Bingham, 1928, p. 21). He was especially intrigued by the sexual indifference of one captive-born, physically mature male chimpanzee and speculated that the animal's lack of sexual responsiveness might have resulted from early social deprivation (!) and the consequent absence of opportunities to observe copulation and competition for females with other males. It is noteworthy that essentially all subsequent studies of the ontogeny of sexual behavior in chimpanzees attempted to resolve this very same issue, i.e., the influence of early social experience on the development of competent sexual behavior.

Despite the differences he observed between sexual behavior of monkeys and chimpanzees at the Abreu estate, Bingham recognized that it would be unwise to draw inferences regarding species-typical patterns of mating from the behavior of adult animals about whom nothing was known of their past history. He considered his observations and the conflicting reports in the literature regarding sexual behavior of nonhuman primates to represent merely cross-sections of their subjects' life histories. He noted that the data available on sexual behavior of

most captive primates were insufficient to permit clarification of behavioral phenomena attributable "to genus, age, past experience, mating opportunities or seasonal circumstances" (p. 20). In particular, he
lamented the lack of information on the development of sexual behavior
in any primate species and emphasized the importance of such information for resolving issues regarding species-typical patterns of mating,
especially for species in which experience and learning were thought
to play an important role. In further support of the ontogenetic approach, he cited its value in resolution of a related controversy among
early anatomists regarding the comparative relationship among primate
genital structures. He noted that "a more satisfactory explanation of
their subject matter (was achieved) when it was arranged in its developmental relations" (Bingham, 1928, p. 9). In his behavioral study,
therefore, Bingham sought to define relatively early experiential contributions to the development of sexual behavior. He conducted his
study over a two-year period with four wild-born chimpanzees, beginning when the animals were estimated to be two to three years of age.
His major interest in innate and acquired aspects of sexual behavior
reflected the influence of the nature-nurture controversy, still active at
that time.

The study of the young chimpanzees consisted initially of daily pairings of male and female partners, there being two of each sex. Eventually, partners were exchanged, group encounters were arranged and
isosexual pairings were studied. Bingham found considerable variability
in the behavior of his subjects and concluded that the development of
sexual behavior in chimpanzees was largely an individual matter. The
animals did not achieve complete patterns of sexual interaction during
the study, but they did exhibit increased genital contact and stimulation
over time, approximating the adult form, but generally lacking penetration. Bingham observed many components of behavior in the
youngsters' social interactions which had no apparent sexual significance when first displayed, but which were observed at later ages in
the context of sexual contacts. He proposed that certain early actions
provide the basis for much adult behavior, including sexual, and that
these actions, originally nonsexual, acquire sexual significance as they
occur in conjunction with increasing sexual responsiveness. Behavior
such as pushing and pulling, originally having no sexual implications,
was later observed in the initiation of copulatory adjustments by the
males. Rushing about and embracing during excitement, and flight
from novel stimuli, represent other patterns whose components were
later rearranged and participated in interactions with a conspicuous
sexual focus. Bingham concluded that the emergence of adequate sexual

behavior in chimpanzees was "a new synthesis of responses. . ." that involved "no new elements of behavior" (p. 78), but, rather, a novel combination or reorganization. He proposed that further research would confirm a relationship of increasing variability in the motor patterns of primate sexual behavior, from monkey to ape to human, and that this reflected, in part, an increased role of early experience and learning in behavioral development. He viewed bodily contact as the basic factor in sexual behavior development, especially at the younger ages, and excitement, a consistent precursor to sexual responses. An important hypothesis that derived from this study asserts that early social experience in these higher primates is a necessary condition for the development of competent sexual behavior, presumably by providing a responsive organism with which to practice and elaborate several primitive response systems. The remaining studies that followed the ontogenetic approach were investigations of the effects on sexual behavior development of several different rearing conditions which varied in the amount of early social experience available to the subjects. It is pertinent to note for later considerations that all four of the chimpanzees studied by Bingham, animals separated from their mothers at two to three years of age, matured and exhibited normal mating patterns and fertility (see Tinklepaugh, 1933; Yerkes and Elder, 1936).

B. Early Social Restriction

Subsequent work on the development of sexual behavior by chimpanzees was delayed until about 10 years after the establishment of the Yerkes Laboratories of Primate Biology at Orange Park, Florida. The study of sexual behavior was but one aspect of the Infant Studies Program which was inaugurated in 1939, under the direction of Henry W. Nissen. Nissen described the objectives of the program in his foreword to Riesen and Kinder's (1952) monograph on chimpanzee postural development as "designed to provide information about the physical, physiological, and behavioral development of chimpanzees, reared from birth under controlled and uniform conditions" (p. *x*). A preliminary report of the study was presented by Nissen (1954), but the complete results were not published until some years after his death (Riesen, 1971).

All the chimpanzees in Nissen's study (seven females and nine males) were born in the Laboratories and separated from their mothers shortly after birth, most of them on the day of birth, and were reared in separate cribs in the nursery for the first two years of life. During this time, they had considerable contact with human caretakers and investigators (e.g., Riesen and Kinder, 1952) and visual and auditory stimulation from each other, but lacked physical contact with conspecifics. The infants were first given social experience with each other during the third year of life, initially in pairs and subsequently in mixed-sex living groups in which they remained until they were 6–7½ years of age. They were then placed in isosexual groups prior to initiating the tests of sexual behavior.

Nissen used three different conditions to assess the chimpanzees' sexual behavior: (1) relatively brief pair-tests with restricted animals only; (2) extended periods of living together in pairs with restricted animals only; and (3) finally, because of the lack of success in the former conditions, testing and living with sexually experienced chimpanzees of the Laboratories. The lack of sexual responsiveness by the chimpanzees surprised Nissen, since wild-born animals reared in the Laboratories copulated and reproduced effectively (Tinklepaugh, 1933; Yerkes and Elder, 1936). The laboratory-born chimpanzees, however, failed to copulate at all during approximately two years of pair-testing, and only one pair among 29 copulated after they were allowed to live together for several months. It may be relevant that the male of this pair was the only animal in the study that was given some limited social experience with older animals prior to the second year of life.

Nissen then placed experimental subjects in adjoining cages or into cages with sexually experienced chimpanzees in hopes that the former would learn from the latter. Again, however, there was no success; the experienced chimpanzees copulated with each other, but the experimental animals essentially ignored them. The experimental animals were finally allowed to live with experienced partners, who successfully "tutored" all but one of the remaining females, but only one additional male. This second successful male was the only experimental animal that began living with other chimpanzees before two years of age.

The results of Nissen's study supported the findings of Bingham (1928) in suggesting that "a social learning process is critical for the full development of chimpanzee reproductive behavior" (Riesen, 1971, p. 16). It was clear that captive-rearing per se was not the critical variable in the deficient sexual behavior of these animals because, as mentioned above, wild-born animals reared in the laboratory from two to three years of age developed appropriate sexual behavior. The difference between these wild-born chimpanzees and the laboratory-born animals was that the latter spent their first two years in physical isolation from conspecifics, whereas the former presumably lived in complex social groups with frequent and varied stimulation from conspecifics. It was concluded, therefore, that conspecific social experience was critical during the period of development from shortly after birth until approximately two years of age for acquisition of species-typical patterns of sociosexual interaction in this species.

C.   Varied Conditions of Early Social Restriction

The last of the longitudinal studies of sexual behavior development of chimpanzees to be considered was part of "a long-term multi-environment differential rearing project" (Davenport and Rogers, 1970, p. 338), which was a logical extension of Nissen's study (Riesen, 1971). The objective of this study was to compare the effects on behavioral development of four different artificial rearing conditions which were designed to provide considerable variation in environmental stimulation. Infant chimpanzees were separated from their mothers within 12 hours after birth and reared in enclosed cubicles for the first two years of life. The four different conditions of restriction ranged from the "extreme" condition in which the cribs were completely devoid of visual stimuli and manipulatable objects to the "social enrichment" condition in which two cribs were placed end-to-end, permitting the infants to see each other and to interact with each other somewhat through bars at the ends of the cribs. Following their removal from the cribs at three years of age, the restricted animals (six females and five males) were housed in pairs or small groups and subjected to a variety of behavioral tests and evaluations (Davenport and Rogers, 1970). Beginning when the animals were seven to 10 years of age, they were tested for sexual behavior in several different conditions, ranging from single pairs of only restricted-reared animals to large groups containing both restricted and wild-born animals (Rogers and Davenport, 1969).

One of the interesting findings of this study was that there appeared to be no difference in the behavior of the restricted chimpanzees as a function of the different conditions of restriction (Davenport and Rogers, 1970). All these restricted animals, however, exhibited more frequent and varied stereotypies than other nursery-reared animals under less restriction, e.g., those of Nissen. Regarding their sexual behavior overall, five of the six restricted females copulated and three of the five males achieved intromissions (of which two ejaculated) (Rogers and Davenport, 1969). In both sexes, successful copulation occurred initially only among pairs in which one member was wild-born. The restricted females neither solicited copulation initially nor did they approach males that solicited them, but they did accept the direct mating attempts of the males. Eventually, these females exhibited sexual presenting that did not differ in frequency from a comparison group of wild-born females. Regarding the restricted males, "Copulations were the result of the skill and persistence of the (experienced) females who frequently effected intromission by trapping the male in a corner and backing onto the erect penis" (Rogers and Davenport, 1969, p. 202). The copulatory behavior of these few successful males did improve somewhat over time, but their intromissions continued to be relatively brief, compared to wild-born males, and were frequently interrupted by the performance of stereotypies.

The results of this study (Rogers and Davenport, 1969) supported those of the two previous ones (Bingham, 1928; Riesen, 1971), confirming the importance of early social experience in the development of sexual behavior of chimpanzees. Some additional data on several other chimpanzees reared in captivity support a similar conclusion (Lemmon, 1971). The chimpanzees appeared to differ from monkeys that had been treated similarly in that some of the chimpanzees were able to overcome the effects of early restriction, whereas the monkeys were not. The authors proposed that this difference might be due to the more variable, less stereotyped pattern of sexual behavior in chimpanzees, as compared to monkeys, and an increased potential for modification as a result of experience. The authors also suggested that the deficits in sexual behavior of the restricted chimpanzees were less severe than those of nursery-reared chimpanzees that received more contact with humans, e.g., those of Nissen (1954). Comparison of their results (Rogers and Davenport, 1969) with Nissen's complete data (Riesen, 1971), however, does not support this conclusion. In both studies, the females were less seriously impaired than the males, and most of them eventually copulated with experienced males, e.g., 6/7 for Nissen and 5/6 for Rogers and Davenport. A somewhat greater proportion of males copulated in the Rogers and Davenport study (3/5) than in Nissen's (4/9), but these differences are not significant.

D. Summary of Results

To summarize the results described above, chimpanzees allowed to remain with their mothers for the first few years of life before separation, and then given peer contact after separation from their mothers, develop competent patterns of sexual behavior (Bingham, 1928; Yerkes and Elder, 1936). Chimpanzees separated from their mothers at birth and reared in social isolation for the first two years of life and then given peer contact exhibit gross deficiencies in their postpubertal sexual behavior, especially during their initial introductions to opposite-sexed partners (Riesen, 1971; Rogers and Davenport, 1969). When their experience is limited to similarly restricted partners, little or no sexual activity is exhibited. On the other hand, when exposed to experienced "tutors," most of the females and some of the males develop reasonable approximations of the species-typical mating pattern. Their behavior in other respects remains aberrant, however, especially in terms of frequent and varied stereotyped activity, e.g., rhythmical rocking or swaying and repetitive movement of individual body parts. The data suggest that while early social experience is important for the development of sexual behavior by chimpanzees, it is not indispensable. Later experience, in the form of patient and persistent tutoring by an opposite-sexed, sexually experienced conspecific, can compensate, to some extent, for the early deficit. The ability of the chimpanzee, in contrast to the rhesus monkey, to profit from this type of later experience may be related to its superior adaptive intelligence (Mason *et al.*, 1968; Davenport and Rogers, 1970).

E.   Theoretical Considerations

Mason (1971) described the behavioral development of monkeys and apes in terms of two major trends, filial and exploitative, the relative influences of which change during the course of development. The filial, or contact-seeking, trend is viewed as being predominant during the early postnatal period and serving an arousal-reducing function. The exploitative trend becomes increasingly important during early childhood, persists during adulthood, and serves an arousal-enhancing function. According to this model, the effect of early social restriction "is viewed as a manifestation of these two major developmental trends, operating on reduced or altered environmental input" (p. 65). When the mother, the primary object toward which early filial responses are directed, is separated from the infant during the earliest stages of the postnatal period and the infant is reared in social isolation, there is a redirection of those responses, with the result being development of self-clinging, rocking, and other stereotyped behaviors. Support for this aspect of the model derives from the finding that essentially all chimpanzees that have been separated from their mothers shortly after birth have developed stereotyped behavior, regardless of the "richness" of the environment into which they were placed (Nissen, 1956). According to this model, separation from the mother and rearing in social isolation after the early neonatal period has little or a lesser effect on filial responses, but a negative effect on exploitative responses, especially those involved in the development of social skills. This prediction of the model receives partial support from the finding that virtually no chimpanzee at the Laboratories, separated from its mother after five to six months of age, has been observed to exhibit behavioral stereotypies (Davenport and Rogers, 1970). The second part of the prediction, related to assessment of the effects of social restriction on exploitative responses, independent of earlier social restriction, has not been tested adequately.

If the model is strictly interpreted, it predicts that allowing the infant to remain with its mother for the first five or six months would permit development of those responses for which contact with the mother is of the greatest inportance, i.e., filial. Since exploitative responses of a social nature depend importantly on peer interaction in chimpanzees (Mason, 1971), placing the separated chimpanzee with peers during this later developmental stage should permit elaboration of those responses and an increased probability of the development of competent sexual behavior at puberty. This is not to imply that the mother plays no role in the development of exploitative responses by the infant, because Mason described quite clearly the several ways in which the mother is involved in this process. The model does suggest, however, that peer interaction after five or six months of age will have a positive influence on development of social responses, whereas prior to that time, it is unlikely to do so. Assessment of this prediction has significance for understanding the development of social, including sexual, behavior of chimpanzees and obvious implications for the breeding of these animals in captivity.

A final note is appropriate regarding the stimulus characteristics of "mother" that contribute to the development of appropriate filial responding, at least as defined by the absence of behavioral stereotypies. Mason and Berkson (1975), in a study of rhesus monkeys, reported that the development of stereotyped behavior as a consequence of early maternal separation was prevented by providing the infants with an artificial mother that moved intermittently and irregularly about the cage. Monkeys treated identically, except for having an artificial mother that did not move, developed typical stereotyped behavior patterns. Whether or not and, if so, to what extent movement of an artificial mother would influence filial responding of infant chimpanzees has not been studied, nor has the question been addressed experimentally of how such early rearing would influence development of exploitative responses during later peer-rearing, as described previously.

The studies of the ontogeny of sexual behavior in chimpanzees provided some insight into the major questions that provoked the original investigation, i.e., those of Bingham (1928). The data supported Bingham's hypothesis regarding the greater amount of time and experience required for chimpanzees, in comparison to monkeys, to develop competent patterns of sexual interaction. On the other hand, although the time scale is reduced in the (rhesus) monkeys, the requirement for appropriate experience is no less in this species than in the chimpanzees (Mason, 1960). The question of variability in copulatory positions, proposed by Bingham to increase from monkey to ape to man, was not addressed adequately from the comparative perspective. The detailed account of his subjects' activities, however, presented by Bingham himself, lends credence to his view. Bingham's speculation or hypothesis that early social deprivation could account for the sexual indifference to an estrous female he observed in a captive-reared adult male chimpanzee was supported by both succeeding investigations (Nissen, 1954; Rogers and Davenport, 1969). The latter investigations, in addition, demonstrated that despite the initial deficits in sexual behavior that developed in adult chimpanzees subjected to early social restriction, these animals were capable of some improvement over time, given appropriate experience with a compatible and persistent conspecific tutor (the chimpanzee version of a sex therapist). Males, in general, were more adversely affected in their sexual behavior than females, in that they were more resistent to improvement. Finally, it should be noted that although the early restriction of chimpanzees discussed above was described as "social," this is not to deny that the development of aberrant behavior may be secondary to changes in the central nervous system, such as those proposed by Prescott (1970). Recent data on socially isolated stumptailed monkeys, for example, indicate that these animals exhibited decreased complexity of dendritic branching in specific areas of the neocortex (Struble and Riesen, 1978). These authors concluded that motoric deprivation accompanied the social isolation of their animals and accounted for the altered neural structures.

III.   ESTRUS AND THE HORMONAL REGULATION OF SEXUAL BEHAVIOR

It was stated at the beginning of this chapter that the controversy regarding the biological relationships among species, instigated by Darwin (1859), had its referents in the study of sexual behavior, especially with respect to the question of a phyletic trend in estrus, i.e., the periodic display or enhancement of sexual receptivity by females. There was one position, as represented by Papanicolaou, cited by Tinklepaugh (1933), which asserted that "the copulative or oestrous period has lost its typical expression in the higher mammals and especially in the primates, which have been deprived of the reaction of a strong sexual desire at a definite time of the cycle" (p. 528). This point of view probably derived from two early reports based on observations of monkeys, especially rhesus monkeys (Hamilton, 1914; Kempf, 1917). Montané (1915), however, reported that the Abreu chimpanzees also mated daily, including the period of gestation, and the caretakers at the estate gave a similar description to Bingham (1928). As noted above, two other reports on chimpanzees presented the contrary view that estrus accompanied female genital swelling in this species (Keith, 1899; Köhler, 1925).

Zuckerman's (1930) analysis of the issue was critical, perceptive, and, for the most part, accurate. He noted that "there is no readily recognizable period of oestrus" (p. 693) in women and that "no definitive statements can as yet be made about the breeding habits of any subhuman primate" (p. 748). Based on his analysis of the literature and his own observations at the London Zoo, he hypothesized that "the matings of the primate are diffused over the entire cycle, paralleling the continued action of the follicular hormone, but varying in frequency according to the varying degrees of activity of that hormone" (p. 748). Zuckerman proposed, in other words, that "estrin," i.e., an estrogen, mediated the estrous behavior of nonhuman primate females, including chimpanzees. "The female gradually passes into an attractive condition, and remains in that state while she becomes desirous. After ovulation she reverts into her unattractive and nondesirous state" (p. 730). He qualified these statements by adding that "even under completely normal conditions, however, the matings of monkeys are not altogether controlled physiologically" because these animals have "a highly efficient sensori-motor apparatus" (p. 731). The latter quotation appears to relate to the primate's relatively superior cognitive abilities, or what Yerkes later referred to as "behavioral adaptiveness" (Yerkes, 1939). It may be that Zuckerman (1930), thereby, antedated Yerkes (1939) in suggesting that the hormonal regulation of sexual behavior across species was inversely related in degree to their taxonomic position, encephalization, and/or intelligence.

A. The Description of Estrus in Chimpanzees

In their classic paper on menstrual cycle patterns of female genital swelling and sexual behavior in chimpanzees, Yerkes and Elder (1936) noted that "to Tinklepaugh belongs the credit of first describing oestrus in the chimpanzee. Previously it had been noted, but never convincingly described" (p. 1). Köhler's monograph (1925) includes one of the brief reports on sexual behavior of chimpanzees, alluded to by Yerkes and Elder, and it contains a rather accurate, although qualitative, description of the corresponding increases in genital swelling and "sexual desire" of the female which occur during the intermenstrual period and the abatement of "her sexual instinct" during menses (p. 303).

Tinklepaugh (1933) conducted his study with a single female, one of the four chimpanzees obtained by Yerkes that were studied originally by Bingham (1928). He recorded genital swelling and sexual behavior, and examined vaginal contents from the female during an 18-month period encompassing the onset of cyclicity (puberty), several cycles prior to conception and the period of gestation. During this time, the female was caged with a wild-born male of approximately the same age, i.e., seven to eight years.

Tinklepaugh reported that "sexual receptivity became evident . . . when the first pronounced swelling of the genitalia appeared" (p. 529), and the female first solicited copulation from the male at that time. The male, apparently also recently matured, exhibited increased interest in the female, especially towards her genital swelling. Mating first occurred during the third period of swelling, and the animals copulated several times a day for eight days. In this cycle and in all subsequent ones for this female, mating took place only during periods of swelling, beginning shortly after the onset of tumescence and terminating at about the time of detumescence. Following detumescence, the female no longer solicited the male sexually, but, rather, attempted to remain near him and to embrace him, which attempts the male rebuffed. Tinklepaugh examined the vaginal contents of the female for sperm in order to eliminate the possibility that the animals copulated at times other than those when they were observed directly. The evidence indicated to him "beyond question the limitation of copulation between these two animals to the period of genital swelling in the female" (p. 530). Tinklepaugh reported further that the female exhibited nine consecutive periods of swelling following the last menstruation before parturition, i.e., following conception. Although these periods of swelling were less regular than the previous ones, they were all accompanied by estrus and, during the first seven when the male was present, copulation occurred daily.

Data had been presented previously which suggested that genital swelling in baboons was controlled by a follicular hormone, then called estrin (Parkes and Zuckerman, 1931). The correspondence in time of the periods of female genital swelling, estrus, and mating of the chimpanzee, a species closely related to the baboon, was consistent with Zuckerman's (1930) view, presented in the preceding, that both the morphological and the behavioral activities in this species were under the regulation of this same hormone. The evidence suggested, in addition, that primates with pronounced genital swelling, such as the baboon and chimpanzee, exhibited estrus in conjunction with increases in the swelling, whereas primates that did not possess such extensive specialized tissue did not exhibit clear evidence of cyclicity in sexual responsiveness, e.g., the rhesus monkey (Hartman, 1928).

B. Menstrual Cycle Studies of Estrus in Chimpanzees

1. *Essential Characteristics of Sexual Behavior*

Following the study by Tinklepaugh (1933) of a single pair of chimpanzees, Yerkes and Elder (1936) conducted a study in which they "sought to discover the essential characteristics, relations, and conditions of mating in chimpanzee" (p. 1), in a formal experiment using several individuals of both sexes. Such a study was possible because of "a special laboratory" that was conceived and constructed specifically to conduct "studies of the biology of the primates" (p. 2), in much the same way that more traditional laboratory species were studied. This laboratory was the southern extension of the Yale Laboratories of Primate Biology, located in Orange Park, Florida (Yerkes, 1932).

Yerkes and Elder (1936) conducted their investigation by testing oppositely sexed pairs of adult chimpanzees for sexual behavior frequently throughout the female's sexual cycle. Female genital swelling was recorded daily as a morphological index of cycle phase, to which the behavioral interactions were ultimately related. Because of the known and hypothesized relationship between female sex hormone concentrations, especially estrogen, and female genital swelling in several species of primates, the pattern of genital swelling of the chimpanzee could also be used as (1) an indirect measure (approximation) of relative estrogen concentrations in the follicular phase, and (2) a predictor of the time of ovulation (Elder, 1938). The pattern of mating in relation to genital swelling, therefore, permitted analysis of sexual interactions in terms of their presumed temporal relationship to these physiological conditions.

One of the most significant findings in this study relates to differences in sexual responsiveness during the sexual cycle between different individuals and different pairs (consorts). Although Yerkes and Elder (1936) asserted in their hypothesis that estrus was characteristic of chimpanzees, they did not believe they recorded female behavior in sufficient detail to support this position. In this initial study, sexual receptivity was inferred from data on copulation and female initiative, both measures the authors believed were confounded by the dominance of the male. They reported that mating overall occurred most frequently during the 10-day midcycle period of maximal genital swelling, but it occurred also at other phases of the cycle. The authors proposed that when the consorts were mature, sexually experienced, and familiar to each other, the female tended to control mating. Under such conditions, copulation was restricted to a relatively brief period of maximal genital swelling, rarely exceeding one-third the total length of the cycle. On the other hand, when the female was immature, inexperienced, fearful, and/or unfamiliar to the male, and the male dominant and sexually assertive, mating could occur much more frequently during the cycle and irrespective of female sexual status. The considerable involvement of social and experiental factors in determining the frequency and distribution of mating in the cycle led the authors to propose that the chimpanzee more closely resembled the human in the regulation of its sexual behavior than any other species studied to that time.

Female initiative, defined by the female moving into the male's cage at the start of the test, was more frequent than male initiative overall and also varied with cycle phase. Female initiative occurred on 85% of the tests conducted during the phase of maximal swelling, but on only 65% of the tests during other phases. The authors inferred from these data that the female was most attracted to the male during maximal swelling and that her more frequent movement into the male's cage during that time was indicative of her heightened sexual receptivity. They proposed, in addition, that female initiative, as defined, also reflected the influence of factors other than sexual ones, e.g., female acquiescence to male persistence or intimidation. They attempted to support their hypothesis regarding estrus in chimpanzees by describing several "case studies" from their intimate knowledge of their subjects' life histories. They acknowledged, however, that future research would be advanced if sexual receptivity were measured more directly and objectively. This study, therefore, provided considerable insight into the sexual relations of chimpanzees in the laboratory, raised several questions in this regard, and suggested modifications in experimental approach to facilitate further investigation.

2. Control of Sexual Responsiveness

The subsequent experiment by Yerkes (1939) was primarily an attempt to describe the conditions and dimensions of estrus in chimpanzees in a more precise manner than before. In this, he felt he had succeeded. He recognized, however, that he had not resolved the problem regarding an objective and reliable method for assessing estrus in chimpanzees. The approach he took was to identify several items of behavior, some subjectively defined, that were related to estrus in the previous study (Yerkes and Elder, 1936), and combine them in an unspecified way to arrive at a single rating on a scale, termed sexual receptivity. Seven items were used in calculating the rating, but three were considered especially valuable indicators. Female initiative, defined by the female's movement into the cage of the male at the start of the test, was one of those indicators. The presentational behavior of the female was a major factor and was rated both in terms of frequency and adequacy for intromission. The third item was the initial response of the male to the female in terms of acceptance or rejection.

Continuing research on the issue of estrus led recently to formalization of a proposal to define several aspects of behavior associated with estrus in terms of the three separate concepts: sexual attractivity, proceptivity, and receptivity (Beach, 1976). According to this proposal, the definition of receptivity is confined to the female's behavior that is necessary and sufficient for fertilization. The degree to which the female's presentational behavior facilitates intromission and ejaculation reflects on her receptivity. Appetitive responses of the females toward the males, such as reflected in the frequency of presentation and sexual initiative, are now defined as proceptive behavior, in recognition of the active role by females in sexual interactions. Appetitive responses of the males towards the females relate to the concept of female sexual attractivity, one measure of which is the frequency of approach. The main reason the current approach is superior to Yerkes' is that evidence suggests the three characteristics are differentially regulated by hormonal and nonhormonal factors. Separate assessment of different characteristics related to estrus permits separate investigation of the causal factors associated with each (Beach, 1976). Interspecies comparisons, considered here, also are facilitated as a result of the increased refinement in definitions. In the presentation of Yerkes' research to follow, his term "sexual receptivity" is replaced by the more general one, "sexual responsiveness," because of the more specific definition now applied to the former term. Analysis of great ape sexual behavior in terns of the concepts of attractivity, proceptivity, and receptivity is presented here, following description of the laboratory and field data related to these concepts.

The second study of chimpanzee sexual behavior by Yerkes (1939) demonstrated again the complexity of behavioral interactions and the significance of individual and social factors in determining the outcome of a mating test. The results, in general, were similar to those of Yerkes and Elder (1936). Mating occurred primarily during the phase of maximal genital swelling which, in some cases, was coextensive with maximal sexual responsiveness. Other evidence, however, suggested that the pattern of behavior recorded in the laboratory might not be characteristic of this species in a more natural setting, and that sexual responsiveness was not equally intense throughout the period of maximal genital swelling. It was proposed that if sexual responsiveness was measured independently of the male's influence or under natural conditions, it would be found to increase gradually during maximal genital swelling to reach a peak during the latter half of swelling, persist for only a few days, and terminate at about the time of detumescence of the swelling. (It was this latter period of maximal genital swelling that was found to be the most probable time of ovulation in chimpanzees (Elder, 1938)). Yerkes derived this hypothetical pattern of sexual responsiveness from the behavior of certain laboratory subjects, whose social relationships in the laboratory resembled his concept of the species-typical pattern of social relations among free-living chimpanzees. The significance of his hypothesis that responsiveness changes during maximal swelling and is most intense during the latter half of swelling is considered here in relation to field data on feral chimpanzees (Tutin, 1975; Tutin and McGinnis, Chapter 9, this volume).

Regarding the question of sexual responsiveness and mating during menstruation, gestation, and lactation, somewhat different results were obtained than previously. Whereas Yerkes and Elder (1936) reported that mating did not occur during any of these physiological states, Yerkes (1939) subsequently found that sexual responsiveness increased and mating did occur during nonbleeding cycles of pregnancy in association with periodic increases in genital swelling. Such swellings were most prevalent during the first two or three months of gestation and were less regular in pattern than menstrual-cycle swellings. Responsiveness during these times was, in general, lower than during genital swelling in the nonpregnant female. This finding, in conjunction with other related behavioral data, suggested to Yerkes that the matings that took place during pregnancy did not necessarily imply heightened female responsiveness. He proposed that the behavior of the male, perhaps conditioned during prior sexual testing, was stimulated by the genital swelling per se, as a cue to sexual status.

As a result of these studies on the chimpanzee, Yerkes concluded that estrus indeed was characteristic of the female chimpanzee and, given suitable methods of investigation, would be shown to be characteristic of all primates, including human females. Regarding the regulation of sexual behavior among mammals, in general, he proposed the broad hypothesis, "The higher the order of behavioral adaptiveness (general intelligence) and the more dominant the male of the species, the wider the range of copulatory responsiveness in the typical sexual cycle and the greater the tendency of the female to respond accommodatingly to his advances irrespective of her sexual status" (Yerkes, 1939, p. 79). Yerkes used the term "male dominance" to refer to the male's role in initiating copulation and his ability to coerce or intimidate the female to copulate in the absence of heightened female responsiveness. Yerkes' hypothesis asserts, therefore, that extension of mating outside the peri-ovulatory period is a function of (1) relatively advanced taxonomic position (i.e., encephalization and, presumably, general intelligence), and (2) male, as opposed to female, sexual initiative. The hypothesis predicts that for species of the same taxon, female initiation of sexual activity results in peri-ovulatory mating, whereas male initiation results in mating irrespective of cycle phase. Other investigators of animal sexual behavior subsequently presented hypotheses, either explicitly (Beach, 1942) or implicitly (Young and Orbison, 1944), that were similar to Yerkes' with respect to proposing an inverse relationship between encephalization and intelligence and the extent of the hormonal regulation of sexual behavior. These latter hypotheses did not include the qualification regarding male dominance, however, a qualification that proved applicable to subsequent analysis of data on sexual behavior of the remaining great ape species, gorilla and orangutan (Nadler, 1977b). It was found, for example, that peri-ovulatory mating was characteristic of gorillas, a species in which the female is primarily responsible for initiating sexual activity. Among orangutans, by contrast, mating occurred irrespective of cycle phase as a result of male assertiveness in sexual initiation (see below).

3. *Sexual Responsiveness and Social Behavior*

The final laboratory study of sexual behavior in chimpanzees was conducted by Young and Orbison (1944), using many of the same animals and similar methods of testing as those used in the earlier studies (Yerkes and Elder, 1936; Yerkes, 1939). This study investigated menstrual cycle patterns of 18 measures of behavior, some of which had been used previously, and additional ones that had not.

For the most part, Young and Orbison (1944) supported the results of the earlier studies. When they tested their behavioral measures for differences between the follicular and luteal phases, they found only seven of the 18 for which the differences were statistically significant. In other words, for 11 of the 18 behavioral measures, differences between individuals and consorts were greater than differences between cycle phases. This supported Yerkes' conclusion that individual and social factors contributed more than hormonal ones in determining the behavioral interactions of chimpanzees in the laboratory. The seven measures for which follicular and luteal phase differences were statistically significant were (1) pre-observation sexual excitement of the male (i.e., whether the penis was erect or flaccid), (2) the female's pre-observational interest in the male (i.e., whether waiting at the door or indifferent), (3) presentation by the female, (4) copulation, (5) nonresponsiveness of the female, (6) nonresponsiveness of the male, and (7) time spent together. All but the nonresponsiveness measures were greater during the follicular phase. Not supported by the statistical analysis in this study were cycle phase differences in female movement to the cage of the male and grooming by the male and female, reported by Yerkes and Elder (1936) and Yerkes (1939), and playfulness of the female, reported by Tinklepaugh (1933). The results of these measures were in the predicted direction, however, and suggested the presence of trends worthy of additional study.

In order to test Yerkes' (1939) conclusion that female responsiveness and copulation increased in frequency to a peak during the second half of the phase of maximal swelling, Young and Orbison (1944) analyzed their data, normalized to the day of detumescence of the genital swelling. [Young and Orbison (1944) used the mean number of presentations by the female during a 30-min test as their measure of sexual receptivity.] Although peaks in the frequency of female presentation were observed during the period of maximal swelling in certain of the consorts, they were balanced out and eliminated when the data for all consorts were combined. The significance of this result is not clear. The timing of ovulation and of the associated hormone patterns in relation to the genital swelling of chimpanzees is not invariable (Elder, 1938; Graham, 1970). Normalization of the behavioral data to genital-swelling detumescence, therefore, could obscure a relationship between hormones and behavior. Since the condition of estrus and its temporal association with female hormone concentrations and ovulation is one of the major issues under consideration in this review, the data on the chimpanzee related to this issue are further considered below.

Following the menstrual cycle study of chimpanzees, Young and Orbison investigated the behavioral effects of ovariectomy, conducted on three of the females. During two periods of testing, eight and 12 months following ovariectomy, the behavioral data recorded were similar to those of the luteal phase of the menstrual cycle and significantly different from those of the follicular phase. These data, therefore, further supported the hypothesis that hormones of gonadal origin exert a measurable influence on sexual interactions of chimpanzees, albeit an influence that, under certain circumstances, is overwhelmed by individual and social factors.

4. Summary of Results

These studies on the chimpanzee, the first of the great apes to be investigated for estrus and related sexual behavior in the laboratory, were consistent in their results regarding the principal issues under investigation. They suggested that (1) cyclic augmentation of sexual responsiveness was characteristic of the female chimpanzee, (2) copulation occurred most frequently during the period of maximal genital swelling, but (3) copulation occurred also at other phases of the cycle and during menstruation, gestation, and lactation, as a consequence of varied individual, social, and environmental influences (including the laboratory conditions of testing). Based on the known and hypothetical patterns of hormone concentrations during the menstrual cycle and the effects of ovariectomy on hormone concentrations, the behavioral data collected under these conditions supported the hypothesis that hormones, especially estrogen, were implicated in the enhancement of sexual responsiveness and copulation of chimpanzees at midcycle. The more specific proposal by Yerkes (1939) that sexual responsiveness undergoes change during maximal genital swelling to reach a peak of just a few days' duration at about the time of swelling detumescence was not supported by the subsequent work of Young and Orbison (1944). That a pattern such as Yerkes proposed might yet be uncovered was not ruled out, however. According to Yerkes (1939) and already mentioned above, such a pattern might only occur under natural conditions or under laboratory conditions in which the male's influence (dominance) was reduced, and the female was given greater choice in the matter of whether or not to mate with a male. Laboratory research adequate to test this hypothesis has not been conducted and may be crucial to a comprehensive assessment of hormonal regulation of sexual behavior in the apes. Recognition and measurement in future research on the chimpanzee of the several aspects of estrus described above should clarify some of the previously reported discrepancies.

C. Menstrual Cycle Studies of Sexual Behavior in Gorillas

1. *Perineal Labial Tumescence (Genital Swelling)*

Laboratory research on sexual behavior of the remaining great apes, gorilla, and orangutan, was initiated relatively recently and followed, for the most part, the methods of investigation used for the chimpanzee. Following up on an early study conducted on a single zoo specimen (Noback, 1939), all female gorillas at the Yerkes Regional Primate Research Center were trained to mount the front of their cages to facilitate measurement of their genital swellings. It was found, in confirmation of Noback (1939), that female gorillas do not have as extensive a genital swelling as chimpanzees, but that the external perineal labia undergo tumescence and detumescence comparable in pattern to the swelling of chimpanzees (Nadler, 1975a). Given the number of gorillas examined in this study, i.e., nine, it was possible to define cycle length more reliably than previously as approximately 31–32 days, in contrast to Noback's estimate of 39–49 days. This finding on periodicity of genital swelling for a group of female gorillas agrees with data on periodicity of sexual behavior in gorillas, reported for captive animals by several authors (see Hess, 1973; Nadler, 1976). It was also found that maximal labial tumescence (LT) of the female gorillas persisted for a shorter period of time than maximal genital swelling of female chimpanzees. In both species, genital swelling is at minimal tumescence at the time of menses, increases to reach maximal tumescence at midcycle, and undergoes detumescence shortly after the midcycle luteinizing hormone (LH) surge, the presumptive time of ovulation (Graham, 1970; Nadler *et al.*, 1979). The major difference between the swelling patterns of female chimpanzees and gorillas is found in the latency to maximal genital swelling in the cycle and the duration of the condition. Maximal genital swelling in chimpanzees occurs within about a week following menses and persists for approximately 10 days before undergoing detumescence (Yerkes and Elder, 1936; Yerkes, 1939). Maximal LT in gorillas develops more gradually and persists for only a few days (median = 1.8 days, mode = 1 day) before undergoing detumescence (Nadler, 1975a). These results suggest that LT of female gorillas may be used as an external marker for monitoring progress of the menstrual cycle in a way comparable to that used by Yerkes and others in research with chimpanzees. [Zuckerman (1930) proposed years earlier that for all primates that exhibited cyclic genital swelling, estrogen would be shown to be the hormone responsible.]

2. Sexual Behavior and Labial Tumescence

A study of sexual behavior in gorillas was conducted in which daily measurements of LT were obtained in conjunction with the daily mating tests (Nadler, 1975b, 1976). The results indicated that the gorillas, like chimpanzees, mated primarily during the period of maximal genital swelling. As noted above, however, maximal LT of gorillas persists for approximately two days, compared to 10 days, on the average, for chimpanzees. The result, therefore, is that whereas chimpanzees mated throughout approximately 30% of the menstrual cycle (e.g., 10 of 35–36 days), the gorillas mated during less than 10% of the cycle (e.g., two of 31–32 days). This finding in gorillas did not support the hypothesis that advanced taxonomic status per se assured independence from hormonal regulation of sexual behavior. On the contrary, a relatively infrequent occurrence and circumscribed period of mating was found, closely associated with a morphological condition (LT) known to reflect elevated estrogen concentrations in several species of primates, e.g., pigtailed macaque (Eaton and Resko, 1974) and chimpanzee (Graham *et al.*, 1972; Reyes *et al.*, 1975). The temporal association between mating and maximal LT in gorillas suggested, therefore, that estrogen played a significant role in sexual behavior of this species. It is worth noting that despite this apparently significant hormonal influence on their behavior, the gorillas also gave evidence of considerable variability in certain aspects of their behavior. For example, in contrast to the stereotyped mounting patterns observed in most species other than man, the gorillas used varied positions during mating, perhaps individually adapted to differences in body size and environmental conditions. One of the most significant findings, however, was that females, rather than males, were the primary initiators of sexual interactions. The males gave no clear indications of soliciting copulation and, indeed, sometimes failed to respond to the female's solicitations, the latter especially during periods of labial detumescence. This finding was of some significance because it supported Yerkes' hypothesis regarding the role of male sexual dominance on patterning of sexual interactions. The gorilla was found to be a species in which the male was not sexually dominant, based on traditional laboratory methods of evaluation and, as predicted, mated only during a relatively circumscribed period which, by analogy to other closely related species, was probably the peri-ovulatory period.

3. Labial Tumescence and Hormones

In order to assess menstrual cycle hormone patterns of female gorillas directly, a study was conducted in which the gonadotropins, prolactin, and gonadal steroids were measured and described in relation to female LT (Nadler *et al.*, 1979). The females were not tested for sexual behavior in this study because of the possibility that mating might alter the hormone patterns. One of the most significant results of this study was the indication that normal menstrual cycle hormone patterns were obtained despite the necessity of restraining and anesthetizing the animals to secure blood samples for assay (periodic rather than daily samples were obtained so as to minimize anesthesia). All cycles exhibited a midcycle LH surge and a luteal phase elevation in progesterone, suggesting normal ovulatory function. The patterns and concentrations of several hormones, moreover, were remarkably similar to those of the human. 17β-Estradiol showed a midcycle elevation that preceded the LH surge and a luteal phase elevation that approximated the time course of progesterone. LT was minimal at menses, reached maximum in association with the midcycle rise in 17β-estradiol, and underwent detumescence approximately two days following the LH surge, when 17β-estradiol was decreasing and progesterone was increasing. Testosterone exhibited a midcycle peak on the day following the LH surge, one or two days after the 17β-estradiol peak, the most probable time of ovulation. The sharply defined midcycle peaks in 17β-estradiol and testosterone, both hormones implicated in facilitation of female sexual behavior, were provocative findings in relation to our earlier data on sharply defined midcycle periods of mating in gorillas, initiated by the females. These data are consistent with the hypothesis that reproduction of gorillas requires a rather precise temporal integration of hormonal and behavioral activities, regulated primarily by the female.

The pattern of mating by gorillas in relation to menstrual cycle concentrations of serum testosterone was of special interest because: (1) testosterone, sometimes referred to as the "libidinal hormone" in both sexes, was implicated in the display of proceptive behavior by female primates (Luttge, 1971); (2) testosterone in female gorillas was found to reach peak levels on the midcycle day after the LH surge, the time when conception is most probable; and (3) the gorilla is a species in which the female was found to be highly proceptive at midcycle. This research, therefore, raised the possibility that proceptive behavior of female gorillas is enhanced by increased concentration of circulating testosterone only at a time in the cycle when mating is reproductively functional.

4. Sexual Behavior and Hormones

A study was conducted on gorillas to test the above hypothesis more directly by measuring sexual behavior and hormone concentrations simultaneously throughout the menstrual cycle in the same female gorillas (Nadler *et al.*, 1978). The females were tested for sexual behavior during two cycles with each of two males, and serum for hormone assays was obtained periodically during the first test cycle only. The hormone patterns obtained in this study were consistent with those obtained in the previous one, suggesting again normal ovulatory function. The behavioral data, however, deviated somewhat from the earlier pattern. The mean number of two copulations per cycle agreed with earlier data (Nadler, 1975b, 1976), but several pairs mated much more frequently, e.g., 11 consecutive days for one pair. Although several of the cycles studied contained peak concentrations of testosterone on the day after the LH surge, female proceptivity and mating frequently occurred at other times during the cycle, and occasionally failed to occur, despite a peak in testosterone. When the frequency of copulations per cycle day was calculated for all pairs and all cycles, the modal number of 10 copulation days occurred on the day following the LH surge (Day LH + 1), and 26 of the 43 copulation days (60%) occurred within the three-day period with Day LH + 1 as the median day. The overall pattern, therefore, was consistent with the hypothesis that copulation was facilitated by elevated levels of testosterone. In order to investigate the source of the variability, the behavioral records of the two males used in the study were analyzed separately. A major difference in the frequency of copulation days per cycle was found, and the male that copulated most frequently (29 days *vs.* 14 days) also copulated over a greater range of cycle days, e.g., 14 days *vs.* five days. Copulation by both males was associated with the three-day period, including Day LH + 1 as the median day, but differed in degree; only 34% of the copulation days of the frequently copulating male occurred within this period, in comparison to 86% for the lesser copulating male. The data may be interpreted in several different ways. They indicate (1) differences in the mating patterns of gorillas are possible when they are tested under certain laboratory conditions; (2) individual differences exist between male gorillas in the frequency and distribution of copulation during the menstrual cycle under the specific laboratory conditions used thus far; (3) gorillas are capable, under certain conditions, of greater independence from hormonal regulation than previously suggested; and therefore, (4) there is overlap among chimpanzees and gorillas with respect to the degree of hormonal regulation of sexual behavior.

5. Summary of Results

Although data analysis for this latter study is not yet complete, there is reason to believe that the male that copulated most frequently also was more aggressive toward the females. If this is confirmed, it would suggest that male dominance in gorillas, like chimpanzees (Yerkes, 1939), is associated with a lesser degree of restriction of copulation to the peri-ovulatory period. Confirmation of a positive relationship between male dominance and frequency of mating in the menstrual cycle of the gorillas, moreover, would help to evaluate the mating patterns of these animals in this laboratory study. Under natural conditions (Schaller, 1963; Harcourt and Stewart, 1978; Harcourt *et al.*, Chapter 10, this volume) and under most conditions in captivity (Hess, 1973; Nadler, 1975b, 1976), it is the female gorilla that initiates and controls mating. The influence of the male in determining the frequency and distribution of copulation during the cycle in the laboratory has not been assessed adequately and, therefore, our ability to describe the influence of endogenous hormones on sexual behavior of gorillas is limited. In nature, a female gorilla is not generally in close, exclusive association with the male and presumably can avoid contact by leaving his immediate vicinity. In the laboratory studies conducted thus far, the female was confined in a single cage with the male for the duration of a test and given no opportunity to avoid contact, apart from climbing to the top of the cage. Since several of the female gorillas did try to avoid the male in this manner, we may conclude that the test situation constrained the behavior of the females to some extent, at least. In the laboratory studies of sexual behavior of both chimpanzees and gorillas, therefore, evidence was obtained which suggested that males exert an influence over the frequency and distribution of mating in the cycle. In both species, moreover, the influence of the male appears to moderate the influence of hormones. In other words, consistent with Yerkes' (1939) hypothesis, the restriction of mating during the cycle of chimpanzees and gorillas was inversely related to the dominance or sexual initiative of the male. Thus, the hypothesis applies both within a species, with respect to differences in the dominance of different individual males, and between species, with respect to the species-typical pattern of mating, as defined by the general pattern for the species, recorded both in the laboratory and under more natural conditions.

D.　Menstrual Cycle Study of Sexual Behavior in Orangutans

For the final species of great ape considered, the orangutan, there is the least amount of information available on its mating in the laboratory. In the single study conducted, four oppositely sexed pairs of orangutans were tested daily for sexual behavior during the menstrual cycle of the female (Nadler, 1977a). Since female orangutans do not exhibit a genital swelling during their cycle, the orangutan cycle was monitored by the periodic occurrence of menses. Urine samples from the females were tested daily for the presence of blood by means of commercially available reagent strips. Behavioral testing of each pair was initiated on the second day of menses and continued through the cycle to the next onset of menses. The results obtained with this species were remarkably different from those obtained on the gorillas and the chimpanzees. During tests of approximately one-hour duration (criterion of 30 min without copulation), three of the four pairs tested mated on every test, i.e., every day of the cycle. The fourth pair mated on 41% of its tests, distributed throughout the cycle. Essentially all copulations, moreover, were initiated by the male in a manner resembling rape. At the beginning of each test, the male pursued the female, who fled and frequently sought to escape by climbing to the top of the cage. The male quickly caught the female, wrestled her, struggling and screaming, to the floor and then forcibly initiated copulation. Once the male achieved intromission and initiated thrusting, the female became passive and permitted the male to position and reposition her frequently throughout the course of copulation. In contrast to chimpanzees and gorillas, in which copulation is generally completed in less than one minute, copulation by the orangutans continued for 14 minutes on the average (range: 1–46 min).

As stated above, each test in the initial phase of this study was terminated 30 min after the completion of copulation. Since there was no periodicity apparent in the sexual behavior of the orangutans under these conditions, a second series of tests was conducted to determine whether multiple copulations would occur differentially with respect to cycle phase in a test of longer duration. This second series of menstrual cycle tests, of five to six hr duration, conducted with only three of the original four pairs (the female of the fourth pair became pregnant during the first cycle in which it was tested), revealed a different perspective on orangutan sexual behavior. During these tests of longer duration, the orangutans mated again during the first hour of testing in the manner described above. In addition, however, they exhibited multiple copulations per day, primarily during the middle phase of the cycle. Especially noteworthy were some few examples of sexual proceptivity by the females, observed during midcycle, despite the fact that the same females had attempted to resist the forcible efforts of the males to copulate with them at other phases of the cycle. Some few examples of cyclic proceptivity have been reported for female orangutans living continuously with males in zoos (see Nadler, 1977a; Maple *et al.*, 1979). These data suggest that orangutans may be less responsive to midcycle fluctuations in female hormone levels than are the other species of great apes. The data are also consistent with the hypothesis that cyclic factors, presumably hormonal, do exert some influence over the behavior of this species of ape.

E. Summary of Menstrual Cycle Studies

The data on orangutans provided additional support for Yerkes' hypothesis regarding male dominance in that of the sexes, the male orangutan was clearly more dominant and aggressive, and copulation occurred irrespective of cycle phase. The sexual behavior of the great apes in laboratory studies, with respect to the degree of restriction to the midcycle, peri-ovulatory period, and the presumed regulation by hormones, could be described, therefore, as a continuum (Nadler, 1977b). The gorilla, at one end, exhibits primarily female sexual initiative (proceptivity) and restricted peri-ovulatory mating; the orangutan, at the other extreme, exhibits male sexual initiative primarily and indiscriminant mating with respect to cycle phase; and the chimpanzee appears to lie intermediate to the former two.

Examination of the laboratory data presented thus far reveals a finding that was common to the three great ape species regarding the influence of the males. Whereas, on the one hand, the typical patterns of mating differed from one species to the other with respect to the frequency and distribution of mating in the cycle, the influence of the male in each species operated in a similar direction, consistent with Yerkes' hypothesis (Table 1). The greater the male's dominance over the female, as reflected by his relatively greater sexual initiative and control over mating, the lesser the restriction of mating to the presumed time of ovulation. In each species, therefore, the female's sexual responsiveness reflected to some degree an interaction with male dominance, as expressed under the particular conditions of laboratory test-

TABLE 1

Relative rankings of the great apes with respect to male and female sexual initiative and peri-ovulatory mating as determined in laboratory studies.

	Chimpanzee	Gorilla	Orangutan
Male sexual initiative	2	3	1
Peri-ovulatory restriction of mating	2	1	3
Female sexual initiative (proceptivity)	2	1	3

Note: The number 1 represents the highest ranking; 3, the lowest.

ing employed. In order to assess more completely the relevance of the laboratory findings for the issue of hormonal regulation of sexual behavior, it is appropriate to examine those findings in relation to data available on these species living under more natural conditions. In such an analysis, the assumption is generally made that the behavior of animals observed in their natural habitats today more closely reflects the selective pressures operating at the time the behavior evolved than the behavior of the same animals in a laboratory environment. Accordingly, the behavior of the animals in the field is considered most appropriate for deducing hypotheses regarding the evolutionary influences on behavior and the adaptive significance of behavior, especially species-typical forms of behavior (Harcourt and Stewart, 1977; Nadler, 1977b; Short, 1977). In a broader sense, the field and laboratory environments represent different sets of conditions, both of which may be analyzed for their respective contributions to the patterns of sexual interaction observed. It is in this latter sense that laboratory research may have special significance for extrapolation of results to the human species.

IV. LABORATORY AND FIELD RESEARCH

A. Chimpanzees

Research on chimpanzees in the field (van Lawick–Goodall, 1968; Tutin and McGinnis, Chapter 9, this volume) and in seminatural group-living conditions in captivity (Kollar *et al.*, 1968; Tutin and McGrew, 1973; Lemmon and Allen, 1978; Coe *et al.*, 1979) support, for the most part, the results obtained by Yerkes and others in the laboratory. Courtship was somewhat more elaborate in the more complex social settings, however, and a somewhat greater percentage of copulations occurred during maximal genital swelling, e.g., 87% (van Lawick–Goodall, 1968) and 83% (Tutin and McGrew, 1973) *vs.* 60% (Yerkes and Elder, 1936) and 73% (Yerkes, 1939). More significant than the above differences and, perhaps, suggestive of their derivation is the comparison between sexual initiation of chimpanzees in the field and in the laboratory environments. Whereas Yerkes found that the female chimpanzee more frequently initiated sexual interactions in the laboratory, field workers subsequently reported that the males primarily initiated mating in the wild. Yerkes would not have been surprised to learn of these differences because he recognized that "the environment of captivity may bring about modifications (in behavior) which only comparable studies of the wild animal will reveal" (Yerkes, 1932, p. 10). In fact, his analysis of chimpanzee sexual behavior provides the basis for the interpretation proposed herein.

Yerkes (1939) concluded that the female chimpanzee is cautious and wary of the male and, when intimidated, mates at any time in the cycle, irrespective of female sexual status. The conditions of laboratory testing are optimal for such male intimidation of the female, since the male is somewhat larger and stronger, and the two animals are confined together in a cage that offers the female no opportunity to avoid or escape the overtures of the male. Yerkes was probably accurate in concluding that the male's prerogatives regarding mating were relatively greater in the laboratory in comparison to nature. The reversal of sexual initiative in the laboratory, as subsequently revealed, therefore, could be interpreted as an accommodation by the female to avoid antagonizing the male and incurring an attack. Given such a condition, in which the female presents to the male relatively more frequently due to factors that are at least partially independent of cyclic influences, one could readily account also for the somewhat greater percentage of copulations in the laboratory studies that occurred outside the period of maximal genital swelling. (It is also likely that determinations of maximal genital swelling in the field are less accurate than those in the laboratory since palpation of the genitalia is required for detection of subtle genital changes. To the extent that maximal genital swelling was recorded more frequently in the field than in the laboratory, the percentages of copulation ascribed to this condition would be increased relative to the laboratory, as reported.)

The interpretation proposed above asserts that the laboratory data reflect a conciliatory action by the female. It is possible that the laboratory conditions were such that the females, rather than being intimidated by the male into mating more frequently, were more highly sexually proceptive in the laboratory for reasons independent of male dominance. The data suggest the contrary, but are insufficient to support a firm conclusion on the matter. The data are inadequate also with respect to the proposal by Yerkes that sexual responsiveness of the female chimpanzee increases during maximal genital swelling to reach a peak of a few days' duration during the latter half of the swelling. Put another way, Yerkes proposed that there was a noticeable change in responsiveness of female chimpanzees during maximal genital swelling. He believed that under natural conditions, mating would only occur during this latter phase of genital swelling, when ovulation was imminent. This belief was based on Yerkes' characterization of chimpanzee social organization in nature, based on limited information from the first field study of chimpanzees (Nissen, 1931). From the evidence available, Yerkes concluded incorrectly that chimpanzees lived in family groups in which males and females were familiar with each other and socially and sexually compatible.

It is interesting to note that the social organization of gorillas in the wild (Schaller, 1963; Harcourt and Stewart, 1977; Harcourt *et al.*, Chapter 10, this volume) more closely resembles Yerkes' concept of chimpanzee social organization than the social organization subsequently described for wild chimpanzees (van Lawick–Goodall, 1968; Tutin and McGinnis, Chapter 9, this volume). The pattern of mating proposed by Yerkes for apes with such a social system, i.e., relatively restricted mating associated with the last few days of genital swelling, is also probably characteristic of gorillas in the wild. Female chimpanzees in nature, however, do not live with males in a family group and do not restrict their mating to the last few days of genital swelling. It was reported initially that female chimpanzees in the field mated frequently with all the adult males in the community throughout the period of genital swelling (van Lawick–Goodall, 1968). It was reported subsequently, however, that some females altered their behavior during the course of genital swelling and formed exclusive consortships with individual males primarily during the latter part of the phase of maximal swelling (Tutin, 1975; Tutin and McGinnis, Chapter 9, this volume). Similar examples of exceptions to promiscuity were reported recently for group-living captive chimpanzees (Coe *et al.*, 1979). The formation of an exclusive consortship among feral chimpanzees, in contrast to the more commonly observed pattern of opportunistic mating, requires the cooperation of the female. The finding that such consortships were formed primarily during the latter part of the phase of maximal genital swelling indicates that feral females altered their behavior from promiscuous mating to consortships at about the time in the cycle that Yerkes (1939) reported increased sexual responsiveness of females in the laboratory. The increased sexual responsiveness described by Yerkes, therefore, may have its counterpart in the natural life of these animals.

Yerkes (1939), as cited previously, proposed that accurate assessment of sexual responsiveness of female chimpanzees required that the influence of the male, judged to be inordinate in the laboratory, be controlled or replaced by some constant stimulus. This type of approach is the hallmark of laboratory research, in that it permits the investigation of individual variables, isolated or partially isolated from other influences that are normally operative. Behavior observed under such laboratory conditions does not necessarily reflect, at a superficial level, the "natural behavior" of the research animal recorded in the field, but may provide data suitable for an increased understanding of certain facets of that behavior. (This is but one aspect of laboratory research and not necessarily its most common function. Problem-oriented research, in which the research animals serve as "models" for less easily studied species, such as our own, is another and probably more common pursuit in the laboratory.) The study of animals in "the field," on the other hand, provides data on behavior under conditions that are presumably similar to those under which the behavior evolved and, therefore, especially appropriate for interpreting the adaptive or functional significance of the behavior. Herein lie the complementary contributions of laboratory and field studies to a comprehensive description of the behavior under investigation, a description not likely to be achieved by either approach alone. Yerkes' proposal that adequate assessment of female sexual responsiveness required experimental control over the male's behavior is still appropriate today, and its relevance to the investigation of the regulation of sexual behavior in the gorilla and orangutan, as well as the chimpanzee, is further considered in the following. Before discussing this matter, it is appropriate first to analyze the laboratory data on sexual behavior of gorillas and orangutans in relation to the comparable data on these species in their natural habitats.

B. Gorillas

In terms of the main parameters of sexual behavior under discussion, i.e., sexual initiative and the frequency and distribution of mating in the cycle, the laboratory data on gorillas, in general, are similar to data from the field (Schaller, 1963; Harcourt and Stewart, 1978; Harcourt *et al.*, Chapter 10, this volume). Since genital swelling and menses were not recorded for wild (parous) female gorillas, it was not possible to relate the mating of these animals to menstrual cycle phase or to the time of ovulation. As in the laboratory, however, mating in the field was initiated by the females and mating occurred for only a few days in succession. The mating of gorillas in the laboratory, therefore, resembled the mating of gorillas in the field with respect to the parameters of reproduction considered; parameters intimately related to fertilization and, hence, species propagation. Some differences between the laboratory and field data, especially with respect to the study of hormones and behavior (Nadler *et al.*, 1978), warrant additional comment.

Whereas mating of gorillas in the initial study (Nadler, 1975b, 1976) was limited to a few days of heightened genital swelling, mating by some pairs was more frequent in the study which included anesthesia and blood withdrawal for hormone assay (Nadler *et al.*, 1978). The data from the hormone-behavior study, therefore, could be considered relatively aberrant, compared to the field data and to the data from the initial study. Several possibilities for interpretation exist. It is possible that the increased mating occurred as a direct but nonspecific effect of restraint, anesthesia, and/or blood withdrawal. The fact that similar behavior was displayed during bleeding and nonbleeding cycles in the pairs concerned suggests, but does not confirm, that the behavioral changes were not directly related to manipulations associated with blood withdrawal. It is also possible that these manipulations acted indirectly on behavior by altering hormone concentrations. Although hormone patterns were similar in the two studies, hormone concentrations were somewhat lower in the hormone-behavior study (Nadler *et al.*, 1978) in comparison to the study on hormones and genital swelling (Nadler *et al.*, 1979). This possibility also seems unlikely for the same reason as above. The animals were anesthetized for blood collection during one cycle only and displayed similar behavior on the subsequent three cycles without such manipulation. Examination of the data from the two hormone studies revealed yet another possibility for the differences in behavior, as well as in hormone concentrations. It was found that the concentration of LH on the day of the LH peak, and the maximal concentration of progesterone during the luteal phase were lower during the months of August to October than during the months of November to March. Since most of the hormone data for the hormone-behavior study were obtained for the months of the former period, they reflect relatively low concentrations, perhaps related to seasonal differences in concentrations of circulating hormones and/or production of hormones. This possibility warrants further investigation. Lowered concentrations of progesterone, implicated for its inhibitory effects on sexual behavior of other primates, could account for enhanced mating (absence of or reduction in hormonal inhibition). The fact that no such enhanced mating was recorded in the initial study (Nadler, 1975b, 1976) during nearly 2000 tests of sexual behavior spanning several years, however, argues against the interpretation that the relatively frequent mating was season-related. The mating of gorillas in nature, moreover, reflects no such variability in the frequency of mating.

A final possibility, suggested above to account for the increased mating of gorillas in the study under consideration, relates to the conditions of testing in the laboratory and the interaction of those conditions with the species-typical pattern of mating and social organization of gorillas, as reflected in the evidence obtained from relatively natural populations living today. It was suggested that social factors, especially the male gorilla's dominance, could account for a change in mating patterns in the laboratory, in the direction of more frequent mating. Preliminary analysis of male gorillas' aggressiveness toward females suggested that mating occurred more frequently and throughout a greater range of cycle days in the pairs that contained the more aggressive of the two males. The relatively less aggressive male mated in a restricted pattern, comparable to the earlier laboratory study and to the field studies. Thus, the increased frequency and distribution of mating in the laboratory menstrual cycle study of gorillas was associated with male dominance and aggressiveness, much as had been reported for chimpanzees tested under comparable conditions in the laboratory. These data suggest, therefore, the need to evaluate the role of male dominance in gorillas, as well as in chimpanzees, in order to gain a more adequate perspective on female sexual responsiveness and the hormonal regulation of sexual behavior in these species.

C. Orangutans

The tests of orangutans in the laboratory contained the most conspicuous examples of male dominance and aggressiveness in the initiation of sexual interactions. In this species, moreover, there was, consistent with Yerkes' hypothesis, the least evidence of cyclicity in mating. Because aggressive and forcible mating, such as that displayed by male orangutans in the laboratory study, is rare, if not unique, among animals, it raises the question of whether such behavior is an artifact related to the conditions of testing. The answer seems to be no and yes. Forcible mating is not merely an artifact of laboratory testing since similar behavior was reported for orangutans in the field. On the other hand, this form of behavior does appear to be stimulated by a specific feature of the testing conditions, a feature which approximates certain conditions that occur sporadically in the field.

Under field conditions, orangutans live in a semisolitary lifestyle (MacKinnon, 1971, 1974; Rodman, 1973; Horr, 1975; Rijksen, 1978; Galdikas, 1979, and Chapter 11, this volume). The most common social unit observed in the field is the mother–infant dyad. Males are believed to live alone throughout their adult lives, with the exception of those times they are consorting and mating with females. The form of social organization among orangutans can be described as a dispersed harem, since several females generally occupy ranges that are subsumed within the larger range of a resident male. The form of forcible mating or rape observed in the laboratory has been observed also in the field and various interpretations have been proffered (see Nadler, 1977b). One area of agreement regarding the rape behavior of male orangutans is that such behavior is exhibited primarily by nomadic subadult males that have not established a home range of their own. It was proposed that forcible mating is not part of the species-typical pattern of reproduction in orangutans, but instead serves a dominance function (Rodman, 1973; Rijksen, 1978). The data suggest that the conditions of laboratory testing were sufficiently similar to certain aspects of the natural environment to provoke behavior on the part of adult males that normally occurs in nature only during an earlier stage of development. A statement by Rijksen (1978), who studied wild orangutans in Sumatra, is relevant. Rijksen (1978) noted that the condition under which most rapes occurred in the wild was "that the female is newly met or met again after a period of separation" (p. 274). The laboratory conditions used thus far for testing orangutan sexual behavior closely replicate the conditions described by Rijksen as common to rape behavior in the field, i.e., relatively brief, daily introductions followed by periods of separation. The pattern of rape observed in the laboratory study, however, was exhibited by four males who were all approximately 20 years of age and considered to be fully adult. The laboratory data are consistent with the field data in certain respects and, in addition, suggest that the conditions under which the orangutan consorts interact, rather than the age or maturational stage of the male, determine whether or not a rape occurs.

The field data do not permit a conclusion regarding the function of rape in orangutans, but they indicate that rape plays some role in the natural life of the species. To the extent that the laboratory conditions replicated conditions in nature that are conducive to the display of forcible mating by male orangutans, and assuming that such mating is not part of the reproductive behavior of orangutans, then it follows that the laboratory conditions used to test the orangutans were, to say the least, less than optimal for assessing female initiative in sexual interactions, i.e., female proceptivity. Despite the apparent bias in the laboratory environment, some few examples of proceptivity were observed during midcycle in two of the four female orangutans tested. These laboratory data and some reports of proceptivity in captive female orangutans living in zoos (see Nadler, 1977a; Maple *et al.*, 1979) indicate that these animals do display such behavior under certain conditions. It was reported that proceptive behavior by orangutans in the wild was observed almost exclusively in adolescent females in consort with males (Galdikas, 1979, see Chapter 11). Whether, under natural conditions, adult females seek out the male during midcycle or at other times, or whether the male seeks out the female periodically to determine her sexual status and to form a consortship, has not been determined.

V.   SEXUAL RESPONSIVENESS IN THE GREAT APES

The laboratory studies of great ape sexual behavior supported the hypothesis that a periodic augmentation of sexual responsiveness is characteristic of the females of all three species. The data indicate further that the display of female responsiveness differed among the species in terms of its intensity and duration in the menstrual cycle under the conditions of testing used in the laboratory. Interpretation of the interspecies differences is incomplete because of inadequate data on several issues which require clarification.

One problem regarding the issue of estrus as investigated under laboratory conditions, the problem of the male's contribution to the behavior, was addressed by Yerkes. Consistent with his results on chimpanzees were the related findings in both gorillas and orangutans that female sexual behavior reflected, to some degree, the assertiveness or aggressiveness of the males in initiating sexual activity, i.e., the greater the sexual initiative of the male, the less restricted was mating in the cycle. The female's contribution to the initiation of sexual activity is apparently compromised by the conditions of testing, and accurate assessment of sexual responsiveness, apart from the male's influence (intimidation), is not possible. Yerkes (1939) proposed that accurate assessment of estrus by chimpanzees in the laboratory required experimental control over the behavior of the male. A similar proposal is suggested as a requirement for adequately assessing this phenomenon in female gorillas and orangutans. The laboratory data permit the general statement expressed above regarding the display of estrus in all three species, but were insufficient to define the specific characteristics of estrus, such as its relative strength and duration in the menstrual cycles of the different species. It remains to be determined whether or not sexual responsiveness of the female great apes differs in these dimensions among the species, assessed independently of male intimidation.

VI.   HORMONES AND SEXUAL BEHAVIOR OF THE GREAT APES

The temporal association in the menstrual cycle between patterns of sexual behavior and concentrations of estrogen in the case of female

                                                    Ronald D. Nadler

chimpanzees (Graham *et al.*, 1972; Reyes *et al.*, 1975) and estrogen and testosterone in gorillas (Nadler *et al.*, 1978, 1979) suggests, but does not confirm, their influences. Both these hormones have been implicated in the sexual responsiveness of other primates and mammals, including humans (Luttge, 1971).

Comparison of menstrual cycle patterns in the concentrations of 17β-estradiol in female chimpanzees and gorillas does not reveal any conspicuous differences related to the different patterns of genital swelling and sexual behavior in these species. The absence of differences in 17β-estradiol concentrations could be interpreted as indicating that differences in genital swelling and sexual behavior of female chimpanzees and gorillas are the result of differences in the hormonal responsiveness of the target tissues that mediate these activities. Because there are relatively few cycles for both species available for comparison, however, the absence of a relationship based on these limited data is not conclusive. Additional menstrual cycle hormone data are required for chimpanzees, especially with respect to testosterone concentrations in the cycle, before competent interspecies comparisons can be made. Menstrual cycle patterns of urinary hormones were reported for female orangutans (Collins *et al.*, 1975), but the few cycles described do not permit interpretation of interspecies relationships for this species either. Additional data on menstrual cycle hormone patterns of orangutans, especially, are required to advance the description of hormone-behavior relationships among the great apes.

Data on plasma and serum concentrations of 17β-estradiol and testosterone in gorillas indicate that both these hormones reach peak concentrations during the midcycle period of maximal genital swelling. The timing of the peaks was different, however, 17β-estradiol reaching its peak on the day before or the day of the LH surge, testosterone on the day after. Given this temporal dissociation of the hormone peaks in gorillas and their species-typical pattern of mating only once or twice per cycle, assessment of differential behavioral responsiveness in relation to the peaks is possible. Characterization of estrogen and testosterone concentrations during the menstrual cycle of chimpanzees and orangutans would indicate whether dissociation occurs in the timing of peak hormone concentrations similar to that found in gorillas, or whether different patterns exist which are related to the differences in mating patterns recorded in the laboratory. The availability of such hormone data on all three species, in conjunction with behavioral data from tests in which male dominance *vis-à-vis* the female is controlled, would provide the basis for more specific assessments of the hormone regulation of sexual responsiveness of the female great apes.

VII. SEXUAL ATTRACTIVITY, PROCEPTIVITY, AND RECEPTIVITY OF THE GREAT APES IN THE LABORATORY

The laboratory data discussed above were not presented by the authors in relation to the concepts of sexual attractivity, proceptivity, and receptivity, as currently defined (Beach, 1976). Some of the data were presented in sufficient detail, however, to permit tentative interpretations in terms of these concepts. Data considered relevant to the concepts of sexual attractivity, proceptivity, and receptivity in the great apes are cited in the following. Figures 1–3 contain curves derived from these data that reflect approximations of relative strength and duration of the concepts in the respective menstrual cycles of the apes, as determined in the laboratory.

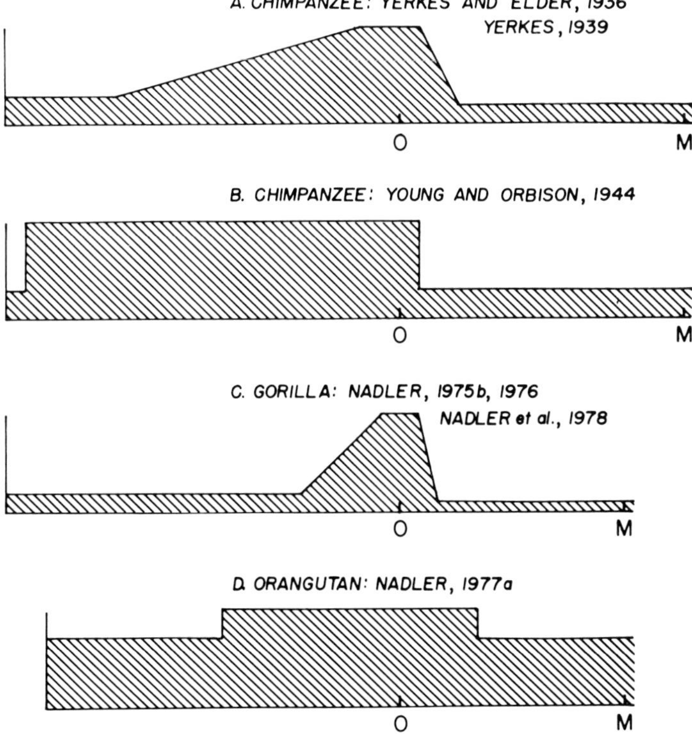

Fig. 1. Menstrual cycle patterns of sexual attractivity in the great apes. Data sources for deriving each pattern are indicated above, center. The ordinate represents relative intensity; the abscissa, days of the menstrual cycle. Zero (0) indicates the day of the LH surge; M, the first day of menses.

A. Sexual Attractivity

This concept reflects the "stimulus value" (Beach, 1976) of the female and is defined by the *male's behavior* toward the female. It should be noted that data related to copulation constitute the main data source available for defining sexual attractivity of the apes. This use of the data related to copulation requires comment, however. The occurrence of copulation is clearly influenced by the female's behavior, i.e., receptivity and proceptivity, as well as the male's. The data on copulation, therefore, represent a confounded measure of sexual attractivity. The copulation data are used in the present analysis for two reasons: (1) completely appropriate data are not available for the apes at this time, and (2) the conditions of testing used to obtain these data enhanced the male's prerogatives for initiating copulation. The data related to copulation, therefore, reflect predominantly the male's behavior toward the female, justifying their use for preliminary estimates of the concept of sexual attractivity in the great apes.

1. Chimpanzees

The data on copulation frequency in chimpanzees in association with cycle phase (Yerkes and Elder, 1936, Table 5; Yerkes, 1939, Table 2) are consistent. They indicate that attractivity is lowest during menses and the adjacent phases, and increases during tumescence to reach maximal levels in association with maximal genital swelling. Attractivity then declines during detumescence to a lower level than in the tumescent phase. The measures of male desire and female acceptability (Yerkes, 1939, Table 2) follow similar patterns. Yerkes and Elder (1936, Table 6) reported that female presentations of a low or intermediate type were followed by copulation 83% of the time, compared to 12% for high presentations or no presentations. Since intermediate and low presentations occur primarily during the tumescent and maximal swelling phases, it follows that females in the tumescent and maximal swelling phases were most attractive to the males. Yerkes and Elder (1936) also reported the data on copulation frequency in relation to cycle days. Analysis of these data indicates general agreement with the preceding data and, in addition, suggests that attractivity increases during the phase of maximal swelling and reaches a peak during the latter half of swelling before declining. In general, therefore, attractivity seems to vary with genital swelling but, in addition, may change during maximal swelling without a concomitant change in swelling (Figure 1A). The pattern of attractivity is also concordant with the pattern of estrogen concentrations during the follicular phase of the chimpanzee menstrual cycle (Reyes *et al.*, 1975), consistent with earlier reports that this hormone increased attractivity of rhesus monkeys (Herbert and Trimble, 1967; Michael and Welegalla, 1968; Johnson and Phoenix, 1976; Keverne, 1976; Wallen and Goy, 1977). The decline in atractivity during the luteal phase is associated with detumescence of the genital swelling, following reduction in estrogen concentration and an increase in progesterone (Graham *et al.*, 1972). The absence of an increase in genital swelling and attractivity during the luteal phase elevation of estrogen in this species suggests that progesterone exerts an inhibitory influence on both measures.

The data of Young and Orbison (1944) that relate to the concept of attractivity are found in Tables 1-A (sexual excitement of male), 1-E and 1-F (frequency of copulation), 1-M (nonresponsiveness of male), and 1-N and 1-O (time spent together). These data indicate that attractivity is greater during the follicular phase than the luteal phase (Figure 1C), but they do not relate to the issue of whether attractivity increases to a peak during the late follicular phase.

2. Gorillas

There were significant differences associated with different degrees of LT for the measures, female success ratio, and percent tests with copulation (Nadler, 1975b, Table 1). The number of copulations per test (Nadler, 1975b, Table 1) were in the predicted direction, but differences related to LT were not significant. These data suggest that attractivity is low when LT is minimal, increases with moderate tumescence, and reaches a maximum in association with maximal LT. Data recently reported indicated that the frequency of copulation days in that study (Nadler *et al.*, 1978) increased sharply two days before detumescence on the day after the estrogen peak, the day of the LH surge. Copulation reached maximal frequency on the day following the LH surge, the day of the testosterone peak, declined slightly the following day, on the day of detumescence, and then declined rapidly. Figure 1C approximates the pattern. Enhanced attractivity recorded shortly after the midcycle estrogen peak, in association with the testosterone peak, may reflect the action of both hormones. Whereas the effects of estrogen on attractivity are considered to be direct, testosterone may exert its effect indirectly by facilitating proceptive behavior. A proceptive (rhesus) female is presumably more attractive than an indifferent one (Beach, 1976; Keverne, 1976). The decline in attractivity is associated with both reduced concentrations of estrogen and elevated progesterone. Since the luteal phase increase in estrogen is not accompanied by an increase in attractivity, the data on the gorilla, like those on the chimpanzee, suggest that the loss of attractivity is primarily related to an inhibitory effect of progesterone.

3. Orangutans

Attractivity of orangutans reflected in the percent cycle days with copulation (Nadler, 1977a) is generally higher through the cycle than for the other apes. The finding that multiple copulations occurred more frequently during the midcycle period suggests that attractivity was somewhat greater during that time (Figure 1D). In all three species, therefore, there is evidence from the laboratory tests of increased attractivity during midcycle and, for the chimpanzee and gorilla, relatively low attractivity during the luteal phase.

B. Sexual Proceptivity

This concept relates to the appetitive *behavior of the female* directed toward the male and, therefore, reflects on the stimulus value of the male (Beach, 1976).

1. Chimpanzees

The data of Yerkes and Elder (1936, Table 6) and Yerkes (1939, Table 4) regarding female initiative portray a similar picture of sexual proceptivity in chimpanzees. Proceptivity is lowest during menses and adjacent phases, elevated during the tumescent phase, maximal at maximal swelling, and somewhat lower during detumescence than during tumescence (Figure 2A). The data on the frequency of presentations (Yerkes, 1939, Table 6) are similar, but the difference between maximal swelling and the other phases is more pronounced. Together, the data on proceptivity are quite similar in pattern to the data on attractivity (Yerkes and Elder, 1936, Table 5; Yerkes, 1939, Table 2). Data related to proceptivity were not presented in relation to cycle days, however, and the possibility that changes occurred during the phase of maximal swelling cannot be assessed. The data from Young and Orbison (1944), analyzed in terms of follicular and luteal phases of the cycle, suggest that proceptivity was greater during the former phase (Figure 2B). The latter authors remarked that some females exhibited peaks in presentational behavior, suggesting that an increase in proceptivity during maximal swelling may occur under certain circumstances.

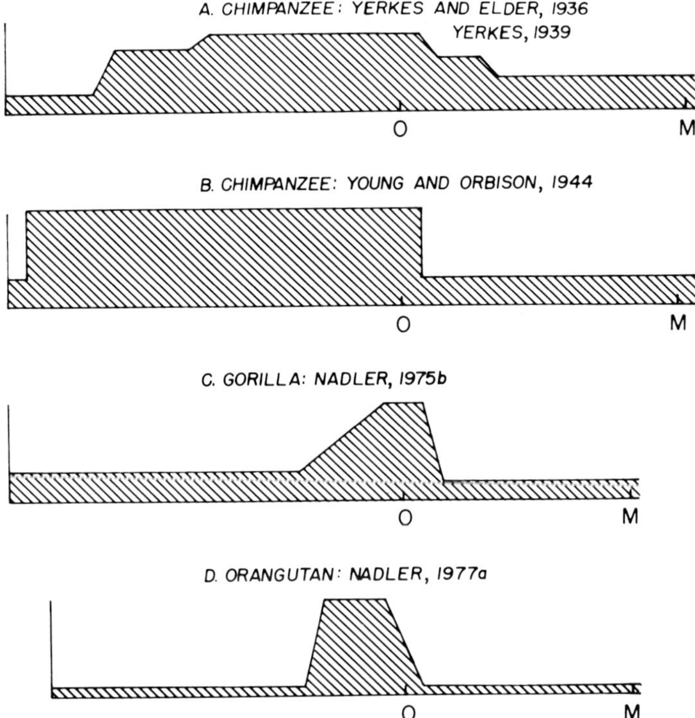

Fig. 2. Menstrual cycle patterns of sexual proceptivity in the great apes. (For explanation, see Fig. 1.)

2. Gorillas

The data on gorillas indicate that the percent tests with presentations and the number of presentations per test increase with increasing LT (Nadler, 1975b, Table 1). The pattern is similar to that for attractivity, but the intensity of proceptivity appears to be relatively greater (Figure 2C). Female gorillas present to males at all degrees of LT (proceptivity), but they are relatively less successful (attractive) during minimal LT.

3. Orangutans

Female orangutans exhibit the least evidence of proceptivity among the apes. Analysis of the four instances of proceptivity reported (Nadler, 1977a) in relation to menses and recently obtained unpublished data suggesting a luteal phase of 12 days' duration in female orangutans reveal the pattern in Figure 2D.

The similarity between the patterns of attractivity and proceptivity for the chimpanzee and the gorilla suggest that some of the same hormones may regulate both conditions. These tentative suggestions clearly require considerably more data before a competent statement can be made regarding the hormonal contributions.

C. Sexual Receptivity

This concept is defined by the *behavior of the female* that permits and contributes to fertilization by a male.

1. Chimpanzees

The data on acceptability of the male (Yerkes, 1939, Table 2) and type of presentation (Table 6) exhibit similar patterns, patterns not very different from those for attractivity and proceptivity (Figure 3A). Receptivity is low during menses and adjacent phases, as indicated by the absence of adequate presentations or any presentations altogether. Appropriate presentations (low or intermediate to the ground) occurred more frequently during tumescence and most frequently at maximal swelling before declining. Since Young and Orbison (1944) did not

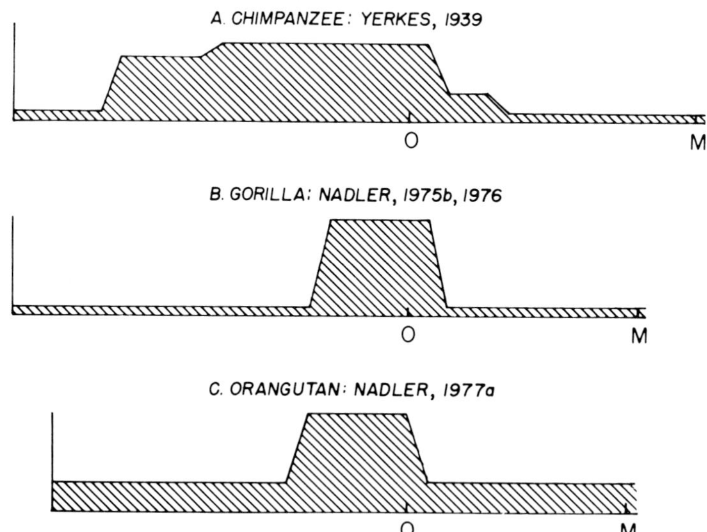

Fig. 3. Menstrual cycle patterns of sexual receptivity in the great apes. (For explanation, see Fig. 1.)

record any measures clearly related to receptivity, no other laboratory data on this measure are available for chimpanzees.

2. Gorilla

Since female gorillas are highly proceptive (and the males relatively nonassertive in initiating sexual activity), receptive behavior is not a conspicuous feature in the mating of this species. Beach (1976) proposed that data related to ejaculation reflected, in part, the attractivity of the female, but such data appear also to reflect receptivity. In order for the male to effect fertilization, the female must assume an appropriate posture and maintain that posture "long enough so that intravaginal ejaculation can occur" (Beach, 1976, p. 125). Ejaculation, therefore, requires receptive behavior by the female and represents indirect evidence of receptivity. In the absence of data directly related to receptivity in gorillas, the indirect evidence on ejaculation is used for a preliminary assessment of this concept.

The data related to ejaculation in gorillas indicate that no ejaculations occurred unless the females were tumescent (Nadler, 1975b, 1976). With respect to the two different degrees of tumescence, one of the measures (percent tests with ejaculation) was relatively higher at maximal LT, but the other two (ejaculations per test and percent ejaculations on a single mount) were similar for the two tumescent conditions. These data on the gorilla, therefore, are not entirely consistent. They do suggest, however, that receptivity is greater during tumescent conditions in comparison to minimal LT (Figure 3B).

3. Orangutans

In order to gain some perspective on receptivity of female orangutans, data from the published study (Nadler, 1977a) were reexamined for evidence of female cooperation in mating. It was found that three of the females offered the least resistance (none) to the males' mating efforts during the midcycle phase. Calculation of the cycle days implicated in this behavioral change suggested the day of the LH surge and the four previous days (Figure 3C).

The preceding discussion is presented primarily to suggest that sexual responsiveness in the great apes is susceptible to analyses similar to those used for other species (Beach, 1976). The analyses presented are largely hypothetical since the data on which they are based are incomplete. They do suggest, however, that despite the differences in the overall mating patterns of the apes, there are several areas of similarity. Although baselines differ with respect to attractivity, all species exhibit evidence of midcycle enhancement. Additional research is required to establish the basis for the variability. All species also show evidence of proceptive behavior. It is interesting to note in this regard that gorilla and orangutan females appear to exhibit similar patterns of proceptivity, despite the considerable difference between sexual initiative of the males under laboratory conditions. Receptivity also seems to follow similar patterns in these species, although different measures of behavior were used to calculate these patterns.

VIII.   CONCLUSIONS

In order to derive more adequate assessments of sexual attractivity, proceptivity, and receptivity in the great apes and, thereby, describe more precisely the regulation of sexual behavior in these species, at least two methodological modifications are indicated: (1) selection and measurement of several behavioral items for each concept that are appropriate for such assessment, and (2) arrangement of test conditions that permit expression of the behavioral items of interest. It was stated above that the data sources used for assessing sexual attractivity of the apes were largely confounded, since they related primarily to copulation. The data relevant to the concepts of proceptivity and receptivity in the apes are meager and require support as well.

A.   Behavioral Items Potentially Applicable to the Concepts of Sexual Attractivity, Proceptivity, and Receptivity in the Great Apes

1.   *Sexual Attractivity*

Items of male behavior appropriate for assessing sexual attractivity of the female include erection, attend, solicit, approach and remain proximate, genital inspect, and groom. Rejection of the female indicates the absence of attractivity. The proportion of female invitations/presentations that are followed by mounting/copulation is a complex measure of attractivity, termed the Female Success Ratio (Michael, 1968) and the Male's Acceptance Ratio (Dixson *et al.*, 1973). An operant response performed by the male to gain access to the female is another measure of the female's attractivity (Michael, 1968).

2. Sexual Proceptivity

Items of female behavior appropriate for assessing sexual proceptivity of the female include approach and remain proximate, solicit/invite/present (frequency), precopulatory vocalization, genital contact (inspect male's genitals, rub own genitals on male), groom, touch/pull/slap, and mount. An operant response performed by the female to gain access to the male may be another measure of proceptivity (Eaton, 1973).

3. Sexual Receptivity

Items of female behavior appropriate for assessing sexual receptivity of the female include present (adequacy for intromission and ejaculation), rejection, and resistance/nonresistance. The proportion of male mount/copulatory attempts that are followed by mounting/copulation is a complex measure of receptivity, termed the Male Success Ratio (Michael, 1968) and the Female's Acceptance Ratio (Dixson *et al.*, 1973). Another version of the female's acceptance ratio is the proportion of male mount/copulatory attempts that are followed by female presenting (Trimble and Herbert, 1968).

It should be noted that not all of the behavioral items listed above apply to each species of ape. Erection prior to copulation, for example, is visible only in the male chimpanzee, among the apes. In general, determination of which of the items listed applies to a given species and whether there are additional items that should be measured is an empirical matter, to be established in future research.

B. Test Conditions Conducive to Assessment of Sexual Responsiveness in the Great Apes

It was stated in several places above that accurate assessment of sexual responsiveness in the great apes, especially female proceptivity, required an approach that reduced the male's influence on the female's behavior. A simple procedure for accomplishing the desired effect is available for use with the gorillas and orangutans, based on the considerable degree of sexual dimorphism in these species. The approach taken is to reduce the size of the doorway separating two compartments of a test cage, such that only the female can pass through it. Preliminary data on the effect of this relatively simple modification of the experiment in orangutans indicate a marked reduction in the frequency of copulation in this species. The females only entered the compartment of the male during a brief period of the midcycle phase, consistent with the pattern of proceptivity described for orangutans in Figure 2D. Since there is little sexual dimorphism in chimpanzees, a procedure similar to that used with monkeys is appropriate (Michael, 1968; Eaton, 1973), in which the door separating the pair can be opened only by the female. Experiments of this type in the great apes will provide another perspective on their sexual behavior in the laboratory and, in conjunction with data on additional behavioral items and menstrual cycle hormone patterns, facilitate more detailed analysis of sexual responsiveness and its hormonal regulation.

ACKNOWLEDGMENTS

The work of the author described herein was supported by NSF Grants GB-30757 and BNS 75-06287, and PHS Grant RR-00165 from NIH. The author gratefully acknowledges the comments and criticisms on an earlier version of this chapter by F. A. Beach, S. L. Washburn, and R. E. Whalen.

REFERENCES

Beach, F. A. (1942). *Psychol. Bull.* **39**, 200–226.
Beach, F. A. (1976). *Horm. Behav.* **7**, 105–138.
Bingham, H. C. (1928). *Comp. Psychol. Monogr.* **5**, 1–165.
Coe, C. L., Connolly, A. C., Kraemer, H. C., and Levine, S. (1979). *Primates* **20**, 571–582.
Collins, D. C., Graham, C. E., and Preedy, J. R. K. (1975). *Endocrinology* **96**, 93–101.
Darwin, C. (1859). "The Origin of the Species." Murray, London.
Davenport, R. K., and Rogers, C. M. (1970). *In* "The Chimpanzee" (G. H. Bourne, ed.), Vol. 3, pp. 337–360. S. Karger, Basel.
Dixson, A. F., Everitt, G. J., Herbert, J., Rugman, S. M., and Scuton, D. M. (1973). *In Symp. Int. Congr. Primat.*, 4th, "Primate Reproductive Behavior" (C. H. Phoenix, ed.), Vol. 2, pp. 36–63. S. Karger, Basel.
Eaton, G. G. (1973). *In Symp. Int. Congr. Primat.*, 4th, "Primate Reproductive Behavior" (C. H. Phoenix, ed.), Vol. 2, pp. 20–35. S. Karger, Basel.
Eaton, G. G., and Resko, J. A. (1974). *J. Comp. Physiol. Psychol.* **86**, 919–925.
Elder, J. H. (1938). *Yale J. Biol. Med.* **10**, 347–364.
Galdikas, B. (1979). *In* "Perspectives on Human Evolution: The Great Apes" (D. A. Hamburg and E. R. McCown, eds.), Vol. 5, pp. 195–233. Benjamin/Cummings, Menlo Park, California.
Graham, C. E. (1970). *In* "The Chimpanzee" (G. H. Bourne, ed.), Vol. 3, pp. 183–220. S. Karger, Basel.
Graham, C. E., Collins, D. C., Robinson, H., and Preedy, J. R. K. (1972). *Endocrinology* **91**, 13–24.
Hamilton, G. V. (1914). *J. Anim. Behav.* **4**, 295–318.
Harcourt, A. H., and Stewart, K. J. (1977). *New Sci.* **76**, 160–162.
Harcourt, A. H., and Stewart, K. J. (1978). *In* "Recent Advances in Primatology—Behaviour" (D. J. Chivers and J. Herbert, eds.), Vol. 1, pp. 611–612. Academic Press, New York.

Hartman, C. (1928). *J. Mammal.* **9**, 181–194.

Herbert, J., and Trimble, M. R. (1967). *Nature (London)* **216**, 165–166.

Hess, J. P. (1973). *In* "Comparative Ecology and Behaviour of Primates" (R. P. Michael and J. H. Crook, eds.), pp. 508–581. Academic Press, New York.

Horr, D. A. (1975). *In* "Primate Behavior: Developments in Field and Laboratory Research" (L. A. Rosenblum, ed.), Vol. 4, pp. 307–323. Academic Press, New York.

Huxley, T. H. (1893). "Man's Place in Nature," Everyman's Library edition. Dent, London, 1921.

Johnson, D. F., and Phoenix, C. H. (1976). *J. Comp. Physiol. Psychol.* **90**, 473–483.

Keith, A. (1899). *Proc. Zool. Soc. (London)* 296–312.

Kempf, E. J. (1917). *Psychoanal. Rev.* **4**, 127–154.

Keverne, E. B. (1976). *In* "Advances in the Study of Behavior" (D. S. Lehrman, R. Hinde, and E. Shaw, eds.), Vol. 7, pp. 155–200. Academic Press, New York.

Köhler, W. (1925). "The Mentality of Apes." Routledge and Kegan Paul, London.

Kollar, E. J., Beckwith, W. C., and Edgerton, R. B. (1968). *J. Nerv. Ment. Dis.* **147**, 444–459.

Lemmon, W. B. (1971). *In* "Medical Primatology" (J. E. Goldsmith and J. Moor-Jankowski, eds.), pp. 432–440. S. Karger, Basel.

Lemmon, W. B., and Allen, M. L. (1978). *Folia Primatol.* **30**, 80–88.

Luttge, W. G. (1971). *Arch. Sex. Behav.* **1**, 61–88.

MacKinnon, J. R. (1971). *Oryx* **11**, 141–191.

MacKinnon, J. R. (1974). *Anim. Behav.* **22**, 3–74.

Maple, T. L., Zucker, E. L., and Dennon, M. B. (1979). *Behav. Processes* **4**, 53–59.

Mason, W. A. (1960). *J. Comp. Physiol. Psychol.* **53**, 582–589.

Mason, W. A. (1971). *In Nebraska Symp. Motivation* (W. J. Arnold and M. M. Page, eds.), pp. 35–67. Univ. of Nebraska Press, Lincoln, Nebraska.

Mason, W. A., and Berkson, G. (1975). *Dev. Psychobiol.* **8**, 197–211.

Mason, W. A., Davenport, R. K., Jr., and Menzel, E. W., Jr. (1968). *In* "Early Experience and Behavior" (G. Newton and S. Levine, eds.), pp. 1–41. Thomas, Springfield, Illinois.

Michael, R. P. (1968). *In* "Endocrinology and Human Behaviour" (R. P. Michael, ed.), pp. 69–93. Oxford Univ. Press, London and New York.

Michael, R. P., and Welegalla, J. (1968). *J. Endocrinol.* **41**, 407–421.

Miller, G. S., Jr. (1928). *J. Mammal.* **9**, 273–293.

Montané, L. (1915). *J. Anim. Behav.* **6**, 330–333.

Nadler, R. D. (1975a). *Anat. Rec.* **181**, 791–798.

Nadler, R. D. (1975b). *Science* **189**, 813–814.

Nadler, R. D. (1976). *Arch. Sex. Behav.* **5**, 487–502.

Nadler, R. D. (1977a). *Arch. Sex. Behav.* **6**, 457–475.

Nadler, R. D. (1977b). *In* "Progress in Ape Research" (G. H. Bourne, ed.), pp. 191–206. Academic Press, New York.

Nadler, R. D., Graham, C. E., and Collins, D. C. (1978). *Ann. Meeting Am. Soc. Primatol., 2nd, Atlanta, Georgia, September 6–9.* Abstract.

Nadler, R. D., Graham, C. E., Collins, D. C., and Gould, K. G. (1979). *Endocrinology* **105**, 290–296.

Nissen, H. W. (1931). *Comp. Psychol. Monogr.* **8**(1), 1–122.

Nissen, H. W. (1954). Genetic, Psychological and Hormonal Factors in the Establishment and Maintenance of the Patterns of Sexual Behavior in Mammals. Unpublished manuscript on file in the Univ. of Kansas Library, F 591.16 Sy 68 1954.

Nissen, H. W. (1956). *Am. Anthrop.* **58**, 407–413.

Noback, C. R. (1939). *Anat. Rec.* **73**, 209–225.

Parkes, A. S., and Zuckerman, S. (1931). *J. Anat.* **65**, 273–276.

Prescott, J. W. (1970). *In* "Medical Primatology" (J. E. Goldsmith and J. Moor-Jankowski, eds.), pp. 356–375. S. Karger, Basel.

Reyes, F. I., Winter, J. S. D., Faiman, C., and Hobson, W. C. (1975). *Endocrinology* **96**, 1447–1455.

Riesen, A. H. (1971). *In* "The Chimpanzee" (G. H. Bourne, ed.), Vol. 4, pp. 1–18. S. Karger, Basel.

Riesen, A. H., and Kinder, E. F. (1952). "The Postural Development of Infant Chimpanzees." Yale Univ. Press, New Haven, Connecticut.

Rijksen, H. D. (1978). "A Field Study on Sumatran Orang Utans (*Pongo pygmaeus abelii*, Lesson, 1827)." Meded. Landbouwhogeschool Wageningen, Wageningen, Nederland.

Rodman, P. S. (1973). *In* "Comparative Ecology and Behaviour of Primates" (R. P. Michael and J. H. Crook, eds.), pp. 171–209. Academic Press, New York.

Rogers, C. M., and Davenport, R. K. (1969). *Dev. Psychol.* **1**, 200–204.

Schaller, G. B. (1963). "The Mountain Gorilla." Univ. of Chicago Press, Chicago, Illinois.

Short, R. V. (1977). *In* "Reproduction and Evolution" (J. H. Calaby and C. H. Tyndale-Biscoe, eds.), pp. 3–19. Griffin Press, Netley, South Australia.

Sokolowsky, A. (1923). *Urol. Cutan. Rev.* **27**, 612–615.

Struble, R. G., and Riesen, A. H. (1978). *Dev. Psychobiol.* **11**, 479–486.

Tinklepaugh, O. L. (1933). *J. Morphol.* **54**, 521–546.

Trimble, M. R., and Herbert, J. (1968). *J. Endocrinol.* **42**, 171–185.

Tutin, C. E. G. (1975). *In* "Contemporary Primatology" (S. Kondo, M. Kawai, and A. Ehara, eds.), pp. 445–449. S. Karger, Basel.

Tutin, C. E. G., and McGrew, W. C. (1973). *Folia Primatol.* **19**, 237–256.

van Lawick-Goodall, J. (1968). *Anim. Behav. Monogr.* **1**(3), 161–311.

Wallen, K., and Goy, R. W. (1977). *Horm. Behav.* **9**, 228–248.

Yerkes, R. M. (1925). "Almost Human." The Century Co., New York.

Yerkes, R. M. (1932). *Comp. Psychol. Monogr.* **8**, 1–33.

Yerkes, R. M. (1939). *Hum. Biol.* **11**, 78–111.

Yerkes, R. M., and Elder, J. H. (1936). *Comp. Psychol. Monogr.* **13**, 1–39.

Young, W. C., and Orbison, W. D. (1944). *J. Comp. Psychol.* **37**, 107–143.

Zuckerman, S. (1930). *Proc. Zool. Soc. London* **45**, 691–754.

A. Immaturity

Quantitative data
between immature
our studies were di
sexual swelling. Ho
Juvenile males freq
play, directing thes
(Fig. 1). Immature n
immature females. F
during play and to
years onwards, but
Male chimpanzee

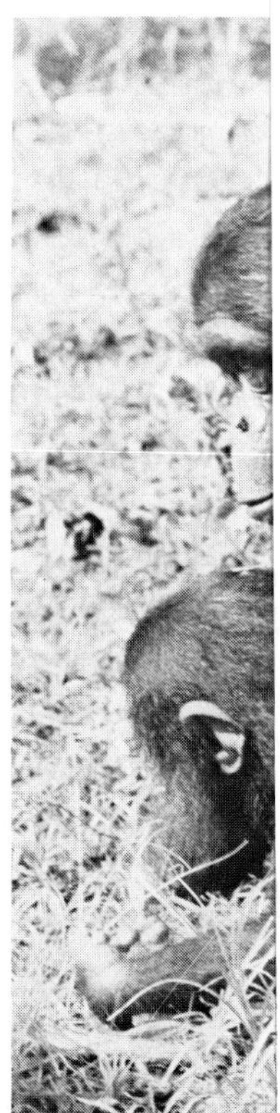

Fig. 1. A juvenile m

Chapter 9

CHIMPANZEE REPRODUCTION IN THE WILD

Caroline E. G. Tutin
Patrick R. McGinnis

I. INTRODUCTION

The chimpanzee differs from the other great apes in its form of social structure and its mating system. The basic unit of chimpanzee social structure is the multimale group, termed a community or unit-group. The only enduring bonds within a community are between mothers and their dependent offspring, whereas other groupings are unstable because of constant fission and fusion. Observations of the chimpanzee

239

Copyright © 1981 by Academic Press, Inc.
All rights of reproduction in any form reserved.
ISBN 0-12-295020-8

population (
Tanzania, re
a large range
(Wrangham,
found that cc
interactions
the ranking
  Chimpanz
competition
(Goodall, 19(
of the chimp
changed this
system of tl
patterns:

  (a) *oppor*
may be mate
  (b) *posses*
lationship w
males from
  (c) *conso*
the other con
to her, both

These three
further dime
munities on
  The data o
National Par
Tutin made
between 197
females show
trated on re
phase of the
corded. The
was fully acc
majority of t
Additional d
same commu
of the Gomb

cocious sexual behavior; the full patterns of courtship and copulation, directed to adult females, appear long before puberty. During 1200 hours of observation in Tutin's study, 408 sequences of sexual behavior between eight immature males (aged 15 mo to 9 yr) and nine adult females were recorded. In 318 (78%) of sequences, at least one intromission occurred. During their first year of life male infants showed an interest in the sexual swellings of adult females, and three male infants successfully achieved intromission before they were two years old (Fig. 2). In infantile mountings (15–36 mo), successful intromission was not an essential prerequisite of pelvic thrusting; young males often thrust against the side of a female's swelling. Courtship patterns emerged gradually during late infancy and the juvenile period. Copulatory sequences of immature males were often lengthy (lasting up to 40 min), with alternate bouts of play and courtship; multiple intromissions and combined sequences of attempted and successful intromissions were common.

B. Adolescence

Precise data on the time course of reproductive development in wild female chimpanzees are not yet available. The oldest female whose

Fig. 2. A 16-month-old infant male copulates with an estrous female who is being groomed by the infant's mother.

birth date is known showed the first signs of sexual swelling when she was 10 years old. Observation of the pattern of development in other females of unknown birth date indicate that menarche normally occurs 1–3 years after the first signs of swelling. The sexual swelling reaches full size some months prior to menarche, and once full swelling size is achieved, the females are attractive and receptive to adult males. After menarche there is a long period of adolescent sterility. The length of this period, between menarche and the first known conception was 13, 28, 30, and 34 months for four females at Gombe. This, plus the tentative estimate of age at menarche of 11–13 years, means that a typical female at Gombe would not give birth until she was 13–15 years old. Female chimpanzees raised in captivity give birth for the first time between 10–11 years, as a result of reaching menarche earlier (mean age at menarche 8 yr 6 mo, $N = 18$, range 7 yr–10 yr 9 mo), and having a considerably shorter period of adolescent sterility (mean 11.3 mo, $N = 7$, range 4–17 mo) (Young and Yerkes, 1943; Smith *et al.*, 1975).

The only chimpanzee male of known age at Gombe to have reached puberty was first seen to ejaculate when he was 9 yr 2 mo of age. Male chimpanzees born and/or raised in captivity reach puberty at an estimated mean age of 7 yr 2 mo $\pm$ 9 mo ($N = 25$) (Smith *et al*, 1975).

C. Adulthood and Reproduction

The lifetime reproductive potential of wild chimpanzee females is extremely limited. The mean interval between births of surviving offspring at Gombe was 5 yr 10 mo ($N = 8$, range 4 yr 5 mo–7 yr 7 mo). Following the death of a suckling infant, conception occurred after 4–8 mo ($N = 5$), resulting in the birth of another infant 12–16 mo after the death of the first. The reproductive life of a female begins with the first pregnancy at 13–15 years of age and continues until death which might occur at 40–45 yr. This gives 25–32 reproductively active years during which a maximum of five or six offspring who survive to weaning can be produced. Almost 30% of detected pregnancies* did not result in live-births, and thus one might expect a typical female to have in her lifetime seven pregnancies, with five resulting in the birth of a living offspring.

* There is no external indicator of pregnancy in the chimpanzee. Sexual cycles continue after conception (Section II, D) but become increasingly irregular and eccentric as pregnancy continues. Thus, it is almost always possible to predict pregnancy 3–4 months after conception. However, as some pregnancies go undetected in the early stages, the estimate of the number of terminated pregnancies must be considered as conservative.

For males, sperm production, and hence reproductive capacity, appears to be continuous from puberty to death, giving a reproductive lifespan of about 30 years. In the Kasakela community, there were about 12 adult females at any time, each of whom would be capable of seven pregnancies in 30 years. This means that during a male's reproductive life 84 detectable conceptions might occur (but prenatal mortality would reduce this to about 60 live-births). In this study population that has approximately equal numbers of males and females, the mean number of expected impregnations per male would be seven, and any individual increase on this could only be achieved at the expense of other males in the community.

D. Reproductive Cycles

1. Sexual Swelling, Ovulation, Estrus, and Copulation

The cycle of sexual swelling of female chimpanzees has been described in detail by Graham (1970, 1973, Chapter 1, this volume), and the cycle lengths of females at Gombe correspond with those reported for chimpanzees in captivity. For 46 cycles shown by nine (three nulliparous and six parous) nonpregnant chimpanzee females at Gombe, the median length was 36 days (range 25–84 days). Cycles were measured from the first day of maximal tumescence as menstruation could not be reliably scored under field conditions. The swelling-cycle phase of maximal tumescence had a median length of 9.6 days ($N = 6$ females, 37 cycles, range 7–17 days). Ovulation usually occurred on the last day of maximal tumescence $\pm$ one day (see Graham, Chapter 1, this volume).

A copulation was defined as an interaction between a male and a female that included at least one intromission. Some copulations included multiple intromissions, and to be classified as separate copulations, intromissions had to be separated by at least 10 minutes of noninteraction between the two individuals.

Data from Tutin's study show that during maximal tumescence, copulation occurred at a mean rate of 0.52 per male hour* (1105 copulations in 2112 male hrs), while during other cycle phases copulation occurred at a mean rate of 0.06 per male hour (32 copulations in 552 male hrs). The cycle phase of maximal tumescence can be called the estrous period in accordance with the strictly behavioral definition proposed by Rowell

---

* Male hour = (number of males observed with female) $\times$ Σ (number of hours each male with female).

(1972). During estrus, the mean copulation rates of seven females averaged over all males, ranged from 0.31–0.99 per male hr. There was no correlation between a female's age or parity and mean hourly copulation rates. Individual males copulated with estrous females at mean rates varying from 0.03–1.14 per hr. There was no overall correlation between male copulation rates and age (Spearman Rank Correlation Coefficient, $r_s = -0.12$, $N = 18$, n.s.), but if infant males are excluded, a statistically significant negative correlation exists ($r_s = -0.94$, $N = 12$, $p < .001$). The highest hourly rates were shown by juvenile and adolescent males; adult males showed lower rates, and the lowest rates were shown by two old males.

2.   Conception

Nulliparous females had a median of 19 cycles ($N = 4$, range 10–27) before the first observed pregnancy (Gombe Stream Research Centre, long-term records). There is no external indicator of conception in the chimpanzee, but the cycle during which conception occurred could be identified on a *post hoc* basis if the pregnancy resulted in a live-birth. Gestation length was assumed to be 228 days, which is the mean duration ($N = 47$, range 202–248 days) reported by Nissen and Yerkes (1943). During 18 pregnancies at Gombe, all 10 females observed continued to show some sexual swelling after conception. In all but one pregnancy this included at least one complete sexual-swelling cycle, i.e., a swelling in which maximal tumescence was reached and maintained for several days. The median number of complete postconception swelling cycles was two ($N = 18$, range 0–6), and all the females also showed irregular sexual swellings that did not resemble the pattern of the menstrual cycle. In most cases the sexual swelling was absent during the last two to four months of pregnancy, but one female showed a complete cycle only one month before giving birth.

In humans, the period of amenorrhea between parturition and the reappearance of menstruation is longer in lactating than nonlactating women (Stix, 1940; Sharman, 1951). The duration of acyclicity in wild chimpanzees can be assessed by the inhibition of sexual swelling, even though menstruation cannot be reliably detected in free-ranging chimpanzees.

For nine females at Gombe the median latency from parturition to maximal tumescence was 43 mos ($N = 10$ intervals, range 11–81 mos). Only one female became pregnant in the first postpartum cycle, and the median number of cycles between parturition and the next conception was 3.6 ($N = 7$ females, 12 intervals, range 1–11 cycles). Most of the females were still suckling their infants when their sexual cycling resumed. Weaning in wild chimpanzees is a long process, and although a mother may begin to restrict her infant's access to the nipple during the infant's second year of life, weaning is not usually completed until after the mother has become pregnant again (Clark, 1977). During Tutin's study, two females who resumed sexual cycles while still lactating were mated very infrequently by adult males in their first postpartum cycles with mean copulation frequencies of 0.12 and 0.22 per male hour. In contrast, one female who resumed sexual cycles two months after the death of her 2½-year-old son, proved extremely attractive to adult males, and a high mean copulation rate (0.43 per male hour) was observed even before her swelling had reached maximal tumescence, when the mean copulation rate rose to .83 per male hour.

III. SEXUAL BEHAVIOR

A. Courtship and Copulation

While the mating system of wild chimpanzees comprises three distinct patterns, the basic behavior patterns shown in courtship and copulation are common to all three. Tutin recorded detailed behavioral data for 1084 copulations involving seven females and 18 males in 106 dyadic combinations. A total of 46 behavioral elements occurred in copulatory sequences, although few were unique to this context. Behavior patterns shown in combinations typical of copulatory sequences but which occurred in interactions *not* including intromission were recorded as incomplete copulatory sequences.

The distribution of the 46 elements of copulatory behavior with respect to timing in the sequence and sex, indicates the nature of the interaction. Table 1 shows that there were few female-specific patterns while there were 12 patterns exclusive to the male that were shown prior to intromission. These elements constitute a complex of courtship behavior. The elements shown by both sexes prior to intromission are concerned mainly with spacing and visual orientation with respect to the other interactant. The group of nine nonsex-specific elements, which were shown either before or after intromission, are spacing and orientation patterns, plus grooming, play, and reassurance patterns.

When directed to a female in estrus, male courtship usually lasted less than one minute before intromission. The mean number of behavioral elements in a random sample of 200 male courtships was 4.6, and the overall range in the 1084 courtships observed was 2–87 elements

TABLE 1

Number of different behavioral elements performed by each sex during the copulatory sequence (N = 1084).

Copulatory Phase	Male	Female	Both Sexes	Total
Pre-intromission phase only	12	3	8	23
Intromission and post-intromission phase only	4	4	2	10
Both phases	4	0	9	13
Total	20	7	19	46

Fig. 3. An adult male shakes a branch at an estrous female who is feeding in a tree.

(each element was counted each time it occurred). The typical sequence of events began when the male, with penis erect, advanced to within sight of the female (if he had not previously been so). He then directed a variety of courtship elements towards the female, the most common being branch shaking (Fig. 3), hair erection, male invitation* (Fig. 4), gazing, and hand-on-branch. These continued until the female responded, normally by glancing, approaching, and presenting to the male. The male rarely showed courtship elements once the female

* The male sits with legs partially flexed and spread apart to reveal his erect penis.

Fig. 4. An adolescent male sits in the male invitation posture as an estrous female approaches in response to his courtship.

had begun to move towards him. Once the female presented to the male, he effected intromission (Fig. 5) and made a mean of 8.8 pelvic thrusts ($N = 1084$, range 3–30) before the female terminated intromission by moving away (Fig. 6). In all cases copulation was dorsoventral, and intromission lasted for a mean duration of 7 sec ($N = 45$, range 3–15 sec). Frequently, one or both participants would inspect their own genitals after intromission, focusing their attention on congealed semen. Social grooming, especially male-to-female, was a common postintromission behavior pattern.

Most male courtship occurred when the pair was further than 1.5 m from each other. Several male courtship patterns are also shown in aggressive displays, especially the autonomic pattern of hair erection, and this may place the female in an approach–avoid conflict. This conflict may be minimized if the male does not approach directly, but remains some distance from the female, allowing her to approach him (McGinnis, 1973). Many male courtship elements appear to be derived from postures and behaviors shown during copulation (McGinnis, 1973). A number of elements of courtship have auditory components (e.g., branch shaking, branch jerking, stamping), or visual components (e.g., hair erection, hunching, male invitation, penile adduction), which

Fig. 5. A copulation between an adolescent male and an estrous female.

Fig. 6. The female leaves, ending intromission.

serve to attract the female's attention and to communicate to her his readiness to copulate.

The presence of a penile erection alone does not constitute courtship, as penile erections occur in nonsexual contexts of high arousal. Only three copulations involving adult males were observed in which the male had a penile erection but showed no other courtship elements prior to the female's approaching and presenting (see also McGinnis, 1973). On 12 other such occasions maximally tumescent females approached and presented, but intromission did not follow. A total of 209 courtship sequences that did not lead to intromission were observed during Tutin's study. In the vast majority of these (86%), the avoidance or nonresponsiveness of the female terminated the interaction. In another 10% of cases the interaction was terminated by extraneous events such as intragroup aggression and intercommunity encounters or predations, which distracted one or both members of the pair. In the remaining nine cases (4%), the male terminated courtship by not responding to the female's approach and presentation.

B. Interference

The response of adult males to the copulations of others depended on the mating pattern. During consortships no other adult males were present. Males showing possessive behavior sometimes responded aggressively to the courtship of others (see Section IV,A,2), but during opportunistic mating overt signs of competition between males for access to receptive females rarely emerged (see Section IV,A,1). Adolescent and adult females were rarely seen to respond to the copulations of unrelated females, but van Lawick-Goodall (1968, p. 242) reported that an old female and her adolescent daughter frequently approached each other's matings and made physical contact, interfering in a way typical of infants and juveniles (see the following).

The most common response to copulation by immature individuals was interference (see also van Lawick-Goodall, 1968), which was observed 341 times in 1137 copulations during Tutin's study. In many cases, the male courtship elements that attract the female's attention also appear to alert youngsters, who respond by advancing to the male, often following the female. Interfering youngsters were never seen to make contact with either male or female prior to intromission, but usually did so immediately when intromission began (Fig. 7). In the 341 observed interferences, the interferer made contact with the copulating male in 72% of the cases and the female in 28% of the cases.

Fig. 7. A juvenile male (center) interferes in the copulation of an adult male and an adult female (Photograph by Gombe Stream Research Centre).

In 91% of the cases where contact was made with the copulating female, the male was also contacted. In these cases the contact with the female appeared to be incidental and an intermediate step (often literally) towards contacting the male. Orientation was toward the male in over 90% of cases, and the interfering individual most often made contact with the male by touching his face (83% of cases). Usually the interferer left as intromission ended and did not interact further with the copulating pair (92% of cases). The interferer's behavioral patterns were submissive and fearful; aggressive patterns were absent.

The responses of 16 immature individuals (eight males and eight females, aged 2–11 yr) to others' copulations were seen on at least 10 occasions. The mothers of six individuals (two males and four females) showed cycles of sexual swelling, affording an opportunity to observe a youngster's reaction to copulations involving the mother. The data (Tutin, 1979a) show that while immature individuals of both sexes interfered in a proportion of their mothers' copulations, females restricted their interference to copulations of related individuals, but male juveniles showed frequent interference in copulations of unrelated chimpanzees. Intense interference first appeared toward copulations of the mother during her first postpartum cycle, providing the infant was older than 18 months when this occurred. For males, the pattern once established toward the mothers' copulations generalized to copulations of unrelated females. Interference by infants and juveniles was tolerated by copulating males who, in 14% of cases, directed reassurance gestures to the interferer. The interference shown by a young male, Goblin, was tolerated while he was a juvenile. Goblin continued to interfere in copulations after he had reached puberty, but gradually the response of the copulating males changed and they became increasingly aggressive towards him, either threatening or attacking him if he approached their copulations. This change in the response of copulating males led to a decrease in frequency, and Goblin's interference eventually ceased.

The motivation and function of interference by immature chimpanzees have been discussed in detail elsewhere (Tutin, 1979a). Interference is governed by the interferer's age, sex, and degree of relatedness to the copulating female. All immature individuals interfere in a proportion of their mothers' copulations. It appears that the mother's first postpartum sexual swelling often coincides with her increased restriction of her infant's suckling. It seems possible that interference at this time is related to the trauma of weaning. Juvenile males also interfere in the copulations of unrelated females, but the behavior decreases in frequency with increasing age and disappears during adolescence as the copulating males cease to be tolerant. This interference by juvenile and young adolescent males may be related to the possessive behavior shown by some adult males (see Section IV,A,2), i.e., an attempt to monopolize the female. It seems that during adolescence, young males become forced to suppress the behavior until they achieve a dominance status that permits its effective expression.

IV.   COMPETITION AND PARTNER PREFERENCES

A.   Intermale Competition

Competition between male chimpanzees in the wild in nonsexual contexts leads to a dominance ordering among the adult males of a community (Bygott, 1979). The order is generally, but not always, linear and is relatively stable over time. Changes in the dominance order are characterized by intense and prolonged conflict between the individuals concerned. Agonistic alliances are sometimes formed between adult male siblings, but they were never observed to form between unrelated males. The adult males of a community did, however, cooperate in attacks of neighboring communities. These intercommunity encounters are often extremely aggressive, and three of them at Gombe resulted in the deaths of adult males (Goodall *et al.*, 1979). Each encounter involved a group of males from one community meeting a lone male from a neighboring community who they attacked savagely and injured fatally.

Intermale competition in sexual contexts differed among the mating patterns, which are treated separately in the following sections.

1. Opportunistic Mating

The majority (73%) of the 1137 copulations observed by Tutin occurred during opportunistic mating. In this pattern estrous females associated with groups of two to 12 independent males*, and detectable signs of competition between males for access to females rarely emerged. Adult males normally showed no response to a copulation even if it occurred less than five meters away. Several examples of apparent social facilitation occurred when a nearby male developed a penile erection during another's copulation and then courted and mated with the female immediately afterwards. This occurred on 37% (98 of 265) of the occasions that an adult male was within 5 meters of a copulating pair. It seemed that social facilitation was even more likely to occur within the two pairs of adult male siblings; brothers were observed to copulate with the same female in quick succession on 64% (27 of 42) of the occasions that they were within 5 meters of each other when one copulated with a female.

On 11 occasions two or more males simultaneously courted the same female. In 10 cases the female presented to the closer of the two males, and the other male showed no aggressive response, but resumed his courtship once the first copulation had finished. The eleventh case occurred early one morning when Pallas (a receptive adult female) was simultaneously courted by Satan (a young adult male), Hugo (an old male), and Goblin (an adolescent male) as she descended from her nest of the night before. After hesitating for a moment, Pallas approached and presented to Hugo who was the closest to her, but before Hugo could proceed, Satan with hair erect charged toward them; Pallas screamed and avoided him. Hugo then turned and attacked Goblin, and Satan joined him, kicking and dragging the screaming Goblin

---

* Young males were classified as independent if they were known to have spent at least 24 hours consecutively away from their mother. It was considered that the movements and associations of younger males were determined by their mothers. For the majority of the period of data collection, there were 12 independent males in the Kasakela community.

downhill. After the attack, Satan and then Hugo copulated with Pallas. This was the only case in which aggression was observed during opportunistic mating.

In opportunistic mating groups, adult males showed little or no overt competition, and in some cases they showed affiliative behavior toward one another in gaining access to receptive females. Figan, the most dominant male, with penis erect twice approached a receptive female who was being groomed by another adult male. In both cases Figan sat and groomed the male for several minutes before moving away slightly to court the female.

2. *Possessive Behavior*

Possessive behavior is not a clear-cut category of behavior but rather, a useful blanket term to cover a variety of behavior patterns exhibited by a male to a female who is experiencing cycles of sexual swelling. In most cases, possessive behavior is shown by a male to an estrous female; more rarely the female will not be in the receptive phase of her cycle. A male was described as acting possessively towards a female if he showed persistent special attention to her beyond the bounds of normal courtship of longer than an hour at a time. This involved the male actively attempting to maintain close proximity with the female, achieved by following, leading, or repeatedly gazing at and waiting for her. Another complex of behaviors frequently shown by possessive males was interruption of copulations between the chosen female and other males. This was achieved either actively by charging aggressively at the intruding male, or passively through merely remaining spatially close to the female.

During the 16 months of Tutin's study, there were 36 incidents of possessive behavior involving seven females and seven males in 16 dyadic and two triadic combinations. The two triadic combinations involved a pair of adult male siblings who, on both occasions, cooperated in the guarding of the female. Not all incidents were observed from beginning to end, but those that were ($N = 20$) ranged in length from one hour to five days. In 78% of incidents the female was in the maximally tumescent phase of her swelling cycle. Table 2 shows the distribution of possessive behavior with respect to the days of the female's swelling cycle. Following Graham (Chapter 1, this volume), the last day of maximal tumescence, i.e., the probable day of ovulation, is designated as Day 0. For the Gombe chimpanzees the typical duration of maximal tumescence was from Day -9 to Day 0. The highest frequencies of possessive behavior occurred on the presumed day of ovulation and the two preceding days (i.e., Day -2 to Day 0).

TABLE 2

Distribution of possessive behavior with respect to the female swelling cycle.

	Day of Female Swelling Cycle (0 = Day of Ovulation)																	
	-18 to -13	-12	-11	-10	-9	-8	-7	-6	-5	-4	-3	-2	-1	0	$+1$	$+2$	$+3$	$+4$ to $+18$
No. of observation periods of at least 2 hr.	8	3	4	7	12	8	7	12	14	11	12	8	7	7	5	3	3	23
No. of times possessive behavior occurred	0	1	2	1	5	1	4	4	5	8	6	8	6	7	2	0	0	2
% of observation days on which possessive behavior occurred	0	33	50	14	42	13	57	33	36	73	50	100	86	100	40	0	0	9

Note: As possessive incidents varied in length from 1 hour to 5 days, their frequency was difficult to quantify. Possessive behavior was scored as occurring or not occurring during each of 153 observation periods on separate days. Possessive behavior was observed on 62 days, the percentage of observation days on which it occurred for each day of the female cycle is shown. The data for all nine females are pooled.

The possessive male actively interrupted copulations or copulation attempts by other males in 21 of the 36 possessive incidents observed. In eight incidents no opportunity to interrupt copulations arose, either because other males present showed no sexual interest in the female, or because no other males were encountered. In the remaining seven incidents, copulations involving other males occurred but the possessive male made no attempt to interrupt them. Possessive males successfully interrupted 70 copulations directed toward the possessed female and unsuccessfully tried to interrupt seven others. In 69 of the 70 interruptions the result of the interaction was related to the males' dominance rankings as established on the basis of observations of the outcome of agonistic, nonsexual, and dyadic encounters; that is, each male successfully interrupted copulations of males ranking lower than himself, but not vice versa. In six of the seven unsuccessful interruption attempts, the copulating male was of higher dominance rank than the possessive male. Possessive behavior enables a male to reduce competition for a receptive female by preventing all males of lower dominance rank than himself from copulating with her.

For Figan, the most dominant male in the community (i.e., the alpha male), possessive behavior was an efficient reproductive strategy that resulted in exclusive copulatory access to an estrous female by interrupting the copulatory attempts of all other males. The effectiveness of this strategy was demonstrated during a two-month period when four females exhibited cycles of sexual swelling. The four females collectively showed seven cycles, and during five of these Figan acted possessively toward the female during maximal tumescence. The two exceptions, when Figan did not show possessive behavior to a cycling female, each occurred when two females were simultaneously in estrus. In each of these cases in which Figan was forced to make a choice, he directed possessive behavior toward the older of the two females.

In seven possessive incidents involving males other than Figan, no attempt was made to interrupt copulations, even if the copulating males were less dominant than the possessive male. This suggests that possessive behavior by the alpha male may serve different functions from that of other males. For the alpha male, possessiveness is, in itself, a valuable reproductive strategy. He is able to maintain exclusive sexual access to the female by actively interrupting or passively preventing other males' attempts to copulate with her. For other males, possessiveness appears, at least in some cases, to be a precursor to the formation of consortships. By remaining spatially close to the female in a group, a male is able to take immediate advantage of any opportunity to initiate a consortship.

3. Consortships

Consorting behavior is characterized by many of the behavior patterns shown during possessive incidents and is also usually directed to females showing cycles of sexual swelling. The male maintains close proximity with the female but extends the leading behavior shown in possessiveness to remove the female from the social group. Once alone, the pair, plus any dependent offspring of the female, cease all loud vocalizations and avoid encounters with other chimpanzees. This avoidance often results in the pair moving to the edge or even outside the normal community range (see also McGinnis, 1979).

In 16 months, during which nine females showed 42 cycles of sexual swelling, there were 15 consortships involving five males and seven females in 12 dyadic combinations. The median length of the consortships was 7 days (range 3 hr–28 days), and for 12 of the 15 consortships, the female was maximally tumescent at least for part of the time. In 10 of the 13 cases where details of the female's swelling cycle were known, Day 0 (ovulation day) fell during the consortship. Nine of the 15 consortships were observed for all or part of their duration. The remaining six (which ranged in length from 7–28 days) were assumed to exist on the basis of the strong circumstantial evidence of prolonged, coincident absence from all observation records of a single male and female. Four unsuccessful attempts to form consortships were also observed.

Maximally tumescent females were usually accompanied by most, or all, of the community's adult males, and so obstacles existed for a single male in initiating a consortship. The strategy used in all observed initiations ($N = 11$; seven successful, four unsuccessful) was an opportunistic one. When the adult males in a group moved away, the consorting male either immediately led the female away in a different direction or remained close to her, grooming or feeding for a while before leading her in a different direction from that taken by the other males. While the male made the first move in initiation, a consortship only developed if the female cooperated by remaining silent and following the male.

During eight of the 11 observed consort initiations, vocalizations from other chimpanzees were heard, and each time the female made a response which involved either stopping and orienting to the calls, vocalizing, or changing her direction of movement to approach the calling males. In five cases the female remained silent and, after the calls, continued to follow the consort male. In the other three cases the female's response to the vocalizations terminated the consort attempt.

Vocalizations from other chimpanzees appeared to be a crucial event in consort initiation, as their occurrence made the consort male's intentions clear to the female at a point when she could chose whether to terminate or to continue the relationship. In all three cases where the female's response to vocalizations from other males terminated the consort attempt, the consort male made no attempt to prevent the female moving toward the other males ($N = 2$), or to evade if the other males approached ($N = 1$).

Once alone, the consorting pair traveled rapidly to the edge of, or even beyond, the normal range of the community. After this, the distance traveled daily was small, and the pair intensively utilized their small, temporary range. They ceased all loud vocalizations and responded with silent avoidance to calls from other chimpanzees. Consortships were characterized by high frequencies of social grooming, especially male-to-female, which occupied an average of 14.8% of their waking hours, as opposed to the 5.3% of time males spent grooming maximally tumescent females in social groups. Females groomed males slightly more frequently when in consortships (6.3% of their waking time) than when maximally tumescent in social groups (3.9%).

Consortships have obvious selective advantage for both males and females, in that they give the male exclusive copulatory access to the female and allow the female to exercise choice of mates. On the other hand, there appear to be several disadvantages associated with consorting. Probably the greatest potential danger to the consorting pair is the increased risk of intercommunity encounters, which comes from moving to the edge of their own community's range. As mentioned in Section IV,A, such encounters are characterized by extreme aggression that has resulted in adult male deaths. In addition, adult males have attacked and killed infants of females from neighboring communities (Suzuki, 1971; Bygott, 1972). Thus it seems that an intercommunity encounter might have drastic consequences for a consorting pair, especially if the female is accompanied by an offspring.

Another disadvantage of consorting comes from the risk of aggressive reunions with community males after the consortship ends. Riss (unpublished data) reviewed the significant dominance changes among males of the Kasakela community and found that four attacks that resulted in changes in males' status occurred when they returned to the community after a prolonged absence. Two of the returning males were known to have been consorting with females, and in each case the male who had been absent suffered a severe defeat. Observations of the termination of consortships indicated that both male and female were vulnerable to such aggression, but its consequences were potentially more damaging to the male if it resulted in a loss of dominance status.

4. *Comparison of the Mating Patterns*

Individual differences in participation in the restrictive mating patterns (i.e., possessiveness and consorting) existed, especially among adult males. The frequency of male involvement in the restrictive patterns was found not to be statistically correlated with age (Spearman Rank Correlation Coefficient, r_s = .08, N = 9, n.s.), dominance rank (r_s = .53, N = 9, n.s.), or the amount of agonistic behavior males directed to females in social groups (r_s = .31, N = 9, n.s.). Significant positive correlations emerged between the frequency of male involvement in the restrictive patterns and:

(a) the amount of time spent associating with estrous females in groups (r_s = 0.72, N = 9, p < .05);

(b) the proportion of this time that males spent grooming estrous females (r_s = 0.80, N = 9, p < .01); and

(c) the frequency with which males shared food with females (r_s = 0.68, N = 9, p < .05).

Data were inadequate to allow statistical assessment of individual differences in involvement in consortships and possessive incidents separately. However, it is evident that while dominance rankings of the adult males as a group did not show a statistically significant correlation with participation in the restrictive mating patterns, the alpha male of the community did gain an advantage. Solely the alpha male could monopolize an estrous female by showing possessive behavior; other males were able only to reduce competition from less dominant males by interrupting their copulations. Figan was not observed to form consortships once he became the alpha male, but he showed a high frequency of possessive behavior. His alpha rank allowed him to monopolize females without being exposed to the risks associated with consorting (for further details, see Tutin, 1979b).

During the period from November 1966 to February 1975, there were 16 pregnancies that resulted in the birth of a living offspring in the Kasakela community. For 14 of these cases it was possible to estimate the cycle of sexual swelling during which conception occurred and to establish the female's activities during the presumed conception cycle. Two females had emigrated temporarily and were associating with groups of males of a neighboring community; four were associating with groups of males at the time of conception (one of these females is known to have been involved in a possessive relationship with the alpha male, but no detailed information is available for the other three); and seven females were involved in consortships. In the fourteenth case the calculated conception date fell midway between two phases of maximal tumescence. During the first, the female was involved in a possessive relationship with the alpha male, and during the second she was in consortship with a young adult male. Thus, at least half of the 14 females conceived while consorting.

It is difficult to calculate the relative frequencies of the different mating patterns in a meaningful way, as it is not obvious what to compare. The patterns vary in duration from one hour to 28 days, and opportunistic mating and possessiveness are primarily restricted to the female cycle phase of maximal tumescence, but consortships often begin or end during other cycle phases. Also, some data on consortships were inferred from circumstantial rather than direct evidence, whereas no indirect assumptions could be made about the other mating patterns. Although the sample size is small, it appears that more females conceived during consortships than would be expected from a knowledge of consort frequency. The data on conceptions also show a lack of relationship between frequency of copulation and reproductive success: 73% of the 1137 copulations occurred during opportunistic mating (in 68% of the observation hours); 25% during possessiveness (in 18% of the observations hours); and only 2% during consortships (in 14% of the observation hours).

B. Partner Preferences

Partner preferences, as reflected by differential participation in the restrictive mating patterns and by variation in dyadic copulation rates in opportunistic mating, do exist. The most striking feature to emerge from the data is the strong aversion between primary kin. The chimpanzee, like many other species of nonhuman primates, shows a general behavioral mechanism that reduces inbreeding by the emigration of individuals from their natal group before reproductive age. In chimpanzees it is only females who migrate, and at Gombe it appears to be restricted to adolescent females (Pusey, 1979). If all females emigrated from their natal communities at or before puberty and did not return, brother–sister and father–daughter matings would be precluded, even though the identity of the father is unknown. Migration is not always permanent, as although all young females move at adolescence, some return to their natal communities once they have become pregnant, and remain there during subsequent periods of receptivity. Copulation between mothers and their mature sons is virtually unknown, only one instance having been observed at Gombe during 15 years. Copulations do occur between mothers and their infant sons during the mother's initial postpartum swelling cycles, but these usually occur in the context of reassurance and are seldom preceded by courtship. Copulations between mature males and their sisters are rare, and two adolescent females consistently avoided the courtship of their older brothers. No mother–son nor brother–sister possessive relationships nor consortships have ever been observed.

It is the male who initiates courtship, possessive incidents, and consortships, but females have some degree of choice in whether or not to cooperate. Only two cases of forced copulation occurred in the 1137 copulations observed by Tutin; each involved a female who was attempting to avoid a male's courtship but was trapped high in a tree. One of these cases was the only copulation ever observed between a mature male and his mother. Of the 209 incomplete copulatory sequences observed by Tutin, 180 were terminated by female avoidance or lack of responsiveness. Differences in dyadic copulation rates shown in opportunistic mating, indicated that partner preferences existed. Most of these preferences reflected differences in individual male–female relationships, but adult males showed a general preference for older females. Of the 30 occasions when adult males made a direct choice between two estrous females during opportunistic mating, the older of the two was selected 27 times ($X^2 = 9.6$, $df = 1$, $p < .01$).

Females probably have little influence as to which males direct possessive and consorting behavior toward them, but they do have a choice as to whether or not a consort relationship develops. Once the alpha male directs possessive behavior to a female, it is doubtful whether she has the ability to terminate the relationship, although she can make it more difficult for him to maintain his copulatory monopoly. However, it is unlikely that a female can form a consort relationship with another male if the alpha male is directing possessive behavior toward her.

The male characteristics that correlate positively with differential involvement in the restrictive mating patterns (Section IV,A,4) give some indication of the male qualities preferred by females. The small sample size (seven females and five males) necessitates the pooling of data for analysis, but we tentatively conclude that females generally prefer males who spend the most time with them during maximal tumescence and who show high frequencies of the affiliative behaviors, grooming and food sharing.

To the extent that affiliative behavior is heritable, two mechanisms would tend to maintain or increase its frequency in the population. Both of these are based on the notion that females choose to consort with males with whom they have affiliative bonds. First, if all females showed a consistent preference for males who showed the most affiliative behavior, then these males should enjoy increased reproductive success, at the expense of males who showed lower frequencies of affiliative behaviors. Further, affiliative behavior may provide a survival advantage to the progeny by providing protection in such incidents as intercommunity encounters.

The tendency of affiliative behavior to spread in the population by sexual selection will, of course, be limited by any negative effect this behavior pattern may have on male survival. Spread of affiliative behavior will also be checked to the extent that some males can mate successfully without displaying this pattern; this is typical of alpha males, for instance, who obtain reproductive access to females through the pattern of possessiveness, made possible by their dominance established during intermale competition.

Thus, it appears that the affiliative behavior of males toward females may be propagated through the population by sexual selection, although this tendency is likely to be counterbalanced by the possessive behavior pattern. Consequently, affiliative behavior patterns may serve as either an alternative or an augmentation to high dominance status in the achievement of reproductive success by male chimpanzees.

V. CONCLUSIONS

Chimpanzee females have a limited reproductive potential and a heavy energy investment in each offspring. Males, on the other hand, have a high reproductive potential and make no direct contribution to the rearing of their offspring. In this situation it would be biologically optimal if females exercised choice, selecting the male whose behavior, and ultimately therefore, genetic contribution maximized the probability of the offspring surviving to reproductive age. Males would be expected to be selective only when presented with a choice of estrous females. This expectation was borne out, with adult males showing a preference for older females, as reflected by direct choice in opportunistic mating, and by the alpha male's pattern of possessiveness (Section IV,A,2).

Behavioral mechanisms allow females a degree of genetic choice, and the fact that a high proportion of conceptions occurred during consortships (Section IV,A,4) suggests that these mechanisms operate successfully. Many aspects of female reproductive physiology function in a way that allows a female copulatory experience with all the adult males in the community under conditions when she is unlikely to conceive, namely, the long period of adolescent sterility, the initial postpartum cycles that typically are anovulatory, and the nine days of estrus prior to ovulation. Such copulatory experience, and the interactions associated with it, enable the female to establish, maintain, or renew familiarity with the different males in the community without jeopardy of impregnation. To the extent that females are able to exercise choice, such familiarity may form the basis for selection of a copulatory mate at a time when the female is likely to conceive.

The data presented here all come from male-female interactions during times when the females were showing cycles of sexual swelling. This represents only a fraction of the female's adult life, and it seems certain that male-female interactions during pregnancy and the long periods of lactational amenorrhea also influence the individual relationships on which female choice is based. The flexible mating system of the Gombe population allows females to exercise choice and gives a reproductive advantage to the alpha male of the community. More longitudinal data are required before the effects of these features of the mating system on individual reproductive success can be established.

ACKNOWLEDGMENTS

We wish to thank the Government of Tanzania for permission to work in the Gombe National Park, and Dr. J. Goodall for supervision at the Gombe Stream Research Centre. Caroline Tutin gratefully acknowledges supervision from Professor A. W. G. Manning and financial support from a Royal Society Leverhulme Studentship, a Science Research Council Studentship, and a Science Research Council Postdoctoral Fellowship. Patrick McGinnis gratefully acknowledges financial support from The National Geographic Society, Washington, D.C. Our thanks go to Dr. W. C. McGrew, Dr. A. H. Harcourt, and Professor R. V. Short for comments on the manuscript.

REFERENCES

Bygott, D. (1972). *Nature (London)* **238,** 410–411.
Bygott, D. (1979). *In* "The Great Apes" (D. A. Hamburg and E. R. McCown, eds.), Vol. 5, pp. 405–439. Benjamin/Cummings Publ., Menlo Park, California.
Clark, C. B. (1977). *In* "Primate Bio-Social Development" (S. Chevalier-Skolnikoff and F. E. Poirier, eds.), pp. 235–260. Garland Publ., New York.
Goodall, J. (1965). *In* "Primate Behavior" (I. DeVore, ed.), pp. 425–473. Holt, New York.
Goodall, J. *et al.,* (1979). *In* "The Great Apes" (D. A. Hamburg and E. R. McCown, eds.), Vol. 5, pp. 13–53. Benjamin/Cummings Publ., Menlo Park, California.
Graham, C. E. (1970). *In* "The Chimpanzee" (G. H. Bourne, ed.), Vol. 3, pp. 183–220. Karger, Basel.

Graham, C. E. (1973). *Folia Primatol.* **19,** 458–468.

McGinnis, P. R. (1973). Ph.D. thesis, Univ. of Cambridge.

McGinnis, P. R. (1979). *In* "The Great Apes" (D. A. Hamburg and E. R. McCown, eds.), Vol. 5, pp. 429–439. Benjamin/Cummings Publ., Menlo Park, California.

Nishida, T. (1968). *Primates* **9,** 167–224.

Nissen, H. W., and Yerkes, R. M. (1943). *Anat. Rec.* **86,** 567–578.

Pusey, A. (1979). *In* "The Great Apes" (D. A. Hamburg and E. R. McCown, eds.), Vol. 5, pp. 465–479. Benjamin/Cummings Publ., Menlo Park, California.

Rowell, T. E. (1972). *In* "Advances in the Study of Behavior" (D. S. Lehrman, R. A. Hinde, and E. Shaw, eds.), Vol. 4, pp. 69–105. Academic Press, New York.

Sharman, A. (1951). *Fertil. Steril.* **2,** 371–393.

Smith, A. H., Butler, T. M., and Pace, N. (1975). *Folia Primatol.* **24,** 25–59.

Stix, R. K. (1940). *Millbank Memorial Fund Q.* **18,** 239–256.

Suzuki, A. (1971). *J. Anthrop. Soc. Nippon* **79,** 30–48.

Tutin, C. E. G. (1975a). *In* "Contemporary Primatology" (S. Kondo, M. Kawai, and A. Ehara, eds.), pp. 445–449. Karger, Basel.

Tutin, C. E. G. (1975b). Ph.D. thesis, Univ. of Edinburgh.

Tutin, C. E. G. (1979a). *Anim. Behav.* **27,** 845–854.

Tutin, C. E. G. (1979b). *Behav. Ecol. Sociobiol.* **6,** 29–38.

van Lawick-Goodall, J. (1968). *Anim. Behav. Monogr.* **1**(3), 161–311.

Wrangham, R. W. (1975). Ph.d. thesis, Univ. of Cambridge.

Young, W. C., and Yerkes, R. M. (1943). *Endocrinology* **33,** 121–154.

Chapter 10

GORILLA REPRODUCTION IN THE WILD

Alexander H. Harcourt
Kelly J. Stewart
Dian Fossey

INTRODUCTION

Our work on the mountain gorilla (*Gorilla gorilla beringei*) was conducted in the Virunga Volcanoes of Rwanda and Zaire (approximately 1° south latitude, 30° east longitude) while based at the Karisoke Research Centre in Rwanda. This study covers 40 months, between

265

Copyright © 1981 by Academic Press, Inc.
All rights of reproduction in any form reserved.
ISBN 0-12-295020-8

September 1972 and March 1977, and in this time we observed 69 copulations, although we recorded over 100. (Some were only heard.) Three-quarters of these observations occurred in just nine months, and in the last six months we recorded none. Thus, copulations in gorilla groups appear to be rare events. Since no other field observer has reported seeing more than two copulations by wild gorillas anywhere in Africa, this report refers only to the Virunga gorilla population. However, studies of captive gorillas (*G.g. gorilla*, the majority of which were obtained in West Africa) indicate such close similarities to the Virunga population that many of our findings may have general application to both subspecies.

Nearly all records came from two fully habituated groups that contained a leading male of more than 18 years of age, one or two subordinate males of 8 to 12½ years, between two and six adult females (<8 yr), and four to eight immature animals. The number of adult females changed because of deaths and movements of individuals into and out of the groups, while changes in the number of immature animals occurred through births, maturation, and emigration from the groups. Immature gorillas are classified as infants up to 3 years, juveniles from 3 to 6 years, and subadults from 6 to 8 years. Males mature later than females and are given a further category before full adulthood of "blackback," which lasts from 8 to 11 years of age. The ages of all immature animals and of many of the adults are known to within six months or less; in fact, the age of some of the immature individuals is known to the day. Data on these animals before September 1972 came from the Research Centre records, collected mostly by Dr. Dian Fossey.

Observation conditions were such that we could sometimes hear copulations that we could not see, or we could identify only one of the partners. Thus, the number of records of copulations differ, depending upon which particular aspect of the subject we are considering at the time.

I. REPRODUCTIVE HISTORY

A. Immaturity

In more than 2000 hours of observation, we saw only 39 mountings that involved immature animals, most of which occurred during play. All immature age–sex classes except infant females (though not all immature individuals) were involved; the youngest mounted animal was 21 months old, and the youngest mounter was 37 months. Thirty-one percent of mounts occurred between two immature individuals ($N = 12$), and, except for one case, the younger partner was always the one mounted. All but one of the other observations were of adults mounting immatures ($N = 26$); 85% of these involved a blackback ($N = 22$), 73% of them being due to a single blackback. The final observation was of a subadult male thrusting against a parous estrous female's shoulder and side, this being one of only two instances of a younger animal mounting an older one.

B. Adolescence

Menstrual bleeding was never observed, so adolescence for females in the wild is taken as the time of the first observed sexual swelling. For three females whose age was known accurately, this occurred at 6 yr, 6 mo; 7 yr, 7 mo; and 7 yr, 11 mo. For a fourth it was between 7 yr, 9 mo and 8 yr, 6 mo. Only two of the females were observed consistently for more than six months after their first swelling, and for both it was five months before they were seen to mate with a fully adult male. These two females did not give birth until 19 and 27 mo after their first observed swelling. Reliable published durations of gestation range from 237 to 270 days, with a median of 252 days ($N = 11$) (Ardito, 1976; Hardin *et al.*, 1969; Mallinson *et al.*, 1976; Nadler, 1975c; Tijskens, 1971). Therefore, the interval from first observed sexual cycle to conception (adolescent sterility) in these two subjects was 10½ and 18½ months.

Although male gorillas in captivity might breed when around 8 years old (Schaller, 1963), we estimate that in the wild, where they face competition from older animals, breeding will not start until the males are at least 15 years of age. Thus, they have an effective period of adolescent sterility of 7 yr, compared to the females' 1–2 yr.

C. Adulthood and Reproduction

The ages of four females at the time of their first parturition is known: they were 9 yr, 6 mo; 9 yr, 10 mo; 10 yr, 0 mo; and 10 yr, 10 mo. The youngest might be a little outside the normal range because she was a female who stayed to produce her first offspring in her natal group. Usually females emigrate from the natal group during adolescence, and most do not stay with the male whom they first join (Harcourt *et al.*,

1976; Harcourt, 1978). Such movements can obviously delay conception, making it probable that 10 years, or a little over, is nearer the normal age of a female when she first gives birth.

Intervals between births when the previous offspring survived have varied from 3.0 yr to 4 yr, 11 mo, with a median of 3 yr, 10 mo (N = 5 females, eight intervals). Death of an offspring causes a shorter interval, as is common in many species. Thus, the median time between death of a previous offspring and birth of a subsequent one is 1.0 yr, with a range from 1.0 yr to 1 yr, 3 mo (N = 3 females, four intervals). The interval seemed not to depend on the infant's age at death, since the range of intervals (4 mo) was far less than the range of ages of death (17 mo).

Only four of 13 parturient females have had all their offspring survive, and one of them was primiparous. In agreement with findings for captive gorillas (Nadler, 1975b), primiparous females in the wild were less successful at raising offspring than were multiparae: four of six known firstborn offspring died, whereas only 6 of 17 known later-born ones died. These figures give a total mortality rate of around 40%. It must be stressed that this rate applies only to the groups in the well-protected study area. In other areas of the park, or elsewhere in Africa where there is more disturbance or where food is less abundant, the mortality rate is expected to be higher.

We calculate the number of surviving offspring produced per year per female ranges from 0.14 to 0.32, with a median of 0.21 (N = 7 females). In other words, the median female produces a viable offspring every 4¾ years. This value is greater than the previously quoted one of 3 yr, 10 mo, because it includes females that produced less than two offspring during the study period. Given the first offspring produced at 10 or 11 years of age, and a reproductive life of 20 to 25 years (assuming 30–35 years as the normal age of death), a female might produce four to five offspring surviving to adulthood. In the 11 years the study groups have been under observation, females have produced up to three viable offspring.

Successful gorilla males control a harem, amd so have a higher potential reproductive output than do the females. The males of the two main study groups are known to have produced, respectively, four (possibly six) and eight (possibly 10) surviving offspring. A male's reproductive life is possibly five to 10 years shorter than a female's (see Section I, B), but with three females in his group, his potential output is six to 13 surviving offspring, or about twice the female's output. (Three females per group is the median for the section of the park in which the study area lies; N = 20 groups.) Again, this value refers only to the relatively well-protected study area in the western half of the park, in which food appears to be abundant. Elsewhere, the animals' reproductive output will be far lower. For example, in the eastern section of the Virungas, census results show a median of two females per group with the rate of production of offspring being 0.125 per female per year, or almost half that in the study area.

D. Reproductive Cycles

1. *Sexual Swellings, Ovulation, Estrus, and Copulation*

In a number of primate species, ovulation has been found to occur within one day of the last day of maximal swelling (Graham, Chapter 1, this volume). Sadly, this finding is usually of little use to the observer of wild gorillas. Although under special conditions in captivity sexual swellings on parous females can be seen and measured (Nadler, 1975a), in the wild they are normally visible only on nulliparous females. Thus, it was impossible in our study to relate estrous behavior or copulation to the timing of ovulation. In addition, menstruation is very rarely seen, with the result that in this paper we refer to estrous cycles, not menstrual cycles, and by cycle length we mean the interval between periods of estrous behavior and copulation.

a. Cycle Lengths We have information on the lengths of estrous cycles of only two nonpregnant adult females, and for one of these there is only one record—a length of 38 days. The other female's cycle lengths ranged from 27 to 39 days with a median of 28 days ($N = 5$), which is a shorter "average" than reported from captivity (Nadler, 1975a).

It appears that adolescent females do not cycle every month. We have observed intervals of two and six months between first and second swellings in two adolescent females. Excluding such long intervals, the median cycle length of four nulliparous females was 27 days, although considerable individual variability occurred; for example, the cycle lengths of one subject ranged from 13 to 32 days.

Estrous behavior during pregnancy is not unusual among primates (Rowell, 1972) and has been reported for captive gorillas (references in Stewart, 1977). Four of five females in our study came into estrus while pregnant and were mounted by the leading male. The length of 10 inter-estrus intervals in 12 females was established; one interval in one female was 20 days, while nine in the second female were very irregular, ranging from 4 to 27 days with a median of 10 days. The last estrous period for this female was only one week before parturition. Pregnant gorillas in captivity also have been reported to show irregular cycles (Hess, 1973; Nadler, 1975a), and Stewart (1977) suggested that observation of erratic intervals might be considered an indication of pregnancy.

b.   The Estrous Period   Considering six adult females first, the lengths of their estrous periods ranged from one to three days (median = 1 day; N = 32 periods; 26 of these observations were based on three females in which five or more periods were observed). The observed one-day median duration of estrus is considered to be an underestimate because of discontinuous observations in some instances; the true median duration of estrus is expected to be about two days. There are very few data for adolescent females but what there are indicate that their estrous periods might usually be longer than the adults', possibly three days instead of two.

Adult females differed markedly in their rates of copulation on days that they mated. Counting only days when observers were with the group for more than one hour, there was a total range of 0.16–1.39 copulations per hour, somewhat less than recorded for one pair of Hess' captive gorillas (Hess, 1973). The range of the females' medians was 0.25–0.85 copulation per hour with an overall median for all females of 0.30 (N = 8 females), or 0.34 for females with at least five records (N = 4). Because only once was more than one female in estrus on the same day, copulation rates for males were very similar to those for females; the two leading males' medians were 0.36 (N = 7 records) and 0.37 (N = 38).

There are, however, three possible sources of error in these rates. First, most observation time was concentrated around noon, but the gorillas might have mated most frequently at dawn, as do chimpanzees (Chapter 9, this volume). Second, most observation time was during rest periods, yet most copulations occurred during travel/feeding periods (Section II, C). Third, some observation periods were less than three hours long, yet one copulation per three hours was the median rate.

2.   Conception

The limited data on duration of adolescent sterility, and irregularity of adolescent cycles, prevents estimation of the number of swellings prior to first conception. For parous females the number of estrous periods between birth and subsequent conception appears to be low. Three females watched over several months before estimated conception date experienced only one estrous period. A fourth female probably had four. (However, neither this female nor one other was seen to mate in the estimated month of conception.) A further indication of the number of estrous periods needed for conception is given by the time between death of an offspring and subsequent birth (See Section I, C). A median interval of one year between infant death and the next birth, and a gestation period of 8½ months indicates an average of three estrous periods before conception; however, this number of cycles may be further reduced according to the unknown duration of acyclicity after death of the infant. Here, considerations suggest that three or less cycles are required for the successful impregnation of a fertile female.

A fifth female, not mentioned in the preceding paragraph because she was atypical, had six to 10 estrous periods before conception of her second offspring. She had unusually long copulations with the silverback, her *median* duration being 0.6 min longer than the longest recorded copulation of any other female. After cycling regularly over a period of at least five months, this female emigrated from the group leaving behind her four-year-old daughter. She joined another male and conceived from him four months later. (Her previous mate also subsequently proved himself to be fertile.) Since this female was very solicitous of her daughter, and since her breasts were unusually large for the age of her offspring, a possible explanation of her infertility lies in prolonged lactation allowing cycling but not implantation.

The existence of lactational amenorrhea is indicated by the fact that the interval between death of one offspring and birth of the next is shorter than the interval between living offspring (Section I, C). The average period of amenorrhea when the offspring survives can be estimated if date of conception is known and if three months from first estrus to conception are assumed (see above). The result is a range from 25 to 47 mo, with a median of 35 mo ($N = 6$ females, nine periods). The relation between postpartum amenorrhea and lactation for our study animals is not precisely known. Infants have been seen on the nipple regularly up to at least three years of age, although it is possible that from six months most of their diet is composed of plant food.

II. SEXUAL BEHAVIOR

Four adult males and eight adult females contributed to the results described in this section. However, for some pairs so few observations of copulation were made that their results are not entirely reliable. Thus, as well as presenting data for all pairs, we give them separately for pairs that had at least five records of the behavior under consideration. Depending on the behavior, one to three males and three to five females were partners in pairs that had five records or more.

A. Courtship

Females normally initiated copulations by solicitingly approaching the male (Table 1a), although very occasionally the male followed the female around and initiated mating. Once near the male after her characteristically slow and hesitant soliciting approach, the female would stand facing him with her body slightly sideways-on, as if waiting for a signal, when she would abruptly turn and back into the normal dorso-ventral copulatory position. The signal appeared to be the male's opening or raising of his arms as he prepared to hold the female around her waist in order to mount. Sometimes the male seemed to ignore the female until she was within a few feet of him, but it was rare that a solicitation was totally disregarded. At no time during their proximity were adults ever seen to touch or to sniff one another's genital regions, although immature animals of both sexes not infrequently sniffed an estrous female's rump. The mildly aggressive component of male courtship behavior that is seen in captivity (Hess, 1973; Nadler, 1976) was never observed in the wild. Most captive gorillas, including Hess's and Nadler's, are western lowland gorillas (*G.g. gorilla*), so whether the described difference is a subspecific difference or an artifact of confinement is not known.

B. Copulation

All 69 observed copulations were dorso-ventral, although twice one pair briefly assumed the ventro-ventral position before turning to copulate normally. During copulation the male sat upright and held the female around the waist. She squatted, sometimes with her hands on the ground, and sometimes with one or both hands holding the male's hands. Facial expressions during copulations were exactly as Hess (1973) observed in captivity, the female with compressed lips and the male with pursed lips. During copulations the female would occasionally look over her shoulder at the male (as in Fig. 15 in Hess, 1973), and sometimes wave her head from side to side with her eyes closed, giving copulatory whimpers as she did so.

Normally mounting was followed by about a ½ to ¾ min interval of adjusting movements, or no movement at all, before thrusting began.

TABLE 1

Comparison of male and female roles in courtship and copulation.

	(a) Approaches to within 2 m and 5 m of partner					(b) Thrusts		(c) Vocalizations					(d) Break contact		(e) Departures			
	2 m			5 m				Frequency			Duration				2 m		5 m	
	♂	=	♀	♂	♀	♂	♀	♂	=	♀	♂	♀	♂	♀	♂	♀	♂	♀
All pairs	1	1	7	2	5	9	1	2	3	5	3	7	2	8	1	5	1	7
Pairs with $N \geq 5$[a]	0	0	3	0	2	4	1	1	1	2	1	3	1	3	0	3	0	3

Note: The column headings ♂, =, ♀ represent the number of pairs in which the behavior was displayed with greater frequency or duration by the male (♂), the female (♀), or equally by both (=).

[a] Pairs in which the behavior was observed at least five times.

It was impossible to tell when intromissions occurred, so we timed copulations as the duration of the copulatory position. This ranged from 15 sec to 19 ¾ min with the median pair's median being 96 sec (Table 2a), almost double the duration reported by Hess (1973) in a captive gorilla pair. Males and females have been observed to thrust and vocalize. However, in only half of the pairs did the female thrust, whereas she vocalized in most instances. In fact, females tended to vocalize during more copulations than did males (Table 1c), a finding that contrasts with Nadler's observations on young adult captive gorillas (Nadler, 1976). The nature of thrusting and vocalizing differed between the sexes; whereas males executed rapid pelvic thrusts, females performed deep, slow body thrusts. The females' normal copulatory vocalizations were rapid pulsating whimpers, but the males usually gave long grumbles that lasted 5 to 20 sec, or pulsating pants, similar to the whimpers, but harsher and more breathy; occasionally males whimpered. Duration of bouts of thrusting was about half a minute for both sexes (Table 2b), while periods of vocalization usually lasted around 20 sec (Table 2c). Males normally exhibited more bouts of thrusting that lasted longer compared with females (Table 1b), but females usually vocalized longer and more often than males (Table 1c). Thrusting and vocalizing increased towards the end of the copulation and then ceased abruptly as the mount was broken, usually by the female (Table 1d). Occasionally the male gave a deep sigh at this point, but we never saw any sure sign of ejaculation. Finally, it was normally the female that left the male, rather than vice versa (Table 1e).

C.   Interference

No adult was ever seen to interfere with the copulations of dominant males, although these males not infrequently stopped the matings of subordinate males (see Section III, A). One factor contributing to the general lack of interference, besides the fact that the dominant male was more capable than others of punishing interference, was the timing of copulations. Seven of all 10 pairs seen copulating, including four of five pairs for which there are at least five records, mated more frequently during travel/feeding periods than rest periods. Under these conditions of travel, a second male was rarely close enough to a copulating dominant male to have an opportunity to interfere.

In contrast, immature gorillas and the adult female's offspring were usually near mating pairs. Although the youngsters occasionally showed interest, only once was active interference seen. In this case the female's

TABLE 2

Duration (in seconds) of copulation, thrusting, and vocalizing.

	(a) Copulations			(b) Thrusts ♂			(b) Thrusts ♀			(c) Vocalizations ♂			(c) Vocalizations ♀		
	R	M	N	R	M	N	R	M	N	R	M	N	R	M	N
All pairs	30–310	96	11	6–52.5	27.5	8	15–31	30	3	1–42.5	17.5	7	1–67.5	15	10
Pairs with $N \geq 5$[a]	75–310	96	6	6–52.5	27.5	6	15–31	30	3	1–27.0	17.5	3	1–42.5	23	4

(*Note:* R = range; M = median; and *N* = number of pairs.)

[a] Pairs in which the behavior was observed at least five times.

offspring exhibited interference similar to that described for chimpanzees (Chapter 9, this volume).

D. Homosexual Behavior

Ten instances of homosexual behavior between females were recorded, involving four different pairs and four individuals. Typically one female would approach another, and sometimes they would embrace. Then either in the ventro-ventral (three mounts) or dorso-ventral (six mounts) position, one would thrust briefly, usually with accompanying copulatory vocalizations from at least one of the partners (seven of eight mounts). They would then dismount and go their separate ways. Parker and Pearson's (1976) suggestion that such homosexual mounting between females in other species was a means whereby the estrous females attracted the dominant male does not apply to these gorillas. The only time the male was attracted, he had mated vigorously with the estrous female less than one hour previously. When he returned, rather than mating with either female, he attacked them. Homosexual mounts in other primate species often seem to be associated with differences in status between the partners and appear to be some sort of affirmation of the difference (see Hanby, 1976). In our study, however, we think that they were purely sexual in context, because on six of the occasions, at least one of the females copulated with the male on the day of the mount and a seventh did so the day before. Further, on five of these six occasions (we could not see the seventh) it was the estrous female that thrusted, the other acting passively.

By contrast, homosexual behavior between males did appear to be agonistic in context, in that it was very similar to the notifying behavior of hamadryas baboons described by Kummer (1968). However, we need more than the three cases that we saw before we can confirm and elaborate on this impression.

III. COMPETITION AND PARTNER PREFERENCES

A. Intermale Competition for Females

Some male gorillas associate with several females for many years, while others remain solitary (Fossey, 1972; Harcourt, 1979a). The implied variation between males in reproductive success suggests competition between them for females (Trivers, 1972).

In the context of competition between gorilla males for females, it is important to separate intra- and extra-group competition. Adult males within a group are probably closely related and have known one another from immaturity (Harcourt, 1979a), while males of different groups are far less likely to be consanguineous. We will consider extra-group competition first.

A fight between strange males can result in many square yards of blood-spattered, flattened vegetation, and wounds that take days, sometimes weeks, to heal (Harcourt and Stewart, 1977). We found that 15 of 19 (79%) encounters between strange males involved violent threat displays. Restricted conditions of observations permitted us to determine that a fight actually occurred in eight out of 16 encounters (50%). These figures contrast with the impression given by Schaller (1963, p. 111–121) that encounters were normally peaceful. However, most of his observations seemed to be encounters between two strange *groups*, whereas nearly half of ours were of encounters between strange *lone males* and groups. Conflicts between lone males and dominant males controlling females are expected to be more violent, because the lone male has more to gain than males who already possess a harem (cf. Hrdy, 1977 for a similar interpretation concerning variation in intensity of contests between langur males).

In contrast with the normal severity of extra-group competition, intra-group competition was mild. The most serious incidence observed was of a leading male hitting a subordinate male on his back, an attack that caused no injury whatsoever. It appears that the leading male in a gorilla group is responsible for all, or nearly all, conceptions: the dominant male allows copulations with infertile females, but attempts to prevent them with fertile ones. Thus, subordinate males mated without hindrance with subadult females, and Schaller (1963, p. 283) saw at least one, possibly two, cases of a leading male permitting a subordinate to mate with a pregnant female. However, subordinate males were sometimes prevented from mating with nulliparous females, and never even attempted to mate with a potentially fertile parous female.

B. Partner Preferences

Most maturing females leave the group in which they were born and join another male. Subsequently, most transfer at least once again, and one female has transferred between males six times (Harcourt *et al.*, 1976). Eventually the female remains with a particular male for a prolonged period, perhaps for life. Possible determinants of her choice of mate have been discussed elsewhere (Harcourt *et al.*, 1976). Females of species that live in a multimale group, such as chimpanzees, can choose a different partner each estrous period.

Gorilla females, by contrast, hardly have any choice at estrus because most groups contain only one fully mature male (Chapter 12, this volume; Schaller, 1963), and if there is more than one mature male present, the evidence indicates that the dominant one gets most of the mating (Section III, A). The significance of mating systems that control female choice of mate is discussed in Chapter 12, this volume.

Although copulations usually lasted less than two minutes (Section II,A,2), females in estrus spent more time near the male than when anestrous (Harcourt, 1979b). This period of closer proximity might be analogous to the sexual consortship of chimpanzees (Chapter 9, this volume). A detailed comparison between chimpanzee consortships and associated partner preferences outside of estrous with the long term relationships evident between gorilla males and females might be of value in elucidating the origin of group living in gorillas, and perhaps that of other primate species as well.

ACKNOWLEDGMENTS

A. H. H. thanks the Medical Research Council and Leverhulme Trustees for the studentships that financed the fieldwork on which this paper is based. K. J. S. thanks Mr. & Mrs. James Stewart for financial support. D. F. thanks the National Geographic Society for continuing support. We are all very grateful to the governments of Rwanda and Zaire for permission to work in their country, and thanks are also due to Annette Walker for her meticulous typing. Finally, the manuscript benefited from criticism by Dr. C. E. Graham.

REFERENCES

Ardito, G. (1976). *J. Hum. Evol.* **5,** 213–222.
Fossey, D. (1972). *In* "Marvels of Animal Behavior" (T. Allen, ed.), pp. 28–29. National Geographic Society, Washington, D.C.
Hanby, J. (1976). *In* "Perspectives in Ethology" (P. P. G. Bateson and P. H. Klopfer, eds.), pp. 1–67. Plenum Press, New York.
Harcourt, A. H. (1978). *Z. Tierpsychol.* **48,** 401–420.
Harcourt, A. H. (1979a). *In* "The Great Apes" (D. A. Hamburg and E. R. McGown, eds.), Benjamin/Cummings Publ., Menlo Park, California.
Harcourt, A. H. (1979b). *Anim. Behav.* **27,** 325–342.
Harcourt, A. H., and Stewart, K. J. (1977). *New Sci.* **76,** 160–162.
Harcourt, A. H., Stewart, K. J., and Fossey, D. (1976). *Nature (London)* **263,** 226–227.
Hardin, C. J., Danford, D., and Skeldon, P. C. (1969). *Int. Zoo Yearb.* **9,** 84–88.
Hess, J. P. (1973). *In* "Comparative Ecology and Behaviour of Primates" (R. P. Michael and J. H. Crook, eds.), pp. 507–581. Academic Press, New York.

Hrdy, S. B. (1977). *In* "The Langurs of Abu." Harvard Univ. Press, Cambridge, Massachusetts.

Kummer, H. (1968). *In* "Social Organisation of Hamadryas Baboons." Univ. of Chicago Press, Chicago, Illinois.

Mallinson, J. J. C., Coffey, P., and Usher-Smith, J. (1976). *Int. Zoo Yearb.* **16,** 189–194.

Nadler, R. D. (1975a). *Anat. Rec.* **181,** 791–798.

Nadler, R. D. (1975b). *Proc. Symp. Congr. Int. Primatol. Soc., 5th* 207–216. Japan Science Press, Tokyo.

Nadler, R. D. (1975c). *Int. Zoo Yearb.* **15,** 134–137.

Nadler, R. D. (1976). *Arch. Sex. Behav.* **5,** 487–502.

Parker, G. A., and Pearson, R. G (1976). *J. Natur. Hist.* **10,** 241–245.

Rowell, T. E. (1972). *Adv. Study Behav.* **4,** 69–105.

Schaller, G. B. (1963). "The Mountain Gorilla: Ecology and Behavior." Univ. of Chicago Press, Chicago, Illinois.

Stewart, K. J. (1977). *Primates* **18,** 965–976.

Tijskens, J. (1971). *Int. Zoo Yearb.* **11,** 181–183.

Trivers, R. L. (1972). *In* "Sexual Selection and the Descent of Man" (B. Campbell, ed.), pp. 136–179. Heinemann, London.

Chapter 11

ORANGUTAN REPRODUCTION IN THE WILD

Biruté M. F. Galdikas

INTRODUCTION

The following is based primarily on data gathered during the first four years (November 1971–November 1975) of a continuing study on wild orangutans in a 35-km^2 study area (2° 45–48′ south latitude; 111° 57′–112° 1′ west longitude) at the Tanjung Puting Reserve of Kalimantan Tengah (Central Indonesian Borneo). During this time, Rod Brindamour

281

Copyright © 1981 by Academic Press, Inc.
All rights of reproduction in any form reserved.
ISBN 0-12-295020-8

and I, along with one trained local assistant, independently amassed 6804 hours of direct observations on target orangutans. Most of this observation time (70%) consisted of whole days and involved 12,361 animal-hours. Observations were conducted primarily on 58 individually recognized and named animals. As in other regions of Borneo and Sumatra (Horr, 1972; MacKinnon, 1971, 1974; Rodman, 1973; Rijksen, 1975), the basic units of the orangutan population studied at Tanjung Puting consist of 12 solitary adult males, 14 lone adult females (usually accompanied by one or two dependent offspring), and 18 independent immature animals, occasionally in association with their natal unit. Immature animals are classified as infants during the first four years of life, as juveniles from four until eight years, and females as adolescents from nine to 15 years, while males are adolescents from nine to approximately 12 years. Males go through an additional stage termed "subadulthood" (when as large or larger than adult females) before reaching full maturity at about 18 years of age. While females remain in the same general vicinity of the forest, males are more mobile. Although two mature males were "resident" in small areas for long periods, not one single adult or subadult male stayed withim the study area throughout the entire four years.

Since copulations are infrequent, all observed occurrences are included in the discussion. Eighteen different wild pairs copulated or attempted to copulate 52 times. When copulation occurred, it was rather conspicuous. Although details were sometimes difficult to observe, the close proximity of two individuals, assorted vocalizations, and movements drew the attention of the observer, making it unlikely that many copulations by animals under observation were missed.

In addition to collecting reproductive data on wild orangutans, we made incidental daily observations of a released excaptive adult female who normally stayed within 100 m of our camp and who depended largely on us for food. However, unless otherwise specified, all data presented here concern wild orangutans. I will use data collected on the excaptive female only when wild data are inadequate.

Among Tanjung Puting orangutans, most copulations take place within the context of consortship between a male and a female. Consortships occur at least for several days, presumably at the female's midcycle, and are characterized by close association between the pair. Consortships may involve a pair of adults or an adult and an immature. While consort copulations involving adult males are cooperative, consort copulations with subadult males frequently involve resistance on the part of the female. Copulations between animals not in consortship (usually an adult and immature or two immatures) also frequently involve an element of rape.

I. REPRODUCTIVE HISTORY

A. Subadulthood and Adolescence

Orangutan females have not been reported to undergo any readily visible physical changes, such as genital swelling, correlated with estrus. Although on a few occasions at Tanjung Puting, the genitalia of consorting females appeared pinker than usual, this was not a reliable guide to estrus, nor was it possible to monitor menstruation. Thus, the onset of female cycles at adolescence could only be judged from behavior. Although juvenile females behaved proceptively towards subadult males, only adolescent females copulated. We were able to make a reasonably accurate estimate of one small adolescent female's age as about seven years when we initially encountered her (1972). She first approached adult males proceptively when approximately 11 years old (1973) but did not give birth until approximately 15 years of age (1977). Another adolescent female cycled for at least one year (as determined from consort periods) before becoming pregnant. Observations of these two females indicate that periods of adolescent sterility exceed one year.

In captivity, orangutan males as young as eight years of age have fathered offspring (Graham–Jones and Hill, 1962). In the wild there was no way to detect the onset of spermatogenesis. Although male infants and juveniles were observed masturbating (in one case, between a mother's toes), subadult males begin to copulate at about 10 years of age when secondary sexual characteristics appear.

B. Adulthood and Reproduction

Although only one female's age at the time of her first parturition is known, it may be that other females initially give birth at about the age of 12 years also.

Birth intervals between the successive offspring of individual females exceed five years and probably average six or seven years. In 1971, there were in the study area 11 adult females, of which four were carrying small infants (all were either approximately one year of age or less), and two large adolescent females. (Although a total of 14 adult females were observed, one individually recognized female was infrequently encountered, so I exclude her from the discussion.) Five adult females and the two younger females gave birth between November 1971 and July 1973. No births were recorded from July 1973 to the end of the study period (November 1975). Two adult females, one of whom appeared very aged, did not give birth at all. By November 1975, none of the 11 females who had given birth in the preceding five years had yet borne new infants or were visibly pregnant. However, weaning of two older infants had begun.

Infant mortality rates seem low. With the exception of a sick infant whose mother also died, no wild female lost an infant. (However, the excaptive female lost her firstborn undersized infant after three months but successfully raised her second past that age.) Assuming a female's reproductive life lasts approximately 25 years (from 15 years to 35–40 years of age), and that she produces three or four offspring every six or seven years, we can estimate that a female produces three or four offspring over a lifetime. Two offspring is the most that any female in the study area is known to have produced, despite the fact that one female became a grandmother when her oldest-known offspring gave birth in 1977. However, at this time the grandmother herself was consorting and showed no signs of old age.

It is not known how many offspring a male produces over a lifetime, although the potential number is much larger than for a female. However, no male is yet known to have fathered more than one offspring. Each of the two resident males probably fathered at least one offspring (one born in 1973; the other in 1976), and a large subadult male who regularly frequented the study area may have fathered a third infant (also born in 1976).

C.   Reproductive Cycles

1.   Estrus

In captivity, orangutan females exhibit a 30-day menstrual cycle (Nadler, 1977). Although Nadler reported that captive orangutan pairs mated either on a daily basis or relatively frequently, regardless of the female's cycle or response, he noted that female proceptive behavior was witnessed only during midcycle.

Consortships between adults are distinguished from all other heterosexual contacts by close association for at least three days. No other heterosexual association involving two adults lasts longer than part of one day, although occasional immature heterosexual associations last weeks. The majority of copulations (64%) occur within consortship. Since females initiate many consort periods and invariably terminate them, we assume that consort periods correlate with estrus. Periodicity, determined wholly by behavioral criteria, seems similar to that of captive orangutans, recurring on an approximately monthly basis. While six consort periods involving one adolescent female were witnessed, we did not observe many consecutive monthly periods. This may have been due to sampling limitations or to the fact that this female was not cycling every month. When consecutive consort periods were observed, she resumed consortship at intervals of 44, 26, and 22 days; copulations were observed 51 and 31 days apart. This is not surprising given the general irregularity of adolescent cycles in higher primates.

Observations on one adult female indicated that in her case there were 29 days between copulations on two consecutive consort periods.

In seven instances we witnessed consort periods (involving three different pairs) from beginning to end (adult male–adult female (5); adult male–adult female (1); subadult male–adult female (1)). Consort periods (probably corresponding with estrus) arranged in duration from three to eight consecutive days (mean − 5.4 days). One to six copulations occurred per complete consort period (mean = 2.9 copulations). Copulations occurred on one to six days (mean = 2.4 days).

2. Physical Changes Correlated with Pregnancy

Although orangutan females do not show perineal swellings at estrus, pregnant females exhibit small but marked swellings (Schultz, 1969) and whitening of the perineal region, as well as enlargement of the nipples and eventually the whole abdominal region.

At Tanjung Puting every female who subsequently gave birth displayed swelling and whitening of the labial tissues for at least some months prior to birth. Initially, these changes were slight but became more pronounced with successive months so that, with experience, an observer could gauge rather accurately the stage of any pregnancy. It was difficult to determine exactly when this swelling began since we were never able to get closer than 5 m to a pregnant wild female. Further, females differed considerably in the amount of hair surrounding the genitalia, so that swellings became obvious on one female long before they did on another.

However, while close inspection and handling of wild females was not feasible, we were able to examine the perineal region of the released excaptive female two weeks after she terminated consort. At close range, slight swelling of the genital region was evident although it may not have been noticed from any distance. In her case, swelling probably began a few days after conception. This may occur with wild females as well. Females exhibiting the swollen genitals indicative of pregnancy did not copulate.

3. Conception

As noted earlier, adolescent females do not cycle regularly, and conception may occur only after long periods in excess of one year following the first observed copulation. Adult females with large juveniles or no offspring seem to conceive within one or two estrous periods. One wild adult female, whose large juvenile was no longer suckling, consorted twice in two consecutive months and probably conceived during the second period. However, another adult female, whose small juvenile was being weaned, first consorted with an adult male and then six months later with a large subadult at which time she may have conceived. This may have been a case of partial lactational amenorrhea (see Harcourt *et al.*, Chapter 10, this volume).

The case of the excaptive female is also instructive. After her first infant died, she copulated 2½ months later. At this time she consorted for five days with a wild adult male who probably impregnated her since she exhibited swollen genitals soon afterwards. She did not consort or copulate thereafter.

Conceptions seem to occur only in consortships. The fact that no female produced offspring as a result of nonconsort associations is probably significant. Two such females who had copulated outside consortship eventually gave birth, but the timing was such that in neither case could the nonconsort copulations have been responsible. However, four infants were born within a year after consort periods terminated. Only in one case did a consorting female not eventually produce offspring, and she was exceptionally aged.

II. SEXUAL BEHAVIOR

A. Copulations

It was usually possible to see intromission but was considerably more difficult to see ejaculate. We usually determined whether ejaculation occurred on the basis of behavior. If, after a period of intense continuous thrusting, a male stopped and showed no inclination to resume, we assumed that ejaculation had taken place. We assumed it had not taken place if the female broke away (while the male was still thrusting) and then rapidly moved off, with the male following in an agitated manner. We estimate that ejaculation occurred during 71% of copulations.

Copulations invariably took place in the trees and could occur at any time of day, with 54% occurring before noon. As reported by Mac-

Kinnon (1971, 1974) and Nadler (1977), copulations were generally ventro-ventral. Usually one or both individuals hung from an overhead branch, and the female often reclined against another branch. The male was usually lower than the female. Five copulations took place in nests. Thrusting lasted from three to 28 minutes (mean = 10.8 minutes) in cases where ejaculation occurred. Thrusting bouts as short as 10 seconds and as long as one minute were observed when copulation was incomplete. In the case of adult males, copulation was almost invariably (with one exception) preceded by part of a long call (Section III,D) and by oral contact with the female genitalia. When a male initiated copulation, he approached the female and spread her legs, manually or orally manipulated her genitalia, and then positioned her pelvis so that intromission could occur. He usually continued grasping the female's pelvis during copulation.

Males, in particular adults, often turned their heads away from the female while copulating and rolled their eyes upwards in an unmistakable expression that is observed only during mating, as has also been reported in captive males (Maple, 1978, personal communication). This expression can be described as a "faraway look in the eyes." Females sometimes fidgeted and looked around; one adolescent female continued eating during copulation, reaching and plucking fruits as the male was thrusting. The same female, on other occasions, reached down to her genitalia as the male thrust and then sucked and licked her fingers. While males (subadult and adult) frequently grumbled (or loudly purred) during copulation, females were silent throughout, unless being raped.

Age classes copulate at different frequencies. Both subadult males and adolescent females copulated or attempted copulations at higher rates than adults of their respective sex (Table 1). Subadult males also displayed a much higher percentage of incomplete copulations than mature males (Table 2).

B. Rape

Both MacKinnon (1971, 1974) and Rijksen (1975) described orangutan males "raping" females. In a captive study, Nadler (1977) found all copulations forcibly initiated by males.

Although most copulations at Tanjung Puting were cooperative, some involved elements of "rape." Rape occurred when a male attempted to copulate or copulated with a female who resisted his efforts to position her for intromission. A female's struggles ranged in intensity and

TABLE 1

Copulation rates per age/sex class as a function of each Independent Observation (IO)[a]

Age/Sex Class	Number of Independent Observations	Complete or Attempted Copulations		Completed Copulations	
		Number Observed	Frequency/ IO	Number Observed	Frequency/ IO
Adult males	329	30	.09	24	.07
Subadult males	162	22	.14	13	.08
Adult females	455	20	.04	13	.03
Adolescent females	190	32	.17	24	.13

[a] An independent observation is defined as a contact with one orangutan unit on one day, regardless of the length of contact or the number of units encountered on that day.

TABLE 2

Frequencies of attempted and completed copulations for each age/sex class.

Copulation	Adult Male Number	%	Subadult Male Number	%	Adult Female Number	%	Adolescent Female Number	%
Attempted; no intromission	0	0	8	36.4	5	25.0	3	9.4
Attempted with intromission, but no ejaculation	6	20.0	1	4.6	2	10.0	5	15.6
Completed	24	80.0	13	59.1	13	65.0	24	75.0
Totals	30	100%	22	100%	20	100%	32	100%

duration all the way from brief tussles with squealing and some pushing and slapping at the male's hands to protracted, violent fights in which the female struggled throughout the length of the copulation, emitted loud rape grunts,[1] and bit the male whenever she could. As will be described, most forced copulations took place in nonconsort contexts. The briefest and least intense struggles involved adolescent females

[1] Rape grunts consist of a loud, distinctive vocalization heard in no other context. This seems to be a different vocalization from the rape vocalizations described by MacKinnon (1971) at the Ula Segama. The Tanjung Puting rape grunt consists of a very low gutteral sound, almost like growling.

and subadult males. Relatively brief and low intensity struggles characterized forced consort copulations between subadult males and adult females. Fierce and protracted struggles took place between adult females and males outside consortship. Only one nonconsort copulation between a subadult male and adult female did not involve a fierce struggle. Rape was never observed between consorting adult males and females.

Rijksen (1975) suggests that rape involves dominance considerations rather than reproduction. However, male behavior during rapes at Tanjung Puting seemed sexual rather than aggressive. Males never furiously bit females or even appeared angry. Rather, male energies seemed to be concentrated solely on holding females to prevent their escape. As soon as raping males stopped thrusting, they invariably released the female. Nor did any female sustain injuries or wounds as a result of rape. While males and females fought and chased individuals of their own sex, relationships between the sexes, aside from rape, were remarkably free from aggression. No severe male attacks on females were ever witnessed as has been reported, for instance, among chimpanzees (van Lawick-Goodall, 1968). In total, aside from rape, only two aggressive heterosexual incidents were seen.

C. Proceptive Behavior

During the four years of study covered by this report, female proceptive behavior was observed 62 times and occurred in two patterns: adolescent females (95%) directed responses exclusively towards adult males, and juvenile females (5%) towards subadult males. Adult females were observed exhibiting strong proceptive behavior only after the initial four-year study ended. The majority of proceptive responses (90%) occurred during consortship.

Only 23 instances (37%) of female proceptive behavior culminated in intromission; however, every instance of adult male–adolescent female copulation was preceded by female proceptve behavior. Therefore, although most adolescent female displays are unsuccessful, it seems they are an essential prelude to copulation with adult males. Juvenile females did not mate after behaving proceptively even when subadult males responded. In one case, the male reached out and attempted to hold a juvenile female, but she quickly moved aside.

In captivity female orangutan proceptive behavior involves females pressing genitals against the male, making thrusting movements, or rubbing genitals on a male's head (Nadler, 1977). One proceptive captive adult female behaved more aggressively. She repeatedly approached, pulled, and slapped the male immediately following introduction into his cage.

At Tanjung Puting, female orangutan proceptive behavior consisted of approaching and sitting or standing very close to the male, grooming him (for as long as nine minutes), touching or mouthing his genitalia, and touching his face, stomach, back, or hands. Solicitation posture, apparently analogous to the presenting of rear quarters common to many monkeys and apes, involved the female dangling her perineal region in the male's face from above while facing him.

As on three occasions adolescent females initiated contact with an adult male by approaching, sitting directly above him, and then breaking and dropping branches. One adolescent female also tried pushing a snag (a branchless dead tree) to the ground and urinated on the male. One adult female initiated consortship by repeatedly shaking a vine in an adult male's face, slapping his stomach, and tweaking his penis.

There are no male courtship displays as such. However, the adult male's long call (MacKinnon, 1971) may play a role in initiating proceptive behavior. Forty-three percent of adolescent female proceptive behavior directly following males beginning the long call. In some cases females were several trees away but moved rapidly and directly towards calling males. During consortship adolescent females responded to roughly one-third of their consort's calls by moving towards him (if they were some distance away) or by exhibiting proceptive responses (if close to him).

III. COMPETITION AND PARTNER PREFERENCES

A. Consortship

There are two types of sexual contacts: consort and nonconsort, and some pairs establish consort relationships upon meeting. Consorts involved three different combinations of individuals: an adult male with an adult female (3); adult male with an adolescent female (7); or a subadult male with an adult female (2), a total of 12 observed involving six different pairs. Four periods involved specific pairs consorting once. (It is difficult to say whether these actually represented single nonrecurring instances or were an artifact of sampling.) Eight consorts represented two recurring relationships: one between a resident adult male and adult female seen in two consecutive months, and a second between the other resident male and an adolescent female, which recurred six times over 12 months. While observations may have been incomplete, this consortship resumed approximately monthly throughout the last five months of the 12-month period (June, July, September, and October[2] 1975). During the one month (August 1975) when we saw no consortship, we were unable to locate either member of the pair during the week that we predicted consortship would recur.

Ten consort periods (83%) and both recurring consortships involved adult males with the three adult males participating in 88% of consort copulations. Two of these males were residents, while the status of the third was uncertain. Adult males spent far more time and a greater percentage of their total contact time consorting than did other age/sex classes (Galdikas, 1978). Approximately 80% of the time that adult males were in contact with other orangutans involved consortships.

Whereas males initiated sexual activity with adult females, copulations with adolescent females invariably followed female proceptive behavior. Of four consort copulations between adult females and sub-adult males, three involved females struggling and uttering rape grunts, whereas no forced copulations occurred in consortships involving adult males. Approximately 25% of copulations during consortship represented incomplete bouts, in which either the male or female moved away before either intromission or ejaculation occurred.

B. Consort Formation and Termination

Initial contact between consorting pairs was observed eight times. In two cases involving adult pairs and in three other instances, females were initially accompanied by subadult males whom the adult male immediately replaced. In one instance involving an adult pair, the adult male approached the adult female to initiate consortship. In the second case, the adult female moved towards the adult male, seemingly as a response to his long call. The initiation of adult consortship could be recognized by close association of partners, with such proximity primarily maintained by the male.

Age is an important factor in terms of selection of consort partners. Immature orangutans, whatever their sex, initiate and maintain consortships with adults. However, while subadult males usually followed

---

[2] Due to the pair's close association, I consider two days in October 1975, when we observed the adolescent female and adult male traveling together, as resumption of their consortship, although no copulations occurred in the relatively brief period we observed them.

adult female consorts within 20 m, adolescent females sometimes wandered great distances from their adult partner, especially in the late afternoon. One adolescent female frequently did not resume contact with her adult consort until the next morning. One afternoon we observed the female travel over 200 m away before nesting for the night.

Females prefer adult males, particularly dominant adult males. Only three males (two of them resident) were approached by females eager to consort. Numerous other males were never so approached even when they were in the area at the same time. Both resident adult males were dominant while the third was also high ranking, although never resident in the study area. However, females exercise a considerable degree of individual preference. The most dominant male was not necessarily the one chosen as a consort partner.

Dominant adult males show age preference for adult females by selecting them as consort partners (as previously described) or by attempting to forcibly copulate (one case). No forcible copulations involving adult males and adolescent females occurred. Differential adult male behavior towards adolescent and adult consort partners was also evident. Adult males closely attended adult females but did not follow when adolescent consorts wandered off.

Further, an adult male rebuffed an adolescent female. After directing proceptive responses, an approximately 11-year-old female followed a large adult male for part of two days (but less than 24 hours total) in an apparent effort to initiate consortship. His eventual negative response (chasing and grabbing) served to terminate the encounter. Whenever the end of a consort period was witnessed ($N = 7$), the female invariably terminated the relationship by leaving the male, frequently after he nested for the night.

C. Copulation in a Nonconsort Context

Brief sexual contacts occurred in which the male and female associated together for less than one day, frequently without travel. These involved 15 instances consisting of 12 individual pairs: adult male and adult female (1), subadult male and adult female (8), and subadult male with an adolescent female (6). While most copulations were isolated pairings (11 instances), one subadult male copulated or attempted copulation with one female in a series of four nonconsort incidents spanning two years.

As the ratios in the last paragraph show, nonconsort copulations primarily involved subadult males (95%) with the vast majority of copulations (84%) forced. Without exception, males initiated sexual activity in nonconsort contexts. As association in this context was brief (mean = 2.75 hr), associations were witnessed in their entirety. The longest lasted 11 hr and 25 min, terminating with both partners nesting. Association was not resumed the next morning. The shortest period lasted approximately one minute. A subadult male raced up to an adolescent female and attempted copulation but was chased off by the female's adult consort after she struggled and squealed. Over half of nonconsort copulations were incomplete, in all cases due to female struggle or avoidance.

Sexual activity in nonconsort contexts occurred in a much more sporadic fashion than consortships. While consortships occurred only in early and mid-1972 and then again after mid-1974, nonconsort copulations were more evenly spaced throughout the four years and involved females at various points of the reproductive cycle. Nonconsort copulations primarily involved females with infants (7), but also a female with a small juvenile (1), a female with a large juvenile (1), and nulliparous adolescent females (6).

D. Aggression between Males

1. The Long Call

The long call is the most frequent orangutan vocalization, and the only one that can be heard consistently over great distances. The apparent function of the long call in initiating female proceptive behavior has already been noted (Section II,C). Equally important, other studies indicate that the long call may play a role in repelling male rivals (MacKinnon, 1971, 1974; Horr, 1972; Rodman, 1973). At Tanjung Puting, long calls seem to mediate male hierarchies (Galdikas, 1978). Even when two or more mature males are within several hundred meters of each other in the forest, contact rarely takes place. Lesser-ranking males stay silent and move out of the paths of more dominant males who continue long calling.

Repeatedly observed prime adult males ($N = 5$) at Tanjung Puting called an average rate of 3.6 calls/day, ranging from 2.2 to 4.3 calls/day for different individuals with dominant males calling more frequently than lower-ranking ones. Prime males also spontaneously crash snags as they move through the forest. One adult male dropped snags at the rate of one per every 18 hours of observation. On a windless day the sound of crashing snags frequently indicated the presence of an adult male.

2. Direct Aggression between Adult Males

Although infrequent, contact between mature males (four instances) invariably involved aggression or marked avoidance responses. Aggression was particularly severe in two cases where adult males came into contact near a female, and most severe when a consort was involved. In the most violent incident, a lengthy combat took place between the adult male consort and an approaching male. In the second incident, an adult female with an infant was present. There was no combat, but one male chased the other on the ground for several hundred meters.

Interactions between lone adult males were not as intense as incidents where females were present. In the two instances where lone males met, one male fled along the ground, but in one case the other male did not chase him while in the second case, the chase did not last very long. Whenever adult male–male contact occurred, one of the males long-called at least once.

3. Direct Aggression between Adult Males and Subadult Males

Interactions between subadult and adult males were 10 times more frequent than contact between adult males and lasted considerably longer. Adult males are far more tolerant of subadult males than of each other. Again, the nature of interactions seems determined by presence or absence of females. When lone adult males encounter lone subadult males, their reactions to each other are mild. The usual subadult male response (80%) to an adult male was to move away in the trees. Lone subadult males never fled from an adult male on the ground. This was not the case when females were present. Adult males, especially consorts, appeared aggressive and belligerent. Subadult males frequently fled on the ground (49%) as adult male consorts moved towards them shaking branches, crashing snags, or vocalizing the long call. However, in some cases (6.5%), this was not enough to drive the subadult male away, and he returned to doggedly follow the consorting couple. Although adult males chased subadults, no physical contact was seen.

Every single consort period observed from beginning to end involved subadult males contacting consorting pairs. Since adult males were always in close proximity to adult female consorts, subadult males never attempted approaching them. However, in one particular instance, subadult males made several attempts to copulate with an adolescent female when whe moved some distance from her adult male consort. The first attempt failed when, following the adolescent's loud squeals, the adult male charged back and chased the subadult male away. Later that day, while the adult male chased yet another subadult male, two subadults successfully mated with the adult male's consort before he returned. (The adolescent consort did not conceive at this time.)

4. Aggression between Males: The Indirect Evidence

Indirect evidence indicates the scope of aggression between mature males better than the instances directly observed. This is due to the fact that mature males are distinguished as a class by their disfigurements. Only three out of 12 adult males were without visible physical anomalies, although in one case the anomalies were evident only after death. These anomalies center on the hands (stiff, misshapen, or broken fingers) and head (missing eyes, torn lips, broken canines), the body parts most vulnerable to injury in orangutan combat. The prevalence of such anomalies suggests that direct conflict must play an important part in the lives of most males who survive to maturity.

IV. CONCLUSIONS: THE ORANGUTAN MATING SYSTEM

Adult male and female "reproductive strategies" in the same species are likely to be different, based on the differential amount of parental investment each sex contributes to the offspring (Trivers, 1972). Orangutans are among the most highly sexually dimorphic of primate species (Napier and Napier, 1967). Large degrees of sexual dimorphism in the animal kingdom are almost invariably associated with polygamous mating systems (Selander, 1972) in which there will be strong competition among (usually) the males (Mayr, 1972), whether for territories, for places at a common breeding ground, or for females directly. In the following section, I will discuss the mating system of Tanjung Puting orangutans.

Since orangutans are not gregarious, the initial problem faced by each individual in terms of reproduction is simply locating a sexual partner. Due to long birth intervals, cycling females are a rare commodity in the forest. At Tanjung Puting, females within the study area were in rough synchrony so that for at least two years, there were no receptive prime adult females. Subadult and adult males are probably capable of producing sperm at all times so that female reproductive capacity becomes a limiting resource for male reproductive success (Trivers,

1972). A male's basic problem in achieving successful reproduction is to locate or attract an estrous female and then, since timing of copulation is crucial in ensuring conception, to maintain contact until ovulation occurs, despite aggression and competition from other males.

Adult males demonstrate two patterns in acquiring females. Some males stay in a limited home range for up to a few years, while others range more widely without staying long in any one area. Residence seems loosely correlated with dominance. Of three males who, at least periodically, shared a common range, the two most dominant males were each resident in the area, first successively and then together. The lowest ranking male only periodically entered the area and was twice observed outside it.

A resident male demonstrates several advantages in terms of male–male competition. He is regularly in close proximity, although not direct contact, with several females whose home ranges overlap his. He is, therefore, in a position to locate them quickly as they come into receptivity, especially since such females might be moving towards him in response to his long call. Perhaps a male's residence also serves to familiarize females with him. Due to their long calling, large size, and proclivity for crashing snags, adult males are far more conspicuous than adult females and are easier to locate. Since adult females are, by contrast, less conspicuous, resident male strategy involves calling regularly and frequently so that females know their location. It is possible that nonresident adult males or wanderers represent the surplus males in a polygamous system or the losers in male–male competition.

It is striking that, as with chimpanzees (Tutin, 1975), conceptions among orangutans occur primarily in the context of consortship between one adult male and female. Thus, the choice of sexual partner remains very much a female prerogative. Although orangutan females can sometimes be raped, they cannot be herded, led, or coerced into joining a male, due to the exigencies of aboreal travel. Hence, a female cannot be monopolized by a male in the expectation that he will have access when she becomes fertile. If the male consorts with an unwilling female, he cannot guide her movements but must resign himself to following her (sometimes straight into the path of a dominant male preferable to her) or must copulate with her in the rape pattern.

Female choice is probably limited to males found in her home range. (Although we found calls could be heard over 2 km away, it is not known whether a receptive female moves outside her normal foraging areas to seek preferred males calling in the distance.) She can select from both resident males and those adult and subadult males who periodically move through her range. This selection process augments the effects of competition between males because females seem to prefer dominant adult males who are usually residents.

This female preference can be explained in evolutionary terms by her very large reproductive investment (Horr, 1972, 1975; Rodman, 1973). Since a female invests at least five years (probably more) in one offspring, evolution is likely to favor mechanisms that ensure selection of the fittest possible mate. Among orangutans, fatherhood entails expenditure of sperm but, more importantly, energy to maintain consortship in the face of competition and aggression from other males. By choosing dominant adult males as consorts, females help ensure that their offspring have at least some of the same qualities that make their consort an effective breeder. Big cheekpads, throat pouches, and large size demonstrate that the male survived to maturity and are probably attributes concomitant with breeding success. At Tanjung Puting, a high degree of male–male competition, combined with female selection of male consort partners, works to exclude many males from an active role in reproduction.

The important role of female choice may explain why, alone among pongids, orangutan females do not display perineal swellings correlated with estrus. Rather, they exhibit marked and well-developed proceptive behaviors. Since the orangutan female plays an important role in selecting, locating, and initiating contact with a consort partner who, in all likelihood, is solitary each time she comes into estrus, there are no selective pressures for the evolution of physiological signals associated with estrus. Unlike chimpanzees, orangutan females are rarely in contact with multimale groups. Orangutan females are never in a situation where vivid visual signals of sexual receptivity might give them an immediate advantage over other females in their relationship with a comnunity of males (for instance, protection from male attack).

Although orangutan females do not show swellings in estrus, they are unique among pongids in that pregnancy is unmistakable even in its early stages. Perineal swellings are visible within weeks, possibly days, after conception. Such swellings may have developed as a mechanism for reducing female attractiveness to males during the delicate time females carry a fetus. Subadult males frequently contact pregnant females and sometimes sniff and check their perineal regions but never attempt mating. Perhaps swellings inhibit biologically useless copulatory attempts with the adaptive result that males go elsewhere to seek estrous sexual partners.

Subadult males are in a particularly untenable position in regards to reproduction. Since they cannot attract females by calling, their reproductive strategy seems limited to (1) searching for receptive females and consorting before females locate adult males, and (2) forcibly cop-ulating with females with a small chance that one may be in the estrous condition. Rape is characteristically a subadult male behavior. However, it must be stressed that subadult males frequently associate with ad-olescent and adult females without attempting copulation. Even among subadults, rape is a secondary "strategy." Subadults attempt to main-tain consortships with females whenever possible. However, the odds against subadult males successfully concluding consortships with prime adult females are probably quite large since the mere appearance of an adult male terminates such relationships.

Subadults also associate frequently and for long periods with anes-trous adolescent females (Galdikas, 1978), but these females, when coming into receptivity, locate adult males as sexual partners. Occa-sionally, the subadult male continues following the adolescent as she consorts. Subadult males do have one advantage over adult males; their presence is tolerated by adult males, while the presence of another adult male close by would not be.

Finally, the fact that subadult males prefer to associate with adoles-cent females rather than with other males or adult females may ulti-mately be related to reproduction. Such associations may have a general developmental function comparable to sexual experimentation between young chimpanzees and young chimpanzee males and adult females (Tutin, 1975). Further, the formation of an affiliative relationship with an adolescent female may facilitate the selection of the same male as a consort partner when both become adult. There was no way to test the validity of this hypothesis during the four-year study. Nonetheless, two nonconsort copulations in which an adolescent female did not struggle involved a subadult male who had previously extensively as-sociated with her in nonsexual contexts. It is likely that most orangutan associations at estrus depend, to a large extent, on relationships es-tablished outside of estrus.

ACKNOWLEDGMENTS

The wild orangutan research on which this paper is based was funded by the Wilkie Brothers Foundation, the L.S.B. Leakey Foundation, the National Geographic Society, the New York Zoological Society, the Van Tienhoven Foundation of Holland, the Herz Foundation, the Jane and Justin Dart Foundation, and the World Wildlife Fund. My husband, Rod Brindamour, and I are very grateful to these organizations and to the people associated with them for making our work possible. Dr. L. S. B. Leakey's initial support was instrumental in enabling us to begin the research. Lembaga Ilmu Penge-tahuan Indonesia (Indonesian Institute of Sciences) and Perlindungan dan Pengawetan Alam (Nature Protection and Wildlife Management) branch of the Forestry Department served as our sponsors in Indonesia. We wish to express our deep gratitude to Mr. Soedjarwo (director-general of the Forestry), Mr. Siswojo Sarodja (director, Bina Sarana Usaha), Mr. Prijono Hardjosento (director, Nature Protection and Wildlife Management), Mr. Walman Sinaga, Mr. Goenari, and Mr. Soegito Tirtomihardjo, as well as Mr. Widajat Eddypranoto, Mr. Rombe, Mr. Bahrun Harun, Mr. Beringan, and Mr. Yusran. We are also very grateful to Dr. Bachtiar Rifai (head, Indonesian Institute of Sciences), Ms. Sjamsiah Achmad (former head, International Bureau), Ms. Moertini (head, International Bureau), Mr. Napitapulu, Mr. J. Bima, and Dr. Rio Rachwartono. The provincial government of Kalimantan Tengah encouraged and supported our work. We are particularly grateful to Mr. R. Sylvanus (former governor) and to Mr. G. T. Binti, as well as to numerous other provincial officials. I also wish to express my deep gratitude to Mrs. Nina Sulaiman and Mr. Sulaiman Sumitakusuma whose gracious hospitality during my recent visit to Jakarta enabled me to write part of this paper.

We would also like to express our very deep appreciation to the following individuals who, in one way or another, have provided much help and support: Dr. Rainer Berger, Dr. Joseph Birdsell, Dr. Bernard Campbell, Mr. Robert Gilka, Dr. Jane van Lawick–Goodall, Ms. Mary Griswold Smith, Dr. David Hamburg, Dr. Barbara Harrisson, Ms. Joanne Hess, Dr. Don Lindburg, Dr. Peter Miller, Mr. Ed and Ms. Sandy Johnson, Ms. Joan Travis, trustees and members of the L. S. B. Leakey Foundation, Dr. J. H. Westermann, Mr. Robert Wilkie, and Mr. Leighton Wilkie.

From May 1974 onward, pairs of biology students from Universitas Nasional, Jakarta Indonesia, served for a period of six months (each) as research assistants and carried out their own field research on the primates of the study area. We thank Mr. Suharto Djojosudarmo, Mr. Jaumat Dulhajah, Mr. Sugardjito, Mr. Endang Sukara, Mr. Barita Oloan Manulang, and Mr. Yatna Supriatna.

Finally, we thank our local staff, Mr. Ahmad (who aids in the wild orangutan research), and his wife Ms. Bahriah, and Mr. Mujiran.

REFERENCES

Galdikas, B. M. F. (1978). Orangutan Adaptation at Tanjung Puting Reserve, Central Borneo. Ph.D. dissertation, Univ. of California, Los Angeles, California.

Graham–Jones, O., and W. C. Osman Hill (1962). *Proc. Zool. Soc. London* **139,** 403–410.

Horr, D. A. (1972). *Borneo Res. Bull.* **4,** 46–50.

Horr, D. A. (1975). *Primate Behav.* **4,** 307–323.

MacKinnon, J. R. (1971). *Oryx* **11,** 141–191.

MacKinnon, J. R. (1974). *Anim. Behav.* **22,** 3–74.

Mayr, E. (1972). *In* "Sexual Selection and the Descent of Man, 1871–1971" (B. G. Campbell, ed.). Aldine, Chicago, Illinois.

Nadler, R. D. (1977). *Arch. Sex. Behav.* **6,** 457–475.

Napier, J. R., and Napier, P. H. (1967). "A Handbook of Living Primates." Academic Press, New York.

Rijksen, H. D. (1975). *In* "Contemporary Primatology" (S. Kondo, M. Kawai, and A. Ehara, eds.) *(5th Int. Congr. Primatol., Nagoya 1974).* Karger, Basel, pp. 373–379.

Rodman, P. S. (1973). *In* "Comparative Ecology and Behaviour of Primates" (J. H. Crook and R. P. Michael, eds.), Academic Press, New York.

Schultz, A. H. (1969). "The Life of Primates." Universe Books, New York.

Selander, R. K. (1972). *In* "Sexual Selection and the Descent of Man, 1871–1971" (B. G. Campbell, ed.). Aldine, Chicago, Illinois.

Trivers, R. L. (1972). *In* "Sexual Selection and the Descent of Man, 1871–1971" (B. G. Campbell, ed.). Aldine, Chicago, Illinois.

Tutin, C. E. G. (1975). Sexual Behaviour and Mating Patterns in a Community of Wild Chimpanzees (Pan troglodytes schweinfurthii). Ph.D. dissertation, Univ. of Edinburgh, Edinburgh, Scotland.

van Lawick-Goodall, J. (1968). *Anim. Behav. Monogr.* **1,** 161–311.

Chapter 12

INTERMALE COMPETITION AND THE REPRODUCTIVE BEHAVIOR OF THE GREAT APES

Alexander H. Harcourt

INTRODUCTION

In the previous three chapters we have seen that the behavior involved in mate acquisition, courtship, and copulation varies widely among the great apes. In this chapter those differences are reviewed, and an attempt is made to provide some insight into the selection pressures that might have produced them. A fuller understanding of

301

Copyright © 1981 by Academic Press, Inc.
All rights of reproduction in any form reserved.
ISBN 0-12-295020-8

the evolutionary causes of differences between species and of the selection pressures maintaining them should help us to appreciate the
multitude of factors that govern the behaviors of animals, and also
should provide us with a basis for extrapolation to man's behavior
(Short, 1979, see Chapter 13, this volume; Wilson, 1971).

A large body of literature exists concerning the evolution of species'
mating systems and the selection pressures that control them (e.g.,
Clutton–Brock and Harvey, 1977; Crook, 1965, 1970; Eisenberg *et al.*,
1972; Emlen and Oring, 1977; Geist, 1974; Jarman, 1974; Lack, 1968;
Orians, 1969; Ralls, 1977; Selander, 1972; Verner and Willson, 1966;
Wrangham, 1979). However, possible relations between a species' mating system and the details of its reproductive behavior have been far
less extensively investigated, although Nadler (1977b) and Short (1979)
have made an important start. Therefore, it is this latter problem that
I would like to examine here by considering some of the selection
pressures that might influence the observed interspecific variation in
the behaviors shown in mate acquisition, courtship, and copulation by
the great apes. As is apparent from the chapter title, much of the
discussion will concern competition between males for mates, rather
than competition between females. This is because such competition
is normally far more intense between males than between females,
since males of most species can increase their reproductive success by
mating with more females, but females usually cannot raise theirs by
mating with more males (Bateman, 1948; Fisher, 1958; Trivers, 1972).
Essentially, females are often a limiting resource for males, but rarely
are males so for females.

Appreciation of the selection pressures that might be operating requires an outline knowledge of the social setting in which the competition is occurring. Therefore, I will start with a summary of the great
apes' social organization. From this basis will follow a discussion of
competition between males for females, and of possible relations between the aggression involved in such competition and the observed
sexual dimorphism in size within Pongidae. Finally, differences among
the apes in specific courtship and copulatory behaviors will be related
to dissimilarities between them in the nature of the competition among
males for mates, with supporting comparative data drawn from
Cercopithecidae.

I. SOCIAL ORGANIZATION

This summary of the apes' social organization has to be brief, and
so ignores such considerations as the variation that appears to exist

between the Bornean and Sumatran orangutans (MacKinnon, 1974, 1979). Of greater concern, however, is that much of our knowledge is based on study of a very few populations of each species. For example, nearly all the information on gorilla behavior in the wild comes from a single population, that of the Virunga Volcanoes of Rwanda and Zaire. Nevertheless, existing knowledge allows us to generalize sufficiently for present purposes.

The apes can be graded in gregariousness from the almost solitary orangutan (*Pongo pygmaeus*) through the community-living chimpanzee (*Pan troglodytes*), with its small parties of continuously changing composition and size, to the gorilla (*Gorilla gorilla*), which lives in bisexual groups stable for months at a time. (The pigmy chimpanzee *Pan paniscus* is not considered here because of insufficient information.)

In orangutan populations the only stable groups (i.e., ones lasting for several months) consist of mothers and their offspring. Otherwise, adult males and females range separately from one another, except, of course, when a female comes into estrus. Even when adults do meet at fruiting trees, for example, there is often hardly any social interaction. It appears that the adult male population can be partly separated into dominant, normally resident males and more vagrant, subordinate males, whose ranges overlap the smaller ranges of a number of females. The marked solitariness and particularly wide dispersal of the adult males seems to be maintained by mutual avoidance, aggression, and retreat of subordinate males (Davenport, 1967; Horr, 1975; Rodman, 1973; MacKinnon, 1974; Rijksen, 1978; Galdikas, 1979, Chapter 11, this volume).

Like orangutans, anestrous chimpanzee females live separately from adult males and from one another, despite considerable overlap of range; their only permanent companions are their dependent offspring, and it is only when they are in estrus that they associate for any appreciable length of time with adult males. Chimpanzee males, again like orangutans, occupy larger ranges than do the females and travel over a number of females' ranges. There the similarity ends, however, because the male chimpanzees share a common territory (hence the term "community"), which they defend cooperatively against males from other communities. In fact, groups of males from one community can be seen traveling on boundary patrols and threatening any males from adjacent territories that they find within their own. Moreover, males usually travel and feed in the vicinity of other males from their own community. However, these parties are unstable, and one male rarely stays with the same partner for more than a few hours at a time (Bygott, 1979; Goodall, 1968, 1975; Goodall *et al.*, 1979; Sugiyama, 1973; and Tutin and McGinnis, Chapter 9, this volume).

The gorilla is very different from the orangutan and chimpanzee. Most individuals in a population live in bisexual groups that contain at least one fully adult male and a number of females and their offspring. The composition of these groups remains stable for weeks or even months, and some adult males and females are known to have been together for more than a decade. Usually only one fully adult male is found in a group, but where there is more than one a clear hierarchy exists, the oldest male almost invariably being the most dominant. Lone males are seen in most populations, but only extremely rarely are lone females found. Over 75% of encounters between strange males are aggressive, and because neither the frequency nor the intensity of the aggression seem related to the site of the encounter, it appears likely that the hostility is related to acquisition of females rather than of range (Caro, 1976; Fossey, 1972, 1974; Harcourt, 1979b; Schaller, 1963).

The pattern of dispersal of young animals is obviously an important facet of a species' social organization. In orangutan populations, both males and females eventually leave their mother and establish ranges of their own (Galdikas, 1979; MacKinnon, 1974; Rijksen, 1978). The community system of the chimpanzee is maintained by the permanent residence of males within the community of their birth; whereas females transfer to neighboring communities, no male has yet been known to do so (Goodall, 1975; Kawanaka and Nishida, 1975; Pusey, 1979). The gorilla in some ways appears intermediate between the orangutan and chimpanzee: both sexes leave their natal group, but while females transfer immediately to lone males or to other breeding units, emigrating males usually wander alone, sometimes for years, until they attract females and can establish a group of their own (Harcourt *et al.*, 1976; Harcourt, 1978). The system of female emigration and transfer seen in the chimpanzee and gorilla contrasts with the pattern of adolescent dispersal of most other group-living primates, where males, not females, are the mobile element of the population (Harcourt, 1978).

For present purposes, the important points to emerge from this account of the apes' social organization are that (1) chimpanzees differ from gorillas and orangutans in the number of males competing for *estrous* females, and (2) because chimpanzee males do not leave the community in which they were born, males within a community are more closely related to one another than they are to males from another community.

II. COMPETITION BETWEEN MALES FOR FEMALES

A. Competition for Females

It is important to distinguish competition between males for females *at the time of estrus* from competition for *permanent access* to females. From the foregoing account of the social organizations of the great apes, it ought to be clear that the males of the three species have differing degrees of access to *estrous* females. In a chimpanzee population, because the males live in communities and those within a community have almost totally overlapping home ranges, a number of males are simultaneously available to and compete for any female in estrous. For example, Tutin and McGinnis (Chapter 9, this volume, Section III, A,1) report that 73% of observed copulations occurred during opportunistic mating periods when groups of two to 12 males would associate with an estrous female and take turns to mate with her. By contrast, the wide dispersal of orangutan males makes it unlikely that more than one dominant adult male will be in the vicinity of an estrous female, and certainly no orangutan female is going to be in the position of the female chimpanzees described by Goodall (1968) as "seldom being accompanied by less than six mature males." Moreover, mature orangutan females in estrus actively appear to choose one particular dominant male in an area, so further reducing the likelihood of more than one male being in the vicinity of an estrous female (Galdikas, Chapter 11, Section III, B; Rijksen, 1978). Gorillas are similar to orangutans insofar as their social organization also minimizes the availability of males to estrous females, and hence minimizes competition between males for females at the time of estrus; there is usually only one fully adult male in a group and where there is more than one, the dominant animal enjoys almost sole access to females in estrus (Chapter 10, Section III).

In all three species, however, the males have to compete for *permanent* possession of females if they are to ensure access at estrus. It is this aspect of competition for long-term access to females that can explain the sometimes intense aggression seen between strange males in all the great apes. In the chimpanzee such competition between strange males seems to be for possession of females' ranges (Bygott, 1979; Wrangham, 1979), in the gorilla for possession of females directly (Section I), and in the orangutan, perhaps for both (Chapter 11, Section II, G).

B.   Aggression between Males

The word "strange" in the preceding sentence must be emphasized in the context of aggression between males for females. It has been said that the contrast between the lack of aggression among chimpanzee males competing for females (for example, their amicably taking turns to mate (Tutin and McGinnis, Chapter 9, Section III, A,1)), with the sometimes violent fights between gorilla males (Harcourt and Stewart, 1977) can explain the differences between the two species in the extent of sexual dimorphism found within them. The reasoning was that because chimpanzee males do not fight fiercely over females, an individual larger than the others gains no obvious benefit from his size in terms of access to females. However, an extra-large gorilla male is at a clear advantage because he is more likely to win a fight (and so a female) than is a smaller one (Harcourt and Stewart, 1977; Leutenegger and Kelly, 1977). But these authors have compared, perhaps misleadingly, the lack of aggression between familiar, consanguineous male chimpanzees, with that between strange, probably unrelated gorilla males, ignoring the fact that *strange* male chimpanzees fight no less fiercely than do strange gorilla or orangutan males. Compare, for example, the account of Goodall *et al.* (1979) of male chimpanzees killing others from another community and taking over their range, with Harcourt and Stewart's (1977) description of the ferocity of fights between strange gorilla males, and with Galdikas' description of the frequency of wounds on adult male orangutans (Chapter 11, Section III, C). There does, however, appear to be a difference between the chimpanzee on the one hand, and the gorilla and orangutan on the other in the nature of aggression between strange males. Fights between strange gorillas and orangutans are contests between individual animals, whereas all the intercommunity fights in the Gombe Stream chimpanzee population were between parties of males from one community and a single male from another (Goodall *et al.,* 1979; Tutin and McGinnis, Chapter 9, Section III, A).

The difference between the comparative lack of aggression within chimpanzee communities and the sometimes extreme aggression between males from separate communities can be explained in evolutionary terms by the fact that males from within a community are more closely related to each other than they are to males from another community. In short, an individual suffers a *smaller* decrease in inclusive fitness (Hamilton, 1964) if a relative copulates with a prospective mate than if a nonrelative does so, and a *greater* decrease if it harms a relative than if it harms a nonrelative. Hence, chimpanzee males tend to be more aggressive to others from outside their community than they are to ones from within it.

C.   Sexual Dimorphism and Aggression between Males

There seems to be little doubt that the degree of sexual dimorphism in the apes and in other primate species in which the male is larger than the female is related to the level of aggression between males for females (Crook, 1972). Particularly strong evidence for this is the fact that the most extreme dimorphism usually occurs in polygynous species, which is where the most intense competition between males for females is seen (Clutton–Brock *et al.*, 1977; Crook, 1972; Jarman, 1974; Ralls, 1977; Selander, 1972). The argument is that the more impressive males win more fights, obtain more females and so leave more offspring, and thus have a selective advantage (Crook, 1972; Geist, 1974; Wilson, 1975). However, the apparent lack of correlation in the primates between the degree of sexual dimorphism in body weight and level of polygyny (as measured by adult sex ratio) found by Clutton–Brock *et al.* (1977) suggests that intermale competition is not the whole answer; other factors are obviously operating.

One of these factors is body size. A strong positive correlation between body weight and degree of sexual dimorphism is found in various orders including the primates (Clutton–Brock *et al.*, 1977; Ralls, 1977; Wiley, 1974). In fact, Clutton–Brock *et al.*, (1977) found that body size accounts for a greater proportion of the variation in sexual dimorphism across the primates than does breeding system. Certainly in the great apes, differences in body size could account for a large part of the variation in degree of sexual dimorphism since the highly sexually dimorphic gorilla and orangutan are larger than the less sexually dimorphic chimpanzee. Nevertheless, a further factor must be involved because the body weight of the gorilla and orangutan differs widely, and yet their degree of sexual dimorphism in size is very similar; and while the female chimpanzee and orangutan have similar body weights, the two species differ widely in their degree of dimorphism. I suggest that these discrepancies can be explained by the higher level of aggression apparent between individual males in orangutan and gorilla populations than between individual chimpanzee males.

Two other possible evolutionary causes of sexual dimorphism are selection for occupation of different niches, as happens in a number of bird species (Selander, 1966, 1972), and selection for large males effective in defense of their females and offspring against predators

(Devore, 1963, 1971). However, neither seem important in the case of the great apes. First, there is little evidence of differential niche usage being of importance among primates (Clutton–Brock *et al.*, 1977), and certainly no evidence for the great apes (Fossey and Harcourt, 1977; Rodman, 1977; Wrangham, 1977). Second, although predator pressure might explain the difference in degree of sexual dimorphism in the gorilla as compared to the chimpanzee, it does not explain the difference between the orangutan and the chimpanzee: orangutan males are hardly ever close enough to a female to protect her or her offspring.

In summary, competition between males for access to females seems to be a key factor selecting for sexual dimorphism in apes, but other factors, especially attributes related to overall body size, are certainly also very important.

III.   REPRODUCTIVE BEHAVIOR

Six major differences in the reproductive behavior of the great apes will be considered. These are (1) the change in the relationships of males and females at estrus; (2) the role of the male in initiation of courtship; (3) the existence of a male courtship display; (4 and 5) the frequency and the duration of copulation; and (6) the duration of estrus. If the three species are ranked for these characteristics, the chimpanzee appears very different from the orangutan and gorilla (Table 1). After describing in more detail these behavior patterns, I will discuss how the differences between the three species, especially the contrast between the chimpanzee and the other two, can be related to differences between them in the number of males simultaneously competing for and available to estrous females. Finally, behavior in one-male and multimale cercopithecoid groups will be compared to seek comparable relationships with the great apes.

The data on the great apes' reproductive behavior on which the discussion is based are taken mostly from the previous three chapters. Throughout this section it must be borne in mind that unless stated otherwise, it is the behavior only of fully mature animals that is discussed.

A.   Reproductive Behavior of the Great Apes

1.   Change in Relationships at Estrus

In chimpanzee populations, obvious changes in relationships at estrus tend to be apparent. Not only do the usually separate sexes come

TABLE 1

Comparison of six components of ape reproduction behavior.

Behavior	Multimale	One-male	
	Chimpanzee	Gorilla	Orangutan
Change in relationships between ♂ and ♀ at estrus	1	3	2
Male's role in initiation of courtship	1	3	2
Presence of ♂ courtship display	1	2.5	2.5
Frequency of copulation (incidents per hour)[a]	1 (0.5)	2 (0.3)	3 (0.045)[b]
Duration of copulation (minutes)[a]	1 (0.125)	2 (1.5)	3 (10.75)
Length of estrus (days)[a]	1 (9.5)	3 (2.0)	2 (5.5)
Totals of ranked scores	6	15.5	14.5

Note: The three species are ranked in expression of the components from most (1) to least (3).

[a] Data in parentheses derived from Chapters 9–11. See text for supporting data for behaviors 1–3.

[b] Calculated value based upon 12-hour day; it is assumed that orangutans were not observed at night.

together, but also the male grooms the female, shares food with her, and when in consort persuades her to follow him away from all other members of the community (Chapter 9, Section III, A,4). Furthermore, other interactions within the community are altered. Males presumably spend a greater time together than usual when congregated around the estrous female, while the relationships of a consorting male with other males in the community are disrupted for the duration of his consortship. Likewise, in orangutan populations some alterations in the pattern of social interactions occur when a female comes into estrus because the normally separate adults come together. However, the alterations do not seem nearly so marked as in a chimpanzee community. In the first place, the animals are so solitary that there can be few established relationships to be disrupted when a male and female do consort. And second, the only obvious alteration in behavior of the males and females, besides the appearance of specifically sexual behaviors, seems to be an increase in their time together, although estrous female orangutans sometimes groom their male partners (Chapter 9, Section II, C; Section III, A). The same is true of the Virunga gorillas, where the only change during estrus is an increase in the time the male and estrous female spend in close proximity to one another (Harcourt, 1979a). No change in the grooming patterns of gorillas was seen when a female came into estrus.

2. *The Male's Role in Initiation of Courtship and the Presence of a Male Courtship Display*

The adult male chimpanzee's active role in initiation of courtship contrasts with the more passive role of the *fully adult* male orangutan and, especially, gorilla. In particular, orangutan and gorilla males show nothing like the male chimpanzee's courtship display. This is an impressive behavior pattern that precedes most copulations and involves obvious display of the erect penis, the pale color of which contrasts with the surrounding black fur. Aggressive components are apparent in the chimpanzee male's courtship display, and while similar sorts of behavior are occasionally performed before copulation by male orangutans and gorillas in captivity (MacKinnon, 1974; Nadler, 1976), their expression is very much more subdued than in the chimpanzee courtship displays.

So far, no mention has been made of the famous "rapes" observed in wild orangutan populations (Galdikas, Chapter 11; MacKinnon, 1974; Rijksen, 1978) and also in captivity (Nadler, 1977a). The reason is that in the wild the majority of rapes are performed by subadult, not adult, males. Galdikas, for example, saw only one rape of an adult female by a dominant adult male (Chapter 11). An important difference appears to exist between captive and wild adult orangutan males in this regard in that in captivity "essentially all copulations were initiated forcefully by the males" (Nadler, 1977a; p. 469). However, later observations, employing a slight change in experimental procedure, now confirm the importance of the female's role in initiation of copulation and courtship (Nadler, personal communication).

3. *Frequency and Length of Copulations during Estrus*

Chimpanzees copulate more frequently during estrus, and their copulations are completed more rapidly than is the case for gorillas and, especially, orangutans. As can be seen in Table 1, the dissimilarity is particularly obvious for the duration of copulations. Chimpanzees complete their copulation in roughly a tenth of the time it takes gorillas, and in a hundredth of the time orangutans take.

4. *Duration of Estrus*

All the contrasts in reproductive behavior that have been discussed so far relate largely to selection pressures on the males. The final behavioral difference that remains for consideration concerns the females: it is the duration of their estrous periods. The chimpanzee's period, at 10 days, is about twice as long as the orangutan's, and about five times as long as the gorilla's (Table 1). In addition, a chimpanzee female has a large and very obvious perineal swelling, while the swelling of the adult female gorilla is normally invisible in the wild, and the orangutan lacks an estrous swelling altogether.

B. Reproductive Behavior and Intermale Competition

Clearly the behavioral patterns shown in courtship and copulation differ greatly between the apes, with the chimpanzee, in particular, contrasting with the gorilla and orangutan. This section is a discussion of those differences in relation to interspecific variation in the selection pressures acting on the two sexes.

1. Intermale Competition

In the multimale chimpanzee community, where a number of males are simultaneously in the vicinity of and competing for any estrous female, the operative selection pressures might be characterized as follows: if a male does not initiate courtship by trying to establish social bonds with the female, the others nearby will do so. If a male does not indicate his vigor by energetic courtship displays and his readiness to mate by his erect penis, a female will choose another male who is so displaying his vigor. If a male copulates infrequently, others who mate often will have more chance to inseminate the female. If a male takes a long time copulating before ejaculation, he stands a greater risk of being interrupted before insemination by the throng of surrounding males than the male who ejaculates rapidly.

For the gorilla and orangutan, it is suggested that the proposed selection pressures are far less pronounced. In what can be thought of as essentially a one-male mating system, the dominant gorilla and orangutan male have far less need to compete by establishing bonds with the estrous female, initiating courtship, or displaying their vigor and readiness to mate, because there are so few adult males present from whom the estrous female can choose. This is especially true of the gorilla, where the male and female would not even be together at estrus unless the female had already chosen the male. Where other adult males are in the vicinity of the female (40% of gorilla groups in the Virunga population had more than one adult male), the established difference in status between the males will, it is suggested, effectively remove any element of choice that might have existed for the female.

In orangutan populations, because the ranges of dominant males overlap, some females in estrus might have a choice of more than one dominant male. Perhaps this slightly greater element of choice for the orangutan female, as compared to the female gorilla, explains the slightly greater initiative shown by the orangutan male than the gorilla male in maintenance of proximity and initiation of copulations. In addition, the slow and infrequent copulations of the gorilla and orangutan can also be explained by the same reasoning: in the one-male mating systems of these two species, the selection pressures enhancing competitive ability for rapid insemination of the estrous female are not nearly so great as in the multimale chimpanzee community.

Therefore, the predominant selection pressures on orangutan and especially, gorilla males are proposed to be those that encourage competition for permanent access to females rather than for access only at the time of estrus; male chimpanzees, on the other hand, compete for access mainly during estrus. In other words, male gorillas and orangutans have been selected primarily for competence in fighting to establish long-term "ownership" of females, whereas male chimpanzees have been selected more for competence in attracting and inseminating estrous females.

The hypothesis just advanced in effect relates differences among the great apes in their reproductive behavior to differences among them in the level of competition between males for females at the time of estrus. The more males present, the more competition there is. If this interpretation of the selection pressures acting to produce the observed interspecific differences is correct, there is no reason why similar contrasts should not be seen within species. One would expect, for example, that subadult and subordinate adult males, who face more competition from dominant males than dominant males do from them, should show more active initiation of courtship and copulation and perform more frequent and rapid matings. An obvious, though extreme, example of such an intraspecific difference matching the described interspecific ones are the "rapes" seen in orangutan populations, the majority of which are performed by subadult males (Chapter 11, Section III, C). Where adult males are involved, I would predict that in most cases they would be subordinate adults. It must be said that insemination is not necessarily the only function of rape behavior. Rijksen (1978), for example, suggests that it might serve as a means by which a young male can assert his dominance over females. As a further possible instance of the expected intraspecific relation between level of competition and nature of reproductive behavior, both Galdikas (Chapter 11, Section II, A) and Tutin and McGinnis (Chapter 9, Section II, D) report an association between age of males and frequency of copulation. Specifically, Galdikas reports that subadults copulate more frequently than adults, and Tutin and McGinnis report that the highest hourly rates were shown by juvenile and adolescent males. Unhappily, the data for the gorilla are simply not sufficient to see if similar relationships exist in that species.

All the contrasts in reproductive behavior that have been discussed so far relate to selection pressures acting largely on males. However, sexual selection involves an element of female choice, as well as direct competition between males. The interspecific differences in the length of estrous periods seen in the apes are perhaps related to this element of female choice. In the multimale chimpanzee community, a long period of estrus and of enlarged sexual swelling may allow a female more time to attract a number of males, thus increasing intermale competition and creating the opportunity for female choice (Harcourt and Stewart, 1977, Nadler, 1977b). Choice may be expressed actively by approaching and consorting with a preferred mate or passively by waiting for one chimpanzee to defeat his opponents (see Cox and Le Boeuf (1977) for a discussion of mechanisms of female choice). By contrast, an estrous gorilla has no choice available. Therefore, estrus is only long enough to ensure copulation during the time of ovulation, and there is no need for a sexual swelling to attract the male because he is rarely more than 50–100 m away.

The dispersal and antagonism of male orangutans permit some opportunity for estrus females to approach preferred mates or to avoid others, although the choice is never large. The dispersal of male orangutans might explain the longer period of estrus of orangutans compared with gorillas; the longer period of estrus in the orangutan may actually be necessary to enable the female to locate a distant male before ovulation occurs. The lack of a sexual swelling in the female orangutan may reflect the fact that the mutual intolerance of males prohibits their congregation around the estrous female. Hence, the expression of female choice takes a different form from the chimpanzee. It may be noted that the absence of sexual swelling could also be related to its ineffectiveness as a visual attractant in the dense rain forest (Nadler, 1977a,b).

2. *Accessibility of Mates*

In this chapter it has been suggested that interspecific variation in reproductive behaviors among the Pongidae can be functionally related to differences among the three species in the number of males competing for estrous females. However, Nadler (1977) related the ob-

served variation in reproductive behavior to interspecific differences in
the "accessibility" of mates, by which he meant their usual proximity.
Although these two points of view might seem similar, in fact they
lead to different predictions. For example, the "accessibility" hypothesis
requires the orangutan male to play a larger role than the male chimpanzee in initiation of copulations, while the "numbers" hypothesis
presented here requires the reverse. Because of contrasts between data
from captivity and the wild, especially with respect to the sexual behavior of orangutans, differences between the two hypotheses at the
moment cannot be resolved satisfactorily. Nevertheless, the "numbers"
hypothesis seems to explain a wider range of phenomena than does
the "accessibility" hypothesis, in regards to both the generality of the
predictions outside the Pongidae (see next section) and to the number
of behaviors interpretable by it. However, this is certainly not to say
that accessibility is unimportant as a factor controlling the details of a
species' reproductive behavior. The difference between the gorilla and
orangutan in length of estrus, for example, seems more readily explained by differences in accessibility of mates than in number of males
competing.

C.   Comparisons with Other Primates

Corresponding differences in reproductive behavior to those found
between the chimpanzee on the one hand and the gorilla and orangutan
on the other are seen between multimale and one-male cercopithecoid
groups. Using the gelada (*Theropithecus gelada*) and patas monkeys
(*Erythrocebus patas*) as examples of one-male mating systems, and *Papio*
and *Macaca* as examples of multimale systems within the Cercopithecidae, comparisons of these species' behavior will be made in the same
order as they were for the Pongidae.

In neither the gelada nor the patas is there much change in relationships between males and females when a female comes into estrus
(Dunbar, 1978a,b; Rowell, 1978). By contrast, in multimale *Papio* and
Macaca groups, males and estrous females not only increase their time
in close proximity, but also the male's grooming of the estrous female
increases (Hausfater, 1975; Rowell, 1968; Saayman, 1970; Seyfarth, 1978;
Southwick *et al.*, 1965), and *Papio* males become more responsible for
maintenance of proximity (Hausfater, 1975; Saayman, 1970; Seyfarth,
1978).

Despite technical differences between studies, it is possible to conclude that gelada and patas females are more active in initiation of

sexual interactions than the males, whereas data from Hausfater (1975), Rowell (1972), and Saayman (1970) suggest that the reverse is the case in multimale *Papio* groups. For example, gelada females were found to initiate 66% of sexual interactions (Dunbar, 1978b), compared to a value of only 34% for *Papio ursinus* females (Saayman, 1970). Differences in rates of copulation also appear to be in the same direction as those found among the great apes; the male of a one-male gelada unit, at 0.55 copulations per hour, has a lower rate than does a male of a multimale *Papio* group, with 1.19 copulations per hour (Dunbar, 1978b; see also Collins, 1978).

Finally, there is a close association between possession of large sexual swellings and membership of multimale groups; swellings are most common in the three genera that typically live in multimale groups (*Papio, Macaca,* and *Cercocebus*), while in species that live in one-male groups or monogamously, swellings tend to be absent (Clutton–Brock and Harvey, 1976).

This comparison of cercopithecoids and apes shows that classification of social organizations into one-male and multimale mating systems reveals similar trends in the two groups. The relation of courtship and copulatory behavior, to social structure and pattern of competition for mates, first noted with great apes, thus seem generalizable to other primates.

IV. SUMMARY

Marked interspecific variation exists among the great apes in their reproductive behaviors. In particular, the chimpanzee shows more male initiation of copulation, more obvious male courtship displays, more frequent and rapid copulations, and longer estrous periods than do the gorilla and orangutan. Variation among the three species in body weight could be part of the explanation for these dissimilarities, as it could for the existing interspecific differences in degree of sexual dimorphism. However, it is suggested here that a more important factor is the contrast between the species in the number of males simultaneously competing for and available to estrous females. This is a major difference separating the multimale mating system of the chimpanzee from the one-male systems of the orangutan and gorilla, and it can account for many of the observed contrasts between the species in the behaviors involved in mate acquisition, courtship, and copulation. In effect, competition for females *in estrus* is more intense in the multimale than the one-male system, whereas competition to maintain long-term access to females is greater in the one-male system. Hence, selection for abilities to maintain permanent possession of females or their ranges, it is argued, is greater in the orangutan and gorilla than in the chimpanzee, while selection for abilities to attract and rapidly inseminate estrous females is greater in the chimpanzee.

CONCLUSION

The fine details of the apes' reproductive behaviors are closely related to the overall social organization of each species. This, in turn, is related to the species' morphology, the nature of the habitat, and the animals' adaptations to it (Clutton–Brock, 1974; Crook, 1965, 1970; Eisenberg *et al.*, 1972; Emlen and Oring, 1977; Geist, 1974; Jarman, 1974; Jolly, 1972; Kaufmann, 1974; Lack, 1968; Orians, 1969; Selander, 1972; Verner and Willson, 1966; Wrangham, 1979). Studies in the wild provide a breadth of knowledge of a species' social organization and, especially, of individuals' adaptations to the environment that is very difficult to gain from captivity. Such studies are, therefore, of vital importance. Without them we cannot appreciate the artificiality of the captive environment and the perturbations it might produce in behavior. Only through studies in the natural environment can the normal social and environmental settings of a species be known, and so the details of their behavior be fully understood.

ACKNOWLEDGMENTS

This chapter was considerably improved by criticism from Dr. J. C. Brown, Dr. T. H. Clutton–Brock, Dr. R. I. M. Dunbar, Dr. C. E. Graham, R. A. Harcourt, Professor R. A. Hinde, and Kelly J. Stewart. I am very grateful to them all for their help. I would also like to thank Professor R. V. Short for initiating so many of the trains of thought presented here.

REFERENCES

Bateman, A. J. (1948). *Heredity* **2,** 349–368.
Bygott, J. D. (1979). *In* "The Great Apes" (D. A. Hamburg and E. McGown eds.), pp. 404–427. Benjamin/Cummings Publ., Menlo Park, California.
Caro, T. M. (1976). *Anim. Behav.* **24,** 889–897.
Clutton–Brock, T. H. (1974). *Nature (London)* **250,** 539–542.
Clutton–Brock, T. H., and Harvey, P. H. (1976). *In* "Growing Points in Ethology" (P. P. G. Bateson and R. A. Hinde, eds.), pp. 195–237. Cambridge Univ. Press, London and New York.
Clutton–Brock, T. H., and Harvey, P. H. (1977). *J. Zool.* **183,** 1–39.
Clutton–Brock, T. H., Harvey, P. H., and Rudder, B. (1977). *Nature (London)* **269,** 797–800.

Collins, A. (1978). *New Sci.* **78,** 12–14.

Cox, C. R., and Le Boeuf, B. J. (1977). *Am. Nat.* **111,** 317–335.

Crook, J. H. (1965). *Symp. Zool. Soc. London* **14,** 181–218.

Crook, J. H. (ed.) (1970). *In* "Social Behaviour in Birds and Mammals," pp. 103–166. Academic Press, London.

Crook, J. H. (1972). *In* "Sexual Selection and the Descent of Man" (B. Campbell, ed.), pp. 231–281. Heinemann, London.

Davenport, R. K. (1967). *Folia Primatol.* **5,** 247–263.

DeVore, I. 1963). *In* "Classification and Human Evolution" (S. L. Washburn, ed.), Vol. 37, pp. 301–319. Viking Fund Publications in Anthropology, New York.

DeVore, I. (9171). *In* "Man and Beast: Comparative Social Behavior" (J. F. Eisenberg and W. S. Dillon, eds.), pp. 297–312. Smithsonian Institute Press, Washington, D.C.

Dunbar, R. I. M. (1978a). *In* "Recent Advances in Primatology" (D. J. Chivers and J. Herbert, eds.), pp. 491–492. Academic Press, London.

Dunbar, R. I. M. (1978b). *Anim. Behav.* **26,** 167–178.

Eisenberg, J. F. Muckenhirn, N. A., and Rudran, R. (1972). *Science* **176,** 863–874.

Emlen, S. T., and Oring, L. W. (1977). *Science* **197,** 215–223.

Fisher, R. A. (1958). "The Genetical Theory of Natural Selection." Dover, New York.

Fossey, D. (1972). "The Marvels of Animal Behavior," pp. 208–229. National Geographic Society, Washington, D.C.

Fossey, D. (1974). *Anim. Behav.* **22,** 568–581.

Fossey, D., and Harcourt, A. H. (1977). *In* "Primate Ecology: Studies of Feeding and Ranging Behaviour in Lemurs, Monkeys and Apes" (T. H. Clutton–Brock, ed.), pp. 415–447. Academic Press, London.

Galdikas, B. M. F. (1979). *In* "The Great Apes" (D. A. Hamburg and E. R. McGown eds.), pp. 194–233. Benjamin/Cummings Publ., Menlo Park, California.

Geist, V. (1974). *Am. Zool.* **14,** 205–220.

Goodall, J. van Lawick (1968). *Anim. Behav. Monogr.* **1,** 161–311.

Goodall, J. van Lawick (1975). *In* "Hominisation und Verhalten" (G. Kurth and I. Eibl-Eibesfeldt, eds.), pp. 74–136. Gustav Fischer Verlag, Stuttgart.

Goodall, J. *et al.* (1979). *In* "The Great Apes" (D. A. Hamburg and E. R. McGown, eds.), pp. 12–53. Benjamin/Cummings Publ., Menlo Park, California.

Hamilton, W. D. (1964). *J. Theor. Biol.* **7,** 1–52.

Harcourt, A. H. (1978). *Z. Tierpsychol.* **48,** 401–420.

Harcourt, A. H. (1979a). *Anim. Behav.* **27,** 325–342.

Harcourt, A. H. (1979b). *In* "The Behavior of Great Apes" (D. A. Hamburg and E. R. McGown, eds.), pp. 186–192. Benjamin/Cummings Publ., Menlo Park, California.

Harcourt, A. H., and Stewart, K. J. (1977). *New Sci.* **76,** 160–162.

Harcourt, A. H., Stewart, K. J., and Fossey, D. (1976). *Nature (London)* **263,** 226–227.

Hausfater, G. (1975). *Contrib. Primatol.* **7,** 1–150.

Horr, D. A. (1975). *In* "Primate Behavior" (L. A. Rosenblum, ed.), Vol. 4, pp. 307–323. Academic Press, New York.

Jarman, P. J. (1974). *Behaviour* **48,** 215–267.

Jolly, A. (1972). "The Evolution of Primate Behavior." Collier Macmillan, London.

Kaufmann, J. H. (1974). *Am. Zool.* **14,** 51–62.

Kawanaka, K., and Nishida, T. (1975). *Symp. Congr. Int. Primatol. Soc., 5th* 173–186. Japan Science Press, Tokyo.

Lack, D. (1968). "Ecological Adaptations for Breeding in Birds." Methuen, London.

Leutenegger, W., and Kelly, J. T. (1977). *Primates* **18,** 117–136.

MacKinnon, J. (1974). *Anim. Behav.* **22,** 3–74.

MacKinnon, J. (1979). *In* "The Great Apes" (D. A. Hamburg and E. R. McGown, eds.), pp. 256–273. Benjamin/Cummings Publ., Menlo Park, California.

Nadler, R. D. (1976). *Arch. Sex. Behav.* **5,** 487–502.

Nadler, R. D. (1977a). *Arch. Sex. Behav.* **6,** 457–475.

Nadler, R. D. (1977b). *In* "Progress in Ape Research" (G. H. Bourne, ed.), pp. 191–206. Academic Press, New York.

Orians, G. H. (1969). *Am. Natur.* **103,** 589–603.

Pusey, A. (1979). *In* "The Great Apes" (D. A. Hamburg and E. R. McGown, eds.), pp. 464–479. Benjamin/Cummings Publ., Menlo Park, California.

Ralls, K. (1977). *Am. Natur.* **111,** 917–938.

Rijksen, H. D. (1978). "A Field Study on Sumatran Orangutans." Veenman and Zonen, Wageningen, Holland.

Rodman, P. S. (1973). *In* "Comparative Ecology and Behaviour of Primates" (R. P. Michael and J. H. Crook, eds.), pp. 171–209. Academic Press, London.

Rodman, P. S. (1977). *In* "Primate Ecology: Studies of Feeding and Ranging Behaviour in Lemurs, Monkeys and Apes" (T. H. Clutton–Brock, ed.), pp. 384–413. Academic Press, London.

Rowell, T. E. (1968). *Anim. Behav.* **16,** 585–588.

Rowell, T. E. (1972). *Adv. Study Behav.* **4,** 69–105.

Rowell, T. E. (1978). *In* "Recent Advances in Primatology" (D. J. Chivers and J. Herbert, eds.), pp. 489–490. Academic Press, London.

Saayman, G. S. (1970). *Folia Primatol.* **12,** 81–110.

Schaller, G. B. (1963). "The Mountain Gorilla: Ecology and Behavior," Univ. of Chicago Press, Chicago, Illinois.

Selander, R. K. (1966). *Condor* **68,** 113–151.

Selander, R. K. (1972). *In* "Sexual Selection and the Descent of Man" (B. Campbell, ed.), pp. 180–230. Heinemann, London.

Seyfarth, R. M. (1978). *Behaviour* **64,** 204–247.

Short, R. V. (1979). *Adv. Study Behav.* **9,** 131–158.

Southwick, C. H., Beg, M. A., and Siddiqi, M. R. (1965). *In* "Primate Behavior: Field Studies of Monkeys and Apes" (I. DeVore, ed.), pp. 111–159. Holt, New York.

Sugiyama, Y. (1973). *In* "Comparative Ecology and Behaviour of Primates" (R. P. Michael and J. H. Crook, eds.), pp. 376–410. Academic Press, London.

Trivers, R. L. (1972). *In* "Sexual Selection and the Descent of Man" (B. Campbell, ed.), pp. 136–179. Heinemann, London.

Verner, J., and Willson, M. F. (1966). *Ecology* **47,** 143–147.

Wiley, R. H. (1974). *Q. Rev. Biol.* **49,** 201–227.

Wilson, E. O. (1971). *In* "Man and Beast: Comparative Social Behavior" (J. F. Eisenberg and W. S. Dillon, eds.), pp. 181–218. Smithsonian Institute Press, Washington, D.C.

Wilson, E. O. (1975). "Sociobiology: The New Synthesis." Belknap Press, Cambridge and London.

Wrangham, R. W. (1977). *In* "Primate Ecology: Studies of Feeding and Ranging Behaviour in Lemurs, Monkeys and Apes" (T. H. Clutton–Brock, ed.), pp. 504–538. Academic Press, London.

Wrangham, R. W. (1979). *Biol. Soc. Life* **18,** 335–368.

Chapter 13

SEXUAL SELECTION IN MAN AND
THE GREAT APES

R. V. Short

I. INTRODUCTION

To a biologist interested in the study of human reproduction, the fascination of the great apes lies not so much in their close similarities to us, but rather in the ways in which their reproductive processes differ from our own. After all, we now know that the genetic distance between man and the great apes, as assessed by the amino acid sequence of blood proteins, is no greater than that observed, for example, between sibling species of the fruit fly *Drosophila* (Bruce and Ayala, 1978). It is an attractive hypothesis to suggest that there must have been extreme conservation of the structural genes during evolution,

319

Copyright © 1981 by Academic Press, Inc.
All rights of reproduction in any form reserved.
ISBN 0-12-295020-8

whereas the regulatory genes, which control the expression of characteristics, are the ones upon which selection has operated (King and Wilson, 1975). Thus, it should come as no surprise to find that the qualitative endocrine changes seen throughout life are so similar in man and the great apes; for example, estriol is the major urinary estrogen metabolite excreted during pregnancy in all these species (Lasley *et al.*, 1980; Wright *et al.*, Chapter 7, this volume) in contradistinction to other primates, and there are also marked similarities in the subunit structure of human and great ape chorionic gonadotrophins (Hodgen, 1979). An analysis of the endocrine events of puberty in children and chimpanzees reveals an almost identical pattern (Winter *et al.*, 1980; Hobson *et al.*, Chapter 4, this volume), to the extent that one wonders whether the tables should not be turned; man might be the best experimental model for research on the great apes, bearing in mind the superabundance of the former species and the endangered status of the latter.

In contrast to these marked structural similarities between man and the great apes, there are pronounced differences in the ways in which the reproductive hormones are put to use. For example, all the great apes show estrus, whereas the human does not. We are the only species in which the breasts develop at puberty; in the great apes, there is no noticeable mammary development until toward the end of the first pregnancy (Short and Drife, 1977). The chimpanzee and to a lesser extent the gorilla show perineal tumescence at time of ovulation, whereas the human and orangutan do not (Short, 1979a). It is these fascinating differences that surely offer us some of the most exciting clues about our own evolution.

The usual justification for great ape research has been that study of our closest living relatives will be particularly relevant to our understanding of ourselves. The statement is in fact true, but perhaps not in the sense in which it is usually meant. For example, it is unrealistic to imagine that we will unravel the endocrinology of the human menstrual cycle by studying that of the chimpanzee. As Alexander Pope said, "The proper study of mankind is man." Humans not only are more available, but also are infinitely more cooperative than apes; human reproductive biology is always likely to be pioneered in the human, and it is seldom that the basic facts will first be revealed in an ape, and subsequently extrapolated to man.

What then are the biomedical justifications for carrying out any research on the great apes? Are we even entitled to keep these endangered species in captivity just to satisfy our idle scientific curiosity, if thereby we accentuate the depletion of stocks in the wild? Although breeding populations may survive in captivity after the great apes have been exterminated in their natural habitats, it is unrealistic to imagine that these captive animals will provide a nucleus for subsequently reestablishing the species in the wild. The basic reason for the animal's disappearance from the wild in the first place is not its poor reproductive performance or its ill health, but human destruction of its habitat and human predation, two almost uncontrollable and irremediable events.

But if the only great apes to survive in the twenty-first century are to be found in zoological gardens and laboratories, we will have lost forever many of the vital behavioral clues that explain why these animals are built the way they are. The great lesson of evolution is that form reflects function. Bequeathed only a caged, or worse still, a stuffed gorilla, we would have no way of determining the adaptive significance of any of its anatomical features, and hence we could not begin to understand why it was so different in appearance from ourselves, while genetically so closely related to us. Surely, one of the justifications for great ape research is that if we can begin to correlate form with function in our closest relatives, then by analogy we can gain new insights into the selective forces that have made humans what they are today. Nowhere is this better illustrated than in the case of sexual selection.

II. WHAT IS SEXUAL SELECTION?

In 1859, Charles Darwin first put forward his ideas in "The Origin of Species by means of Natural Selection, or the Preservation of Favoured Races in the Struggle for Life," but he soon realized that natural selection must mean more than mere survival of the fittest if evolution was to take place; the chosen few must also be fertile, and capable of handing on their attributes to succeeding generations. Thus it was in 1871 that Darwin produced his second great work, "The Descent of Man, and Selection in Relation to Sex." In it, he clearly distinguishes between sexual selection and natural selection:

> Sexual Selection depends on the success of certain individuals over others of the same sex, in relation to the propagation of the species; whilst Natural Selection depends on the success of both sexes, at all ages, in relation to the general conditions of life. The sexual struggle is of two kinds; in the one it is between the individuals of the same sex, generally the males, in order to drive away or kill their rivals, the females remaining passive; whilst in the other, the struggle is likewise between the individuals of the same sex, generally the females, which no longer remain passive, but select the more agreeable partners.

Darwin also appreciated that the effects of sexual selection would depend very much on the mating system of the species in question; in polygynous species, where the males mated with more than one female, the competition between males would therefore be enhanced, leading to an exaggerated development of those secondary sexual characters used in intermale aggressive encounters. These concepts are summarized in Figure 1.

Darwin was obviously fascinated by the comparison between the great apes and man himself, and made frequent reference to this in

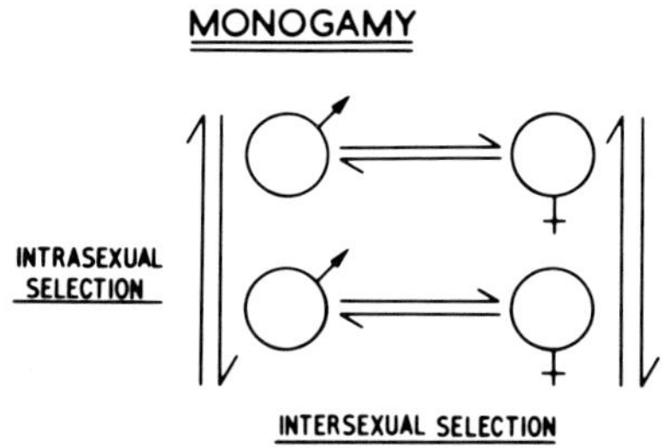

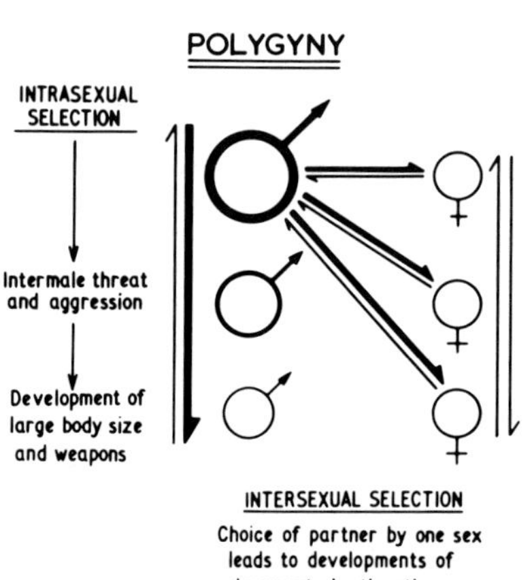

Fig. 1. Effects of sexual selection in a monogamous and a polygynous mating system. In the latter, competition between males is enhanced, leading to sexual dimorphism in body size, and perhaps the development of offensive weapons. Intersexual selection also may lead to the development of adornments aimed at attracting the opposite sex. (Reproduced with permission of Academic Press; from Fig. 1, "Sexual selection and its component parts, somatic and genital selection, as illustrated by man and the great apes," in *Advances in the Study of Behavior* **9**, p. 134.)

"The Descent of Man." He cites Savage and Wyman's (1847) discovery of the gorilla in Gabon, and their secondhand account of its behavior, gleaned from the local natives. This meant much to Darwin, as it was a perfect illustration for him of male behavior in a polygynous species:

> Therefore, looking far enough back in the stream of time, and judging from the social habits of man as he now exists, the most probable view is that he aboriginally lived in small communities, each with a single wife, or if powerful with several, whom he jealously guarded against all other men. Or he may not have been a social animal, and yet have lived with several wives, like the gorilla; for the natives "agree that but one adult male is seen in a band; when the young male grows up, a contest takes place for mastery, and the strongest, by killing and driving out the others, establishes himself as the head of the community.

Darwin was also intrigued by the varying degrees of sexual dimorphism exhibited by man and the great apes:

> The sexes of the chimpanzee are not as different as those of the orang or gorilla . . . As the prodigious difference between the skulls of the two sexes in the orang and gorilla stands in close relation with the development of the immense canine teeth in the males, we may infer that the reduction of the jaws and teeth in the early male progenitors of man must have led to a most striking and favourable change in his appearance.

Darwin's exciting concept of sexual selection lay fallow until taken up by R. A. Fisher in 1930, in his classical work "The Genetical Theory of Natural Selection." Fisher addressed himself to the question of how nature went about producing sexual dimorphisms, whereby the male could differ so markedly from the female in the expression of the physical characteristics used in intermale threat and aggression. Clearly, there were two possible solutions. The simplest would be to locate all the genes regulating male secondary-sexual characteristics in mammals on the sex-specific chromosome, namely, the Y chromosome. Although this would ensure that these characters could only be expressed in the male, it would have the distinct disadvantage that the Y chromosome would have to contain an enormous amount of genetic information, and it is difficult to see how during the course of evolution genes for newly evolving sexually dimorphic characteristics could become Y-linked, since there is very little crossing-over between the Y chromosome and autosomes, or even between the Y and X chromosomes.

The other more subtle alternative, which Fisher pointed out was the one that nature in fact had adopted, was to use the Y chromosome in mammals merely to code for the presence of a testis, and then to use the hormones secreted by the male gonad to activate genes that regulated the expression of sexually dimorphic characteristics. By making these genes *sex limited*, as opposed to *sex linked*, it was thus possible to open up the entire autosomal complement for exploitation in sexual dimorphism.

This concept of sex-limited regulation of sexual dimorphism is far reaching, and yet it seems to have escaped the attention of evolutionary biologists. It tells us, for example, that the structure of the testes themselves is likely to be determined initially by genes located on the *Y* chromosome, whereas any somatic characteristic by which the male differs from the female will probably be hormonally determined. We will see in a moment how this helps us to understand the differences between man and the great apes.

One aspect of sexual selection that has received insufficient attention in the past concerns the differential effects that it may have on the testis and the ovary. Since the gonads are endocrine glands releasing steroid hormones into the systemic circulation, the volume of the endocrine component is likely to vary in relation to the total body mass of the animal. But the principal function of the gonads is to release gametes to the exterior, and here there are fundamental differences in the constraints upon an ovary and a testis. Only a minute proportion of the total population of oocytes in the ovary at birth are ever ovulated; the majority perish through atresia. Since the oocytes of all mammals are remarkably similar in size, major differences between species in body size, life span, or ovulation rate do not necessitate any alteration in the volume of germinal tissue in the ovary. Thus, we might expect the volume of endocrine tissue in the ovary to vary with body size, but the volume of germinal tissue to remain relatively constant and to be completely independent of the mating system.

But the factors governing testicular size are entirely different. Although the size of the individual spermatozoon is relatively constant between mammals, large animals have to produce many more spermatozoa than small ones if only because of the greater volume of the female tract through which the ejaculate becomes distributed. But even more important than this is the frequency of ejaculation. Species that have to ejaculate several times a day will need a far higher rate of sperm production that those that may only ejaculate once or twice a year. The position is complicated still further by gamete selection, which will occur in cases of promiscuity; if more than one male copulates with a female at one estrus, the male who deposits the largest number of spermatozoa in her tract will, other things being equal, be more likely to sire the offspring, and hence will be more favored by selection. In summary, therefore, both the endocrine and gametogenic components of the testis would be expected to vary with body size, and in addition there will be overriding effects of the mating system and copulatory frequency on the volume of gametogenic tissue.

How can the volume of gametogenic tissue in the testis be made to vary independently of that in the ovary? Hayward and Shire (1974) have shown that there is a Y-linked gene (or genes) in the mouse that determines the volume of seminiferous tubular tissue laid down in the prepubertal testis, and this accounts for 41% of the difference in testis size between inbred strains. If a gene is sex-linked in one mammal, it is likely to be sex-linked in all, since there is little crossing over between the X and Y chromosome, or between the sex chromosomes and the autosomes (Ohno, 1967). A Y-linked gene would also respond extremely rapidly to selection pressure, since it is always in the haploid state and is only expressed in the male. Thus, we might expect testis size to have a high heritability; data from the bull show that scrotal circumference, an indirect measure of testis size, in fact has a heritability of 0.67 (Coulter *et al.*, 1976).

What factors regulate the rate of sperm production by the testis? The frequency of ejaculation appears to be without effect on the timing of the spermatogenic cycle, so the only way of increasing sperm production is to increase the volume of seminiferous tubular tissue. Since seminiferous tubule diameters are similar in all mammals, this can be achieved only by increasing the number and/or length of individual tubules. Amann *et al.* (1976) have shown that the rate of sperm production in the rhesus monkey is 23×10^6 sperm per gram of testicular tissue per day, whereas the corresponding figure for the human testis is only 4.4×10^6 (Amann and Howards, 1980). This difference is probably accounted for by the much greater percentage of tubular tissue in the rhesus testis; macaques have the highest testes:body weight ratios of all the primates (Schultz, 1938). That increased testicular size, as a result of an increased mass of tubular tissue, does indeed result in an increased daily rate of sperm production has been convincingly demonstrated in the bull by Hahn *et al.* (1969).

Sexual selection may also tell us something rather interesting about the distinction that we commonly draw between *primary* sexual characteristics, by which we mean the basic differences between the male and female reproductive tracts as laid down in early fetal life, and *secondary* sexual characteristics, which become evident only during puberty and adulthood. After the initial genetically determined event, the development of a testis or an ovary, testicular hormones take over the responsibility for determining the other primary sexual characteristics. Thus, testicular androgens from the Leydig cells stimulate the development of Wolffian duct derivatives, and a peptide derived from the Sertoli cells inhibits the development of Mullerian duct derivatives (Short, 1979b). These early stages of hormonally induced sexual differentiation are beyond the immediate reach of sexual selection, and are a basic ontogenetic feature that is common to all mammals. The secondary sexual characteristics, on the other hand, should be viewed as much more recent adaptations that are a direct consequence of sexual selection. It is probably no accident that the female reproductive tract remains internalized, hidden from view, and hence largely untouched by sexual selection, whereas much of the male reproductive tract has become externalized, so that the penis and scrotum are fully displayed and are often extensively modified in response to sexual selection. A penis is a *sine qua non* for reproduction in male mammals, so it is laid down in early fetal life as part of the basic process of primary sexual differentiation of the male reproductive tract. But the ultimate shape and form that the penis will take in adult life will also depend on the extent to which it is used in display, and thus it may undergo extensive androgen-dependent secondary modification at the time of puberty.

After Fisher, there seems to have been a lull in the development of our ideas about evolution and sexual selection, but there has been a recent reawakening in the guise of sociobiology. Possibly because the great population geneticists of the twentieth century, like Fisher and Sewall Wright, were laboratory-based and mathematically orientated, it took a long time before people like W. D. Hamilton and E. O. Wilson drew attention to the fact that behavior was just as much subject to evolution as any bodily characteristic. Many laboratory-based scientists still have considerable difficulty in appreciating the fact that the adaptive significance of an animal's behavior can only be understood in the context in which that behavior originally evolved—the wild. It is the animal's natural habitat that has been the threshing floor of evolution. In the years to come, we will increasingly appreciate the great debt we owe to that small group of ethologists who have so painstakingly documented the behavior of wild animals. Nowhere is this more apparent than in the case of the great apes, in which the pioneer studies of British, American, and Japanese biologists have given us new and important insights into the reproductive and social behavior of chimpanzees, gorillas, and orangutans in their natural habitats (Hamburg and McCown, 1979; Wrangham, 1979; Short and Weir, 1980). Soon, it will no longer be possible to collect this type of information, and yet it represents the key to our understanding of primate evolution. We are now in the fortunate position of having enough facts available to begin to make sense of what was hitherto an insoluble paradox: genetically and biochemically, man and the great apes are incredibly similar, and yet phenotypically and behaviorally they are so vastly different. The key to this paradox must lie in the reason for the basic behavioral differences. The behavioral differences in their turn have regulated the nature and extent to which sexual selection has operated, and it is sexual selection that has accounted for many of the more spectacular phenotypic differences that separate man and the apes.

III. THE EVOLUTION OF SOCIAL BEHAVIOR IN THE GREAT APES

We now have a reasonable inventory of the social and reproductive behavior of orangutans, chimpanzees, and gorillas in the wild (Hamburg and McCown, 1979; Wrangham, 1979; Short and Weir, 1980), and it is obvious that the mating systems of the three species are very different. Orangutans are solitary animals, and the male and female only associate with one another when the female is in estrus (Galdikas, 1979 and Chapter 11, this volume). Chimpanzees live in large, mixed-sex communities, and the mating system is essentially promiscuous with little competition between males over access to an estrous female; however, consortships can sometimes occur between one male and a particular female for the duration of her estrus (Tutin, 1980; Tutin and McGinnis, Chapter 9, this volume). Gorillas are strictly polygynous, with one mature male constantly traveling with a small group of females and defending them at all times against rival adult males (Harcourt *et al.*, 1980; Harcourt *et al.*, Chapter 10, this volume). So the first question we must ask ourselves is why should there be such extreme behavioral diversity between three such closely related species of great ape?

Wrangham (1979) has advanced a most plausible explanation. He argues that the basic limiting resource throughout evolution has been the availability of food; since the female has the greatest energy investment in reproduction because of the increased nutritional demands of pregnancy and lactation, it has been particularly essential for her to optimize her food supply. The male is a direct competitor with her for this resource, and so she has had to evolve a reproductive strategy that takes account of this fact, maximizing her own food availability at the expense of the male, while permitting him access to her in a way that optimizes her reproductive success. Thus, Wrangham suggests that it is the female rather than the male that holds the key to the evolution of social behavior, at least as far as the great apes are concerned. In practice, the system seems to work as follows.

The orangutan is basically an arboreal frugivore; it has the disadvantage of being very large for an arboreal primate, and hence it needs access to an abundant food supply. Since the density of fruiting trees or those with edible leaves and bark is not particularly high in the thick forest that these animals inhabit, each female needs a large core area to support her nutritional needs. The area is far too large for her to patrol on a regular basis, since arboreal primates are slow movers. A permanent male consort sharing and defending her core area would be a distinct disadvantage for her, so her optimal reproductive strategy has been to live in isolation, seeking the company of the male by proceptive behavior only when she is in estrus. For his part, the male must occupy a large core area (he is twice the size of the female) in a region where females are present; he must advertise his presence to them by long calls, which can be heard at a great distance, since there is almost zero visibility in the forest canopy. He must also try to exclude other adult males from his core area.

The chimpanzee is an omnivore and is both terrestrial and arboreal, living in a much more open habitat where vision becomes important. However, the food availability is relatively low, so each female needs a large core area to support her needs. Although terrestrial locomotion allows her to cover far greater distances in a day, it would still be impossible for the female to patrol her core area, and she cannot afford to share it with a male, who might defend it for her. Thus, like the female orangutan, her optimal reproductive strategy has been to remain apart from the male except when she is in estrus. Shortage of food prevents the male chimpanzee from sharing a female's core area, but at the same time he cannot defend a territory that embraces the core areas of several females, since he could never patrol its vast boundaries. Males therefore have evolved a pattern of collaborative defense of a very large area, encompassing the core area of numerous females. The male troop can travel rapidly to any of the territorial boundaries where sight, sound, or smell suggest an invasion by rival males, or even a strange female. A necessary condition for this social cooperation between males has been a sharing of the sexual favors when any of the females comes into estrus, and so the chimpanzee has a promiscuous mating system.

Finally, we come to the gorilla, which is predominantly terrestrial, and feeds on rather low-quality herbage of which there is usually an abundant supply. Thus, the female does not have to be particularly concerned about competititon for food from other animals, either male or female; there is no need to exclude others from her relatively large core area, of which she only utilizes a small part on any given day. Her optimal reproductive strategy therefore is to establish a permanent consortship with a male, who will defend a moving territory around her at all times. For his part, the male cannot afford to devote his entire attention to a single female, when he can maximize his reproductive success by collecting and defending a moving territory around three or four adult females. If he dies, the group fragments. The mating system is thus a typically polygynous one.

IV. THE EFFECT OF SOCIAL BEHAVIOR ON SEXUAL SELECTION IN THE GREAT APES

A. The Orangutan

The mating system of orangutans is basically a polygynous one, since one male mates with several females, and there is intense intermale competition. This has had a number of interesting consequences for the sexual development of the species (Short, 1979a).

1. Somatic Development

Male orangutans are about twice the size of females. Eckhardt (1975) has reviewed the existing information on the body weights of Bornean and Sumatran orangutans, and has shown that the males weigh 72.8 ± S.D. 17.7 kg (Bornean) or 66.0 ± 17.3 kg (Sumatran), whereas the females of both races weigh 37.4 ± 4.1 kg. Males also show three other anatomical features that are lacking in the female: the development of pronounced lateral facial skin folds, a large laryngeal apparatus (Galdikas, 1979), and a large skull with particularly pronounced canine teeth (Harvey *et al.*, 1978). Since all these sexual dimorphisms develop maximally after puberty, we may safely conclude that they are a consequence of sexual selection. The increased body size of the male is characteristic of all polygynous primates (Clutton–Brock *et al.*, 1977), and this, together with the development of the canine teeth and the sites of attachment for the muscles of the lower jaw, is obviously related to intermale threat and aggression. The facial and laryngeal development is presumably related to the need for males to project their voice over great distances in the forest (they can be heard for distances over 2 km; Galdikas, Chapter 11, this volume), both as a threat to rival males

TABLE 1

Gonad weights as a percentage of total body weight in the great apes and man (adapted from Short, 1979a).

Species	Mating System	Social Structure	Ovaries: Body Weight Ratio (%)	Testes: Body Weight Ratio (%)
Orangutan	Polygynous	Solitary	0.006	0.048
Chimpanzee	Promiscuous	Large mixed-sex troops	0.010	0.269
Gorilla	Polygynous	Small one-male groups	0.012	0.017
Man	Serially monogamous	Family groups	0.014	0.079

and as an advertisement of their presence to any estrous females who can then seek them out. In fact, the facial folds may act like a parabolic reflector, focusing the sound so that its point of origin can be more readily located.

2. Genital Development

The lack of visibility in the forest canopy undoubtedly has minimized the importance of visual cues in the evolution of the orangutan. Thus the female shows no sign of perineal or vulval tumescence at the time of estrus, and no mammary development prior to pregnancy; the male genitalia are also normally completely concealed from view by the long body hair. However, the erect penis is about 4 cm in length, pink in color, and clearly visible. To what extent, if any, it is used in male–female display is uncertain, but its length certainly allows the animal to adopt a wide variety of copulatory positions, necessitated by the fact that copulation usually occurs when both animals are hanging from branches (Short, 1979a).

3. Gonadal Development

Table 1 summarizes the relative sizes of the ovaries and testes in the orangutan, as compared with the other great apes and with man. The female's ovaries are relatively small, and probably no different from those of the other species (Short, 1979a). The testes are extremely small, particularly when compared to testicular weights in other primates (Schultz, 1938). This small size can be attributed to a decreased volume of tubular tissue, since the blood testosterone concentrations (and, by inference, the volume of Leydig tissue) is comparable to that seen in the other apes and man (Short, 1979a).

Although we still lack really good information, it is obvious from the extensive field studies of Galdikas (1979 and Chapter 11, this volume) that copulation in the orangutan is a relatively rare event. The average home range of an adult female is 5–6 square km, and therefore an adult male is likely to be in vocal communication with only a small number of females at any one time. Since Galdikas estimates that the mean birth interval of the females is about 5 to 7 years, this suggests, as in the other great apes, that births are kept well apart by long intervening periods of lactational anestrus. Thus, each of the few adult females within earshot of a given male is likely to come into heat only once or twice every 5 or 6 years. Even when one of these estrous females seeks out a male and establishes a consortship with him, it only lasts for a mean of 5.4 days, and the mean number of copulations per total consort period is only 2.9, spaced over a mean of 2.4 days (Galdikas, Chapter 11, this volume). In the face of such extremely low copulatory frequencies, the spermatogenic capacity of the testis, in terms of the number of spermatozoa produced per day, does not have to be high. Since testis size, the volume of tubular tissue, and spermatogenic capacity are closely related to one another, it is not difficult to see why the orangutan's testes are so small. A moment's reflection will also show why the orangutan's ovaries do not differ in relative size from those of the human or the other great apes.

B.   The Chimpanzee

The chimpanzee is a terrestrial and arboreal omnivore, living in large mixed-sex communities, and basically having a promiscuous mating system. This unique behavior has had a number of interesting consequences for its sexual development (Short, 1979a).

1.   Somatic Development

The degree of sexual dimorphism in body size is the least of all the great apes. In captivity, the mean weight of five males aged 12½–13½ years at the Yerkes Primate Center was 46.6 ± S.D. 5.58 kg, and the mean weight of five females was 41.6 ± 2.82 kg (Gavan, 1971). At the Holloman Air Force Base, the mean weight of 11 mature males was 65.6 ± 5.9 kg, and the mean weight of six mature females was 56.8 ± 2.0 kg (Smith *et al.*, 1975). In the wild, body weights are much lower, and the degree of sexual dimorphism is greater. The mean weight of nine adult males in the Gombe Stream Reserve, Tanzania, was 39.5 kg (range 36.2–43.0), whereas the mean weight of six females was only 29.8 kg (range 26.9–32.9) (Wrangham and Smuts, 1980). Perhaps this is a reflection of the decreased availability of food in the wild, which as Wrangham postulated, will have its most dramatic effect on the females.

The only other somatic characteristic by which males and females obviously differ from one another is in the size of the canine teeth, which are relatively much larger in the male (Harvey *et al.*, 1978); apart from this, the faces of male and female chimpanzees are very similar in appearance.

The reduced extent of sexual dimorphism in somatic characteristics in the chimpanzee is presumably a direct consequence of the promiscuous mating system. If males do not have to compete with one another for access to estrous females, intermale aggression will be considerably diminished. However, since groups of males are still involved in territorial defense against rival males, some dimorphisms obviously persist.

2. Genital Development

The more open habitat in which the chimpanzee lives has undoubtedly increased the importance of visual cues during the course of evolution. Thus, the female shows pronounced vulval and perineal tumescence at the time of estrus; females will normally permit copulation only during the swelling phase of their cycle, when they are most attractive to the males. The swelling is maximally developed in the middle of the menstrual cycle, and lasts for about 10 days (Tutin and McGinnis, Chapter 9, this volume); detumescence occurs about 24 hr after ovulation. Since the gross volume change in the perineum may be as much as 1400 ml during the course of the menstrual cycle (Graham, 1970), it obviously can be seen at a considerable distance by the males. Thus, in contrast to the orangutan, where female attractiveness was unimportant in an animal hidden from sight of the male by the forest canopy, the female chimpanzee has become extremely visually attractive to males and has developed an extended period of estrus. Maybe these features are an adaptation to its promiscuous lifestyle, and perhaps ready access to estrous females has been important for male group cohesion.

The male chimpanzee also shows signs of genital development in response to sexual selection. Although the flaccid penis is normally concealed from view by a prepuce whose orifice is flush with the body wall, the whole genital area is conspicuous in the male because of a bare area of white skin overlying the large testicular swellings. The erect penis is also very large, being 8 cm in length, and its pink color makes it particularly conspicuous. Males almost invariably take the initiative in courtship, and an erect penis seems to be particularly important in the male's courtship display (Tutin and McGinnis, Chapter 9, this volume). A long, thin filiform penis also may be important for intromission, in view of the highly tumescent vulva of the estrous female.

3. Gonadal Development

As can be seen from Table 1, the chimpanzee has by far the largest testes, both relative to body weight and even in terms of absolute size, of all the great apes. This can be explained by the extremely high copulatory frequencies seen in male chimpanzees in the wild.

Tutin (1980) studied a community of 17 male and 24 female chimpanzees in the Gombe Stream Reserve, Tanzania, and observed 1137 copulations in a period of 16 months. Ninety-eight percent of these copulations occurred when an estrous female was in the company of a group of males, and only 2% were during consortships, when a male removed an estrous female from the social group and was alone with her. The mean copulation rates of individual males with all estrous females ranged from 0.03 to 1.14 times per hour.

This is in striking contrast to the extremely low copulatory rates in the orangutan, and can be accounted for by a number of factors. First and foremost, the promiscuous mating system of the chimpanzee makes any of the 24 females in the community potentially available to all the males, whenever a female comes into estrus. Second, the pronounced advertisement and prolonged duration of estrus itself further enhances the copulatory opportunities for the male. It can be calculated that each female will experience estrous cycles for about 15% of her reproductive life, spending about 400 days actually in heat (Tutin, 1980). Thus although the chimpanzee, like the other great apes, has a long mean interbirth interval of 5 years 8 months as a result of lactational anestrus, in a large community of females there are always likely to be some adolescent females coming into estrus, since they experience an average of 19 cycles between menarche and first conception. One or two estrous cycles even occur during pregnancy itself, and there are about four more cycles between the end of one period of lactational anestrus and the next conception. Thus, hardly a month will go by without one of the females in the community coming into estrus.

It is possible to make some rough calculations about the daily sperm production rate in chimpanzees. Given a total testicular weight of 119 g (Schultz, 1938), and using Amann's rhesus monkey sperm production rate of 23×10^6 sperm per gram of testis per day (Amann *et al.*, 1976), the daily output of spermatozoa by the chimpanzee's testes might be about 2737×10^6. Since the content of spermatozoa in one chimpanzee ejaculate is only 603×10^6 (Warner *et al.*, 1974), this means that the testes can produce sufficient spermatozoa for at least four ejaculations a day. Thus, it can be seen that the chimpanzee's reproductive anatomy is perfectly adapted to its unusual pattern of reproductive behavior.

C. The Gorilla

The gorilla has a typical polygynous mating system, but it differs in a number of important respects from that of the orangutan, and this has had repercussions for the gorilla's sexual development.

1. Somatic Development

Although we lack adequate information on the precise body weights of male and female gorillas either in the wild or in captivity, there is no doubt that adult males are at least twice the size of females (Short, 1979a), and this is precisely what one would expect from the mating system. The male is also facially distinct from the female as a result of the massive development of the saggital and nuchal crest (Short, 1980), and the enlarged canine teeth (Harvey *et al.*, 1978). We know that there are occasionally fierce fights between rival males that can lead to severe wounding or even death of one of the antagonists (Harcourt and Stewart, 1977).

2. Genital Development

The female gorilla does exhibit a cycle of vulval tumescence, with maximal development at the time of estrus (Nadler *et al.*, 1979, and Graham, Chapter 1, this volume). However, this is impossible to discern in the wild, even at a distance of a few feet (Harcourt *et al.*, 1980, and Chapter 10, this volume), and thus it is in no way comparable to the florid tumescence of the female chimpanzee. Clutton–Brock and Harvey (1976) have pointed out that female sexual swellings are most highly developed in those primates that live in multimale groups, where they serve to attract a number of males to the estrous female. Since the female gorilla only has access to one adult male, and since it is she who usually initiates copulation in any case by approaching the male in a characteristic manner (Harcourt *et al.*, 1980, and Chapter 10, this volume), she has no need to advertise her receptive state to all.

The male's genitalia are completely concealed from view by the body hair, and even on close inspection it is impossible to discern a true scrotum (Short, 1979a). The penile orifice is flush with the body wall, and even when the prepuce is reflected, the flaccid penis is only about a centimeter long, and its black color blends with that of the surrounding skin. The fully erect penis is only 3 cm in length and is in no way conspicuous. Thus, it seems that the male's genitalia have been left virtually untouched by sexual selection, and it is doubtful whether they play any role in attracting the female. The small size of the erect phallus also may be related to the fact that in the wild, copulation is always dorso-ventral (Harcourt *et al.*, 1980, and Chapter 10, this volume), although in captivity ventro-ventral copulatory attempts are occasionally seen (Nadler, 1980, and Chapter 8, this volume).

3. Gonadal Development

Table 1 emphasizes the fact that the relative size of the gorilla's testes is almost identical to that of the ovaries, a situation that may well be unique among mammals. The testes to body weight ratio is in fact the lowest recorded for any primate (Schultz, 1938; Short, 1979a), and this seems to be due to a reduction in the volume of seminiferous tubular tissue, since both Wislocki (1942) and Hall–Craggs (1962) reported that on histological examination of the testes of wild gorillas, the seminiferous tubules appeared to be small in diameter, few in number, and widely separated by masses of interstitial tissue. The fact that adult male gorillas have blood testosterone concentrations that are comparable to those of normal human males (Short, 1979a, and Wright *et al.*, Chapter 7, this volume) suggests that there has been no comparable reduction in the volume of endocrine tissue in the testis.

These testicular changes can be related directly to the very low copulatory frequency in the gorilla, which is a result of several factors. First, any adult silverbacked male has only 2–6 adult females in his harem. Second, the duration of estrus itself is extremely short; the female will permit copulation for only 1 or 2 days during her month-long cycle. Third, the male's copulatory frequency on the day of estrus is not particularly high; Harcourt *et al.* (1980) recorded rates of 0.32, 0.36, and 0.38 copulations per hour for three different silverbacks, and of course, as in the other great apes, copulations are confined to the daylight hours, as the male and female always make separate nests at night. Finally, the greater part of the female's reproductive life is spent in lactational anestrus, which lasts from 2 to 4 years following each birth (Harcourt *et al.*, 1980). Each period of lactational anestrus is followed by only two or three estrous cycles before the female becomes pregnant again. In addition to these rare occasions on which the nonpregnant adult females come into heat, the adolescent females also undergo an unknown number of irregular estrous cycles prior to first conception, and some females show brief, sporadic periods of estrus throughout gestation (Harcourt *et al.*, 1980).

The general conclusion to be drawn from all this information is that in contrast to the chimpanzee in which intercourse is almost a daily occurrence, in the gorilla it is at most an annual event. This reduced copulatory frequency in the gorilla understandably has had a major impact on certain aspects of sexual selection.

V. WHAT CAN THE GREAT APES TELL US ABOUT MAN?

The first question we must ask of ourselves as a species is, "What is the normal pattern of human social and sexual behavior?" We cannot expect to answer it by studying contemporary human societies, since the extensive modifications of human behavior by civilization have taken place only in the last few thousand years, whereas man has been in existence for several million years. Given that the human generation time is of the order of 25 years, and bearing in mind that only about 80 human generations have elapsed since the birth of Christ, it is most unlikely that there has been any appreciable genetic change in man over the brief period of our evolutionary history during which we have been subjected to the competing and conflicting lifestyles of various civilizations.

There is general agreement among archaeologists and anthropologists that man originally lived the life of a hunter–gatherer; the men hunted game while the women gathered fruit, nuts, roots, and grain. As 99% of our existence was devoted to this type of life, we must look to hunter–gatherer man if we are to explain the selective forces that have made us what we are today. Hunting and gathering may have been an extremely subtle form of existence, since it allowed the male and the female to exploit two completely different food sources within a common habitat without competing with one another. At the same time it made men and women mutually reliant upon one another. This, in its turn, may have been of great importance for the rearing of highly dependent offspring and for transmitting to them all the acquired experience of both parents. But was hunter–gatherer man monogamous, polygynous, or promiscuous?

The most obvious clue comes from our body size. The fact that men are about 20% heavier than women at a given age (Short, 1979a) suggests that we were polygynous, or perhaps promiscuous, but certainly not monogamous. But what form did this polygyny take? Analysis of 185 contemporary human societies uncontaminated by Western culture showed that 74% were basically polygynous, although economic considerations and a shortage of women meant that in practice about half were forced to adopt a monogamous lifestyle (Ford and Beach, 1952). There seems little doubt that as a species we have indulged in strong male–female pair bonding and long-term consortships, which have been reinforced by love and by a relatively high copulatory frequency consequent upon the suppression of cyclical estrus in the female, thus making her constantly attractive to the male and potentially receptive to him at any time from puberty onwards. How can we equate this pair-bonded monogamy with the evidence from body size that we are likely to have been polygynous?

One possible explanation is that we may have indulged in serial monogamy, and that because of the differences in the fertile life span of men and women, this would have resulted in a degree of sexual selection comparable to that seen in more typical polygynous mating systems. The argument runs as follows: There seems little doubt that men almost always choose female partners who are younger than themselves; this is certainly true of contemporary societies. Furthermore, female fertility declines markedly after about the age of 40 and ceases entirely at menopause, whereas male fertility persists throughout life. Indeed, in one contemporary hunter–gatherer society, the !Kung from the Kalahari Desert, female fertility ceases at about the age of 34 (Howell, 1979). If males practiced serial monogamy and always sought out younger, premenopausal wives for whom there would be considerable competition, it is obvious that a successful man would contribute a disproportionately large share of his genes to the succeeding generation, and sexual selection would operate.

It is intriguing to examine some of the other somatic characteristics by which men and women differ from one another. In contrast to all the great apes, there are no clear-cut sexual dimorphisms in the shape or form of the human skull, and gone are the sex differences in dentition. However, there are marked dimorphisms in the soft tissues of the face, with beard growth and baldness being confined to the male. These facial characteristics have presumably served both in the ritualization of threat and aggression between men and also in the enhancement of the attractiveness of the man to the woman. We have long since abandoned the use of our teeth in aggressive encounters, preferring to use the muscular strength of our limbs, enhanced by the deployment of weapons.

The other dimorphic somatic characteristic has been development of the breasts in the woman at the time of puberty, thereby making her attractive to the man a year or more before she actually becomes fertile. But in contrast to the cyclical sexual attractiveness of the tumescent perineum of an adolescent chimpanzee, the human breast is always there and reveals nothing to the onlooker about when the woman is likely to ovulate; breast volume does fluctuate during the menstrual cycle, but it is maximal just prior to menstruation (Milligan *et al.*, 1975). Breast development can therefore be seen as yet another female stratagem for reinforcing the strength of the pair bond.

As far as genital development is concerned, the woman shares with the female orangutan a complete absence of labial tumescence at the time of ovulation. This could be construed in a number of ways; one might take it as evidence that we are not adapted to live in multimale groups like the chimpanzee. Or, perhaps it is a consequence of our upright posture, which effectively conceals the labia from view. Or, perhaps it is just a part of the generalized suppression of all external indices of impending ovulation, which has been a necessary adjunct of making the female constantly attractive to the male.

If the woman has failed to indulge in any genital development, the male more than makes up for it with his large pendulous scrotum and conspicuous penis. The penis not only is visible when flaccid, but also is extremely large when erect, being on the average 13 cm in length. Even the pubic hair in the male seems designed to draw attention to the genitalia, rather than to conceal them as in the orangutan and gorilla. Since it is difficult to imagine that the human genitalia have been used in intermale threat and aggression, it is more plausible to imagine that they have evolved to attract the female. It is interesting to note that penile enlargement must be due to sexual selection since it occurs only at puberty; the genitalia of a prepubertal boy are not so very different from those of a postpubertal gorilla. Perhaps the large size of the erect human penis has enhanced the pleasure of intercourse and the ability of the woman to experience orgasm; it has certainly made a wide variety of copulatory positions anatomically possible.

Finally, we must consider human gonadal development. The human testis is much larger than that of the gorilla or orangutan in both relative and absolute terms, but is much smaller than that of the chimpanzee. There appear to be marked racial differences in testis size: in Caucasians, the mean weight of one testis is 21 g, whereas in Chinese it is only 10 g (Short, 1979b). Although part of this difference may be accounted for by a difference in total body size, it is doubtful whether that could account for all of it, and the finding remains unexplained.

The fact that the human testis is considerably larger than that of the gorilla or orangutan does seem to be related to our much higher copulatory frequency, which is of the order of two to three times per week in younger Western men and which declines with age (Westoff, 1974). From the studies of Freund (1962, 1963), we know that Caucasian men can maintain a normal sperm output at an ejaculatory frequency of 3.5 times/week. It falls to 60×10^6 spermatozoa per ejaculate at a frequency of 8.6 times per week, and to 31×10^6 at a frequency of 16.9 times per week. This is consistent with what we know about human sperm production rates. Taking Amann and Howards' (1980) figures of 4.4×10^6 sperm per gram of testicular tissue per day, this gives a daily sperm production rate for 42 g of testicular tissue of 185×10^6, which is somewhat less than the mean sperm content in a normal ejaculate of 253×10^6. Thus, the human testis is well-suited to meet the spermatogenic demands put upon it, but it seems certain that a man's fertility would be seriously compromised if he attempted to emulate the high copulatory frequencies of a male chimpanzee.

VI. CONCLUSIONS

Much has happened since Savage and Wyman provided the Western world with the first published account of the gorilla in 1847, and Darwin made use of this information to illustrate how an animal's mating system could determine its anatomy through the forces of sexual selection (Darwin, 1871). It has taken over one hundred years to fill in the details of this story for the great apes, and many facts are still missing. But at least we can begin to develop a logical progression of thought. We must look to the sociobiologists for an explanation of the behavioral differences, and to the ethologists for a precise description of those behaviors in the only environment that counts: the wild. Without such information, we cannot begin to make sense of the anatomical and physiological differences that separate man and the apes. Nowhere is this more evident than in the case of the reproductive system. It is the divergence of form and function between male and female that provides us with so many fascinating clues about the nature of the selective forces that have been at work during the course of evolution. It is in this context that the great apes are particularly worthy of study, since they, our glass, can modestly discover to ourselves that which we yet know not of.

REFERENCES

Amann, R. P. and Howards, S. S. (1980). *J. Urol.* **124,** 211–215.

Amann, R. P., Johnson, L., Thompson, D. L., and Pickett, B. W. (1976). *Biol. Reprod.* **15,** 586–592.

Bruce, E. J., and Ayala, F. J. (1978). *Nature (London)* **276,** 264–265.

Clutton-Brock, T. H., and Harvey, P. H. (1976). *In* "Growing Points in Ethology" (P. P. G. Bateson and R. A. Hinde, eds.), pp. 195–237. Cambridge Univ. Press, London and New York.

Clutton-Brock, T. H., Harvey, P. H., and Rudder, B. (1977). *Nature (London)* **269,** 797–800.

Coulter, G. H., Rounsaville, T. R., and Foote, R. H. (1976). *J. Anim. Sci.* **43,** 9–12.

Darwin, C. (1859). "The Origin of Species by means of Natural Selection, or the Preservation of Favoured Races in the Struggle for Life." Murray, London.

Darwin, C. (1871). "The Descent of Man, and Selection in relation to Sex." Murray, London.

Eckhardt, R. V. (1975). *Am. J. Phys. Anthropol.* **42,** 349–350.

Fisher, R. A. (1930). "The Genetical Theory of Natural Selection." Oxford Univ. Press (Clarendon), London and New York.

Ford, C. S., and Beach, F. A. (1952). "Patterns of Sexual Behaviour." Eyre and Spottiswoode, London.

Freund, M. (1962). *J. Reprod. Fertil.* **4,** 143–159.

Freund, M. (1963). *J. Reprod. Fertil.* **6,** 269–286.

Galdikas, B. M. F. (1979). *In* "The Great Apes" (D. A. Hamburg and E. R. McCown, eds.). pp 195–233. Benjamin/Cummings Publ., Menlo Park, California.

Gavan, J. A. (1971). *In* "The Chimpanzee" (G. H. Bourne, ed.), Vol. 3, pp. 46–102. Karger, Basel.

Graham, C. E. (1970). *In* "The Chimpanzee" (G. H. Bourne, ed.), Vol. 3, pp. 183–220. Karger, Basel.

Hahn, J., Foote, R. H., and Seidel, G. E. (1969). *J. Anim. Sci.* **29**, 41–47.

Hall-Craggs, E. C. B. (1962). *Proc. Soc. Zool. London,* **139**, 511–514.

Hamburg, D. A., and McCown, E. R. (1979). "The Great Apes." Benjamin/Cummings Publ., Menlo Park, California.

Harcourt, A. H., and Stewart, K. J. (1977). *New Sci.* **76**, 160–162.

Harcourt, A. H., Fossey, D., Stewart, K. J., and Watts, D. P. (1980). *J. Reprod. Fertil. Suppl.* **28**, 59–70.

Harvey, P. H., Kavanagh, M., and Clutton–Brock, T. H. (1978). *J. Zool. London* **186**, 475–485.

Hayward, P., and Shire, J. G. M. (1974). *Nature (London)* **250**, 499–500.

Hodgen, G. D. (1979). *In* "Animal Models for Research on Contraception and Fertility." (N. Alexander, ed.) pp. 125 136. Harper, New York.

Howell, N. (1979). "The Demography of the Dobe !Kung." Academic Press, New York.

King, M-C, and Wilson, A. C. (1975). *Science* **188**, 107–117.

Lasley, B. L., Hodges, J. K., and Czekala, N. M. (1980). *J. Reprod. Fertil. Suppl.* **28** 121–129.

Milligan, D., Drife, J. O., and Short, R. V. (1975). *Br. Med. J.* **4**, 494–496.

Nadler, R. D. (1980). *J. Reprod. Fertil. Suppl.* **28**, 79–89.

Nadler, R. D., Graham, C. E., Collins, D. C., and Gould, K. G. (1979). *Endocrinology* **105**, 290–296.

Ohno, S. (1967). "Sex Chromosomes and Sex-linked Genes." Springer-Verlag, Berlin and New York.

Savage, T. S., and Wyman, J. (1847). *Boston J. Natur. Hist.* **5**, 417–443.

Schultz, A. H. (1938). *Anat. Rec.* **72**, 387–394.

Short, R. V. (1979a). *Adv. Study Behav.* **9**, 131–158.

Short, R. V. (1979b). *Brit. Med. Bull.* **35**, 121–127.

Short, R. V. (1980). *J. Reprod. Fertil. Suppl.* **28**, 3–11.

Short, R. V., and Drife, J. O. (1977). *Symp. Zool. Soc. London* **41**, 211–230.

Short, R. V., and Weir, B. J., eds. (1980). The great apes of Africa, *J. Reprod. Fertil. Suppl.* **28**.

Smith, A. H., Butler, T. M., and Pace, N. (1975). *Folia Primatol.* **24**, 29–59.

Tutin, C. E. G. (1980). *J. Reprod. Fertil. Suppl.* **28**, 43–57.

Warner, H., Martin, D. E., and Keeling, M. E. (1974). *Ann. Biomed. Eng.* **2**, 419–432.

Westoff, C. F. (1974). *Fam. Plann. Perspect.* **6**, 136–141.

Winter, J. S. D., Faiman, C., Hobson, W. C., and Reyes, F. I. (1980). *J. Reprod. Fertil. Suppl.* **28**, 131–138.

Wislocki, G. B. (1942). *J. Mammal.* **23**, 281–287.

Wrangham, R. W. (1979). *Social Sci. Informat.* **18**, 335–368.

Wrangham, R. W., and Smuts, B. B. (1980). *J. Reprod. Fertil. Suppl.* **28**, 13–31.

Chapter 14

BREEDING GREAT APES IN CAPTIVITY

David E. Martin

I. INTRODUCTION

Breeding of great apes in captivity is both a necessary and a valuable endeavor, and a topic of continuing concern (Teleki and Baldwin, 1975; Martin, 1976). The geographical areas in which these animals naturally live are increasingly being taken over by man. To prevent extinction and to permit further study of these species, the means must be developed to ensure that successful breeding will take place in captive environments, i.e., zoological exhibits and research centers. The great phylogenetic similarity of these animals to man makes them excellent models for the performance of experiments that are ethically inadmis-

343

Copyright © 1981 by Academic Press, Inc.
All rights of reproduction in any form reserved.
ISBN 0-12-295020-8

sible or inappropriate in man, and that can provide man with many insights into human health and disease. However, the use of great apes for such experimentation is justified only if they are available in sufficient numbers for both experimental usage and procreation. For procreation, we increasingly must rely on captive breeding as it is unlikely that a significant number of wild-caught animals will be available in the future.

The accomplishment of a successful breeding program is time-consuming and expensive. Great apes do not reach maturity until between seven and nine years of age, thus requiring a long-term investment before any dividends are reaped in the form of infants. In the unusual instance that a female breeds successfully for a reproductive lifetime of 30 years, one cannot reasonably hope for more than about 15 infants. The size of these animals mandates large and elaborate caging arrangements, and sizable daily expenditures on food and medical care. A well-staffed and well-equipped nursery is essential to guarantee adequate care for premature, sick, abandoned, or injured infants. Detailed menstrual cycle records need to be maintained, and an aggressive and innovative mating program must be initiated to maximize the conception rate.

Despite such difficulties, several zoological gardens and research facilities have had success in breeding these animals; with some species the success has been exceptional. The details of methodology at these several sites already have been reported adequately in primate literature. The purpose of this chapter is to identify and review briefly this literature, to provide a general perspective on breeding activities worldwide, and to discuss certain breeding problems of contemporary and developing importance. Much of the information basic to animal breeding, e.g., menstrual cycle endocrinology, reproductive behavior, and periparturitional physiology, is considered elsewhere in this volume.

II. NATURAL BREEDING OF THE GREAT APES

A. *Pan troglodytes*

1. *Yerkes Regional Primate Research Center, Atlanta, Georgia*

This facility is the oldest great ape breeding center in the world. It began in 1930 when Robert M. Yerkes of Yale University established an experimental station to study primate behavior at Orange Park, Florida, near Jacksonville. He purchased four animals, including two from a collection maintained by Señora Rosalie Abreu at Quinta Palatino in Havana, Cuba. For 35 years, the laboratory at Orange Park flourished with 161 chimpanzee births. During this period, pioneering studies into the reproductive biology of the chimpanzee were initiated by Young and Yerkes (1943) who identified the individual variation in menstrual cycle length, the period of adolescent sterility, and the frequency of abortion in this species. Then, as part of a federally funded project to establish seven regional primate research centers around the United States, the Orange Park facility was relocated on the campus of Emory University in Atlanta. Since August 1965 and through December 1979, another 143 births have put the facility in a class by itself regarding number of births in captivity. This increase in birthrate in Atlanta was in response to increasing demands at Emory for animals for scientific research.

Table 1 indicates that, over the years, almost identical numbers of males and females have been born at the center. There appears to be no seasonality to the chimpanzee birth pattern, as shown in Table 2.

Whereas other chimpanzee colonies have not been in existence long enough to produce second-generation offspring in sizable numbers, this has occurred frequently at the Yerkes Center. In fact, there are several infants who are known fourth-generation offspring. Table 3 summarizes information concerning these, allowing one to calculate an approximate generation time of 13 years. The center presently has 10 live fourth-generation infants, all offspring of four original colony animals.

The long colony history also provides other valuable information for those interested in primate breeding. Of the 287 deliveries, 12 have been multiple, with 11 sets of twins and one set of triplets. The incidence of multiple births in this colony is thus approximately one in 24, which confirms an earlier report for the same colony (Peacock and Rogers, 1959), indicating a considerably higher incidence of twinning than among American human births (approximately 1 in 88; Srb and Owen, 1965). However, only nine out of the 25 multiple-birth chimpanzee infants lived longer than one week. Table 4 summarizes the most relevant colony data regarding these births. One male accounts for the siring of the great majority (17 out of 25) infants, and one female (Flora) for the delivery of 13 of the 25. At least five of the 12 sets of multiple births recorded in the colony were dizygotic conceptions.

During the 48 years of the Yerkes animal colony, there have been 72 chimpanzee mothers, six of whom have delivered 77 of the 304 births recorded as of this writing. Twenty-seven males have accounted for all sirings, with one individual accounting for 50, another 41, and four

David E. Martin

TABLE 1

Outcome of pregnancy at Yerkes Primate Center Chimpanzee Colony.

	Live Births[a]		Stillbirths[c]		Premature Births[b]		Miscarriages[d]		Total Births	
Years	M	F	M	F	M	F	M	F	M	F
1930–1934	6	6	—	—	—	—	—	—	6	6
1935–1939	10	8	—	—	—	—	—	—	10	8
1940–1944	9	14	2	—	—	—	—	—	11	14
1945–1949	16	11	1	2	—	—	1	—	18	13
1950–1954	9	11	2	—	—	—	—	—	11	11
1955–1959	11	11	2	—	2	1	1	—	16	12
1960–1964	5	13	—	2	2	2	—	—	7	17
1965–1969	11	14	2	2	1	1	—	—	14	17
1970–1974	24	17	3	8	2	1	—	—	29	26
1975–1979	30	24	4	—	—	—	—	—	34	24
Totals	131	129	16	14	7	5	2	—	156	148

[a] Includes live-born premature births, i.e., those with a gestation period between 170 and 200 days, inclusive.

[b] Includes conceptuses dead on delivery, and with a gestation period between 170 and 200 days, inclusive.

[c] Includes conceptuses dead on delivery, and with a gestation period longer than 200 days.

[d] Includes conceptuses dead on delivery, and with a gestation period between 60 and 170 days.

additional males each more than 20. In several instances, exact parentage is not known, since the pregnancies were initiated in a compound environment housing more than one male. Serological investigations have allowed determination of some actual fathers (Balner *et al.*, 1978).

The Yerkes colony also has four female and five male type O blood group chimpanzees. Type O is the rarest in the species, and it is fortunate that four of these animals are active breeders.

The colony data also permit rather precise estimations of the age limits within which chimpanzees have conceived. Considering only

TABLE 2

Monthly pattern of chimpanzee births at Yerkes Primate Center.

Month	Orange Park 1930–1965	Atlanta 1965–1978	Total
January	13	15	28
February	14	16	30
March	10	13	23
April	14	7	21
May	22	14	36
June	13	10	23
July	10	13	23
August	11	14	25
September	11	8	19
October	13	6	19
November	17	12	29
December	13	15	28
Totals	161	143	304

those parents with known birthdates, and taking the estimated mean gestational age from natural single-day matings as 227.6 $\pm$ 4 (SEM) days (Martin *et al.*, 1978), several statements can be made. The youngest mother in the colony conceived her first infant at the age of 8 years, 2.5 months (Cuba, born March 24, 1926, conceived Peter, born April 11, 1935), whereas the oldest mother conceived her last infant at 41 years 8 months (Flora, born January 28, 1941, conceived David, born June 24, 1977). The youngest male sired his first offspring at the age of 9 years, 2.5 months (Ked, born May 11, 1947, sired Klem, born May 21, 1957), while the oldest father sired his last at the age of 41 years, 8 months (Hal, born October 7, 1935, sired David, born June 27, 1977). A definite menopause has not been observed in any of the older colony females (Graham, 1979).

Ample documentation of the general operating procedures for animal housing and breeding practices has been provided in the literature (Guilloud, 1969; Bourne, 1972; Keeling and Roberts, 1972; Bourne *et al.*, 1975). Animals have been housed at one of three sites since the facility has been in Atlanta. The majority of the animals are housed in a laboratory environment, paired in an indoor–outdoor cage arrangement separated by a guillotine door. The indoor cages are 2.4 $\times$ 2.4 $\times$ 2.4 m, whereas the outdoor cages are slightly larger (2.4 $\times$ 2.4 $\times$ 3.7 m).

An experimental field station 25 miles away, opened in 1967, has provided an effective semi-free-ranging condition for a breeding nucleus of approximately 15 animals. Although serology has been required to

TABLE 3

Four generations of chimpanzees born at Yerkes Primate Center.

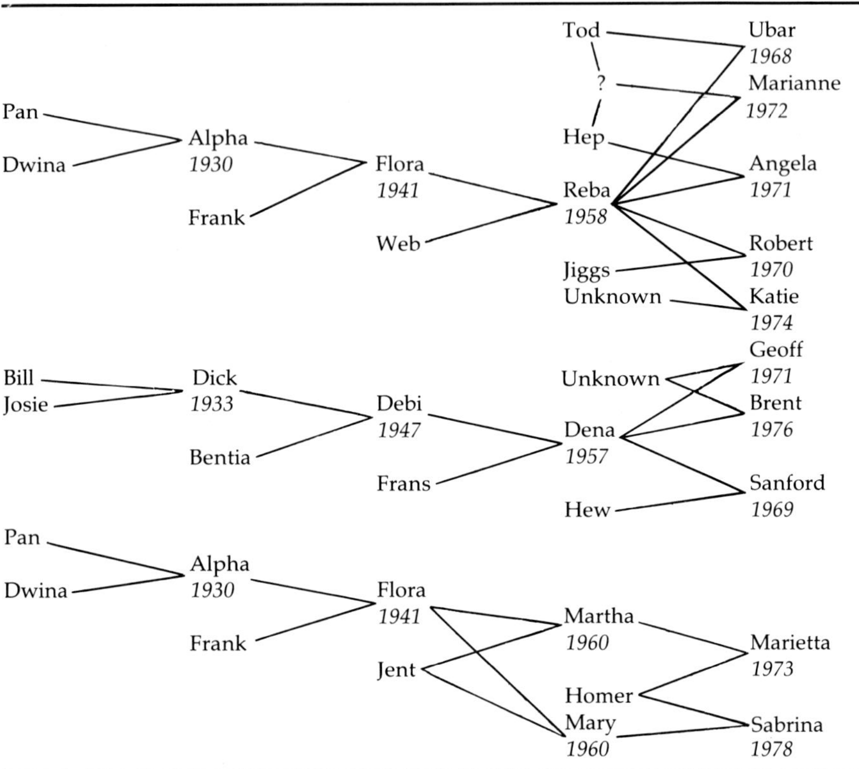

determine the exact parentage of the infants, this environment has proven very conducive for a sizable number of births.

Some experience has been obtained with the difficulties of maintaining a chimpanzee colony in a totally free-ranging environment. In 1972 a small group of animals was put on uninhabited and subtropically vegetated Bear Island off the coast of Georgia; additional animals were added a few years later. Early in 1978, however, the project was terminated due to a combination of unfortunate experiences, including difficulties with observation as a result of heavy foliage, poaching, infant rejection, and deaths from unknown causes.

At the laboratory housing facilities on the Emory University campus, routine procedures are followed to ensure an optimum environment for efficient breeding. Female genital sexual swelling is monitored using a scale of 0 (minimum) to 4 (maximum), and detailed monthly menstrual

TABLE 4

Summary of multiple births at Yerkes Primate Center.[a]

Date of Birth	Sex	Mother	Father	Survival[b]
JUN 26 1933	M F	Mona	Pan	M 182 days; F 457 days
NOV 26 1951	M M	Pati	Hal	(Stillborn)
JUN 06 1953	M M	Helene	Hal	M 4 days; M 111 days
DEC 26 1955	M F	Jenny	Hal	(Premature)
SEP 24 1956	M F	Flora	Jent	Both 3 days
MAY 21 1957	M F	Fanny	Ked	M 64 days; F 2 days
JUL 07 1960	F F	Flora	Jent	Alive in 1980
NOV 25 1962	F F	Jenny	Hal	(Premature)
JUL 05 1967	M F F	Flora	Hal	M premature; F F hours
AUG 22 1968	F F	Banana	Hal	Hours
DEC 09 1969	M M	Flora	Hal	M alive in 1980; M stillborn
MAY 17 1972	F F	Flora	Hal	Alive in 1980

[a] All twins, except for one set of triplets.
[b] Survival times are in months, unless otherwise noted.

cycle charts are maintained for each animal (Graham *et al.*, 1969; Graham, 1970). Animals are generally paired in male–female combinations for one or two months at a time, and if a pregnancy does not ensue, a different match is made.

Experience has shown that not all male–female pairs are compatible, and also that not all animals show mating activity despite demonstrating apparently normal menstrual cyclicity or semen production. Specific research requirements also limit the use of animals for breeding. Hence, at any given time, the actual size of the breeding colony is rather small.

Pregnancy is determined either by rectal palpation or the use of pregnancy tests, or both. Since rectal palpation is generally not sufficient to detect a fetus before the next ensuing ovulation, immunologic pregnancy tests have received considerable emphasis. It has been known for some time (Clegg and Weaver, 1972) that chronic gonadotropin is secreted in sizable quantities and for a prolonged period in the first half of pregnancy. Immunological pregnancy tests, developed for use in humans [Pregnosticon Accusphere Test (Organon Inc., West Orange, NJ) and UCG Test (Wampole Laboratories, Stamford, CT)] react with chimpanzee chorionic gonadotropin.

Woodard *et al.* (1976) have found that a test based on an antiserum produced against a subunit of ovine luteinizing hormone (LH) and originally designed for use in subhuman primates (Hodgen and Ross, 1974), has much greater sensitivity, and is thus more accurate when titers of chorionic gonadotropin are starting to rise (chimpanzee LH does not cross-react in this assay). Use of this Subhuman Primate Pregnancy Test permits detection of pregnancy as early as 15–18 days after sexual-swelling detumescence, which is currently the best external indicator of the presumed time of ovulation (see Graham, Chapter 1, this volume).

The earliest work on the duration of gestation was based on studies of Robert Yerkes and his associates, J. H. Elder and H. W. Nissen, on the time of single fertile inseminations. The reports of these studies recently were re-analyzed (D. E. Martin, 1976) and found to be somewhat contradictory. A paper by James Elder (1938) on the timing of ovulation revealed only seven cases (Table 5) in which a single-day mating occurred and resulted in a term delivery. Data on birthdates from the Yerkes Center files allowed calculation of gestation periods. Surprisingly, Martin calculated a 355-day pregnancy for Mona compared with an average of 227.3 days for the other six. No mention of this discrepancy was made in Elder's paper.

Six months earlier, Yerkes and Elder (1937) had published a paper that apparently for the first time mentioned Mona's pregnancy. Her estimated date of conception was listed as October 31, not July 7, 1935. Elder inadvertently must have included the wrong mating record in his 1938 data summary. Mona's gestation period thus spans 230 days. (Unfortunately, it was incorrectly reported in the Yerkes and Elder paper as 239 days.)

A later paper, by Nissen and Yerkes (1943), summarizing data on 49 births, recorded only six instances of single chimpanzee matings in the Yerkes colony. The data was summarized in Table 5 (data of Nissen and Yerkes), and the reader will note several disparities from the data of Elder in Table 5.

Concerning infants Hal, Cap, and Jule, there was consistent reporting. Fin appears in both papers; however, Nissen and Yerkes mentioned that his case had not been reported earlier. Dina, Ami, and Kola had been removed from the list of confirmed single-day matings. Examination of the original colony records confirmed that these animals were the result of controlled single-day matings. Tom and Helene were found to be the result of prolonged exposure to the male and should not have been included. Art was the result of a single-day mating not previously recorded.

Table 5 (data of Martin, *et al.*) shows all known verified single-day controlled matings resulting in a normal pregnancy. The Hoboh/Banana mating is the result of a single artificial insemination recently performed by our group (Martin *et al.*, 1978). Two additional pregnancies recorded

Single matings resulting in a normal pregnancy.

Data of Elder, 1938

Infant	Birthdate	Father[a]	Mother	Insemination Date	Gestation Period in Days[b]
Hal	10-07-35	Jack	Josie	02-16-35	233
Ami	06-26-36	Jack	Mona	07-07-35	355
Dina	03-27-37	Bokar	Pati	08-02-36	237
Cap	04-13-37	Bokar	Dita	08-16-36	240
Jule	05-08-37	Bokar	Fifi	10-17-36	203
Fin	11-10-37	Bokar	May	03-27-37	228
Kola	12-15-37	Jack	Cuba	05-06-37	223

Data of Nissen and Yerkes, 1943

Infant	Birthdate	Father	Mother	Insemination Date	Gestation Period in Days
Tom/Helene	06-26-33	Pan	Mona	12-02-32	206
Hal	10-07-35	Jack	Josie	02-16-35	233
Cap	04-13-37	Bokar	Dita	08-16-36	240
Jule	05-08-37	Bokar	Fifi	10-17-36	203
Fin	11-10-37	Bokar	May	03-27-37	228
Art	07-20-39	Bokar	Fifi	12-05-38	227

Data of Martin *et al.*, 1978

Case #	Infant	Birthdate	Father	Mother	Insemination Date	Cycle Day[c]	Gestation Period in Days
1	Hal	10-07-35	Jack	Josie	02-16-35	0	233
2	Ami	06-26-36	Jack	Mona	10-31-35	−5	230
3	Dina	03-27-37	Bokar	Pati	08-02-36	−1	237
4	Cap	04-13-37	Bokar	Dita	08-16-36	−4	240
5	Jule	05-08-37	Bokar	Fifi	10-17-36	−2	203
6	Fin	11-10-37	Bokar	May	03-27-37	−4	228
7	Kola	12-15-37	Jack	Cuba	05-06-37	−1	223
8	Art	07-20-39	Bokar	Fifi	12-05-38	0	227
9	Artifee	07-24-76	Hoboh	Banana	11-24-75	−3	244

[a] Obtained from colony records.
[b] Calculated value.
[c] Day 0 = the last day of maximum tumescence.

by Elder in Lia both ended in abortion at four months, but the cycle day of the single matings is known (0 and -3).

No special treatment is given to the animals in the Yerkes colony as they proceed through gestation, and no assistance is provided at parturition unless medically required. From past experience, it is known which multiparous females care adequately for their infants, and recent practice has advocated keeping them together for several months. Nulliparous females are watched very closely, especially if lab-reared, for they often will disregard, reject, or abuse the infant. A large nursery provides for complete care until the animals are several years of age, where they are caged in small groups in indoor–outdoor environments similar to those for the adult breeding pairs.

Currently the chimpanzee colony is relatively stable, with a recent census (October 1, 1980), indicating 134 total animals, of which 115 are at the main center building in Atlanta, two are on loan to other research institutions, and the remainder (17) are at the field station.

2. *Primate Research Institute, New Mexico State University, Holloman AFB, New Mexico*

In 1958, the 6571st Aeromedical Research Laboratory, located at Holloman Air Force Base, New Mexico, began using chimpanzees as experimental animals in aerospace research related to Project Mercury. It was not until 1966, however, that efforts were directed toward an active breeding program aimed at making the center self-sufficient in terms of great ape requirements. A semi-free-ranging environment was constructed by building a water-filled moat around a 30-acre tract of land (van Riper *et al.*, 1967; Kollar *et al.*, 1968). This consortium provided an alternative to the standard procedure of caging animals singly or in pairs.

The consortium's environment proved successful in initiating many pregnancies (Douglas and Butler, 1970). Problems occurred, however, as with any free-ranging colony, in that parentage could not easily be determined. Those individuals who were reproductively normal but unable to coexist socially did not become pregnant, and the animals were not readily accessible for experimental protocols requiring such procedures as blood withdrawal. The consortium idea was abandoned in 1970 when the moat developed a serious leak, and permanent indoor–outdoor cages were utilized exclusively for breeding (Butler, 1972).

Female menstrual cyclicity is recorded on a daily basis. The system used is comparable to the one employed at the Yerkes Regional Primate

Research Center in Atlanta, Georgia. There, the size of swelling is recorded on a 5-point scale.

Excellent support staff is available in the form of veterinary medical personnel, pathology laboratory, clinical chemistry laboratory, and radioimmunoassay facilities. Early in 1974 a modern nursery, staffed by a pediatrician and designed to keep animals up to two years of age with continual care, was opened. Infants are routinely removed from their mothers at birth and hand-reared in the nursery.

In 1971, the animal colony and facilities were leased to Albany Medical College, Albany, New York; the facility was named the International Center of Environmental Safety within the college's Institute of Human and Comparative Toxicology. Then in 1980, the animal colony and facilities were purchased by the New Mexico State University, and the facility was renamed the Primate Research Institute.

The remarkable success of the center in breeding chimpanzees, and the acquisition of additional knowledge about the reproductive physiology of the chimpanzee (Winter *et al.*, 1975; Hobson *et al.*, 1976; Hobson and Fuller, 1977) has allowed development of research programs that require the use of considerable numbers of chimpanzees. Three federal agencies (Food and Drug Administration, National Institutes of Health, and Center for Disease Control) jointly support and use a large group of chimpanzees (Table 6) in a program to develop a vaccine against serum hepatitis B infection. Since the chimpanzee is the only nonhuman primate that is susceptible to hepatitis B, it is used for testing vaccines for research into this disease. Other animals are used in a variety of toxicology and reproductive biology studies.

TABLE 6

Current and five-year estimates of size of Holloman chimpanzee colony.

	1978	1983
Females in breeding colony	48	92
Males in breeding colony	19	19
Births	25	47
Deaths	11	14
Purchases	0	6
Returning from NIH/FDA/CDC research[a]	12	20
Going to NIH/FDA/CDC	20	42
Assigned to other research projects	62	49
On loan to other institutions	80	150
Total animals involved	277	439
Total animals at Holloman facility	197	289

[a] Animals leased for hepatitis research.

Tables 6 and 7 summarize the size and reproductive success of the chimpanzee colony at Holloman. These data suggest that the colony could remain the largest in the world, and that it has enormous breeding potential. There is no detectable seasonality in births.

Mating is accomplished by pairing a male with a female for each menstrual cycle. Between the 15th and 25th days after mating, a blood sample is obtained and assayed by RIA for human LH as described by Faiman *et al.* (1975). Experience with this method indicates that pregnancy can be detected in some females as early as 12 days after mating, but after 15 days it is very reliable for pregnancy determination.

If pregnancy has not occurred, the male/female pairing is allowed to remain for a few more cycles, or another pairing arrangement is established. Thus, for every mating cycle of the breeding females, information ultimately will be available to determine pregnancy rates (number of matings per term infant) and fetal wastage (number of conceptions per term infant). This new program of careful monitoring of reproductive activities will accumulate valuable data within a few years comparable to that currently being obtained at the Primate Center TNO in the Netherlands (see Section II,A,5). When pregnancy has been confirmed, the female is returned to her cage and left undisturbed throughout the course of gestation. Delivery is unassisted unless required by spontaneous complications or experimental protocols.

Detection of the midcycle LH surge by radioimmunoassay can provide the basis for a controlled mating program and timed pregnancy. In two animals a several hour mating period was allowed on the day following the urinary LH surge. One of the pregnancies ended in abortion, and one infant was delivered by cesarean section.

3. *Southwest Foundation for Research and Education (SFRE), San Antonio, Texas*

In 1967, a colony of 18 chimpanzees (10 adults and eight juveniles) was acquired from Africa. It formed the nucleus for a breeding program that has been growing in size and value ever since, although the Southwest Foundation for Research and Education (SFRE) is known primarily for its colony of more than 2000 baboons (principally *Papio cynocephalus* and *Papio hamadryas*).

The chimpanzee colony is maintained in a building completely separate from the baboon and other primate enclosures (Hummer *et al.*, 1969). This building measures 23 m × 18 m, and has both indoor and outdoor adjacent rooms connected by a guillotine door. These room combinations are designed for harem-style grouping of animals, with

TABLE 7

Outcome of pregnancy at Holloman chimpanzee colony.

Year	# Breeding Females	#Stillbirths	#Live Births	Total # Births	Sex of Infants	
					Male	Female
1966	10	0	1	1		
1967	17	2	4	6		
1968	27	2	6	8		
1969	32	4	7	11		
1970	28	1	12	13		
1971[a]	19	0	2	2		
1972[b]	24	1	15	16		
1973	31	0	17	17		
1974	40	2	15	17	12	5
1975	43	6	24	30	18	12
1976	47	1	21	22	10	12
1977	47	4	21	25	11	14
1978	53	7	18	25	13	12

[a] Last year under direction of Holloman AFB.

[b] First year under lease by Albany Medical College.

one ma.e to three or four females. The outdoor cages face south, and a warmed concrete pallet inside provides a comfortable sleeping area. This ensures adequate opportunity for the animals to exercise outdoors even on chilly days, and never to be affected adversely by cool evenings.

Although cage cleaning, feeding, and health monitoring are routine, virtually no outside influences, such as periodic restraint of individual animals for specific research projects, disrupts the activity of the breeding colony. Annual physical examinations are done under anesthesia, and similar restraint of new mothers allows removal of a delivered infant. No urine collections are done for early detection of pregnancy, and no assistance is given with deliveries unless medically justified. The infants are raised in a nursery, and may at that time be selected as candidates for long-term experiments related to viral oncology and multiple sclerosis. When mature, they are returned to the breeding colony, assuming they are in excellent health, and in this way the breeding nucleus increases.

Table 8 summarizes the growth and activity of this breeding colony. Kraemer *et al.* (1975), in assessing the reproductive success of the colony between 1967 and 1975, reported a higher performance, measured in term pregnancies per adult female year (0.49), than that found in the wilds (0.36) of the Gombe National Park (Teleki and Baldwin, 1975). The size of the SFRE colony during those years did not increase accordingly, however, due to the experimental use of embryos for scientific study. From 1973 through 1977, the number of term pregnancies per adult female year has averaged an even higher 0.55, with the number of live births per adult female year averaging 0.43.

In mid-1979, more than 70 chimpanzees previously maintained at the Laboratory for Experimental Medicine and Surgery in Primates (LEMSIP), and involved in studies funded by the National Heart, Lung, and Blood Institute of the National Institutes of Health, were transferred to SFRE, along with the study programs themselves. As a result, the SFRE colony became the second largest in the United States, after the Holloman colony, with the Yerkes collection ranking third in size.

4. Laboratory for Experimental Medicine and Surgery in Primates (LEMSIP), Tuxedo, New York

In 1967 this primate laboratory in New York City became the focus for development of an inter-institutional primate facility (Davis *et al.*, 1969; Davis and Moor–Jankowski, 1971; Moor–Jankowski and Goldsmith, 1971). The laboratory is part of the New York University School of Medicine and is in Sterling Forest, an hour's drive from Manhattan.

TABLE 8

Outcome of pregnancy at Southwest Foundation chimpanzee colony.

Year	Total Population	Total Breeders		Total # Pregnancies	# Stillbirths	# Live Births	Sex of Infants	
		Male	*Female*				*Male*	*Female*
Before 1970	20	5	8	5	0	5	2	3
1970	36	5	13	6	1	5	3	2
1971	39	5	20	13	3	10	4	6
1972	44	5	20	14	0	14	7	7
1973	43	5	21	10	4	6	3	3
1974	48	5	22	12	0	12	8	4
1975	59	6	16	10	1	9	4	5
1976	74	6	16	12[a]	1	9	4	5
1977	83	6	18	10	2	7	2	5
1978	81	6	18	9	1	8	7	1
1979	162	—[b]	—[b]	17	2	15	9	6
1980	170	—[b]	—[b]	14	1	13	9	4

[a] Two abortions reported in 1976.

[b] Information unavailable.

Laboratory research investigations that utilize primates in many areas of medicine and biology are underway, involving both staff members on-site and researchers based at nearby institutions. The New York Academy of Medicine and 14 neighboring medical schools jointly sponsor the facility through their union into the Associated Medical Schools of New York and New Jersey, Inc.

Five chimpanzees were acquired in 1964, and active breeding activities began early in 1970; acquisition of animals from other sources has also contributed to the rapid increase in size of the colony. The impetus for this activity was initiated primarily by research interests in hepatitis, immunobiology, and immunogenetics.

Animals live in cages suspended from the ceiling of the buildings in which they are housed (see Gould, Chapter 15, this volume). Menstrual cyclicity is determined by observed menstrual flow (blood on the perineum or on dry plastic sheets placed beneath the cages). The genital sexual-swelling is graded for its level of tumescence by a rating scale from 0 (maximum detumescence) to $+3$ (maximum tumescence).

Prior to transfer of part of the colony to SFRE, LEMSIP held 30 males of breeding age, although only 18 formed a breeding nucleus. Each of the latter has accounted for from one to eight of the total conceptions. There were 38 females of breeding age, with 27 included in the permanent breeding colony. On the basis of mating trials, the others appear sexually incompetent. Two of the breeding females delivered five infants each, 13 females were primagravids, and eight females each delivered between two and four infants.

Breeding has been accomplished solely by natural mating. Animals are paired according to blood groups and compatible social characteristics. If no conception occurs after approximately six menstrual cycles for the female, then another pairing is made. Polygamous breeding has never been used. Short-term mating encounters are arranged when accurately dated conceptions are desired. Thus far, two such pregnancies have occurred with conception date accurate to ± 1 day.

Pregnancy diagnosis is by a hemaglutination–inhibition test for urinary chorionic gonadotropin. The test is routinely performed at two- to three-day intervals, beginning 15 days after genital sexual detumescence, until either an obvious repeatable positive test is obtained or until the beginning of the next menstrual cycle. Once a pregnancy is confirmed, testing is continued at two-week intervals to confirm the continuance of gestation or to aid in establishing the time of fetal demise in the event abortion occurs.

To date, 27 females have experienced 54 conceptions. Two females were pregnant when they arrived at LEMSIP, and both had successful live births. Of the 54 conceptions, 45 (83.3%) resulted in viable births; seven (13%) were late gestational stillbirths, and two (3.7%) were abortions. Annual conception rates have ranged from one, during the first year of the colony, to 14 or 15 in the past few years. There is no observable seasonal pattern of conceptions.

Following delivery, weak, rejected, or abused infants are immediately transferred to a nursery and are hand-reared. Otherwise, infants are allowed to remain with their mothers from four to six months. During earlier years of the colony, this time of mother–infant rearing was much longer, between 12 and 18 months. When infants reach maturity, provided they are not being utilized for specific research projects, they become part of the breeding colony. Second-generation infants are anticipated in the future, since seven females have recently achieved menarche and are in good health.

5. *Science Park/Veterinary Resources Division (SP/VRD), University of Texas System Cancer Center, Bastrop, Texas*

This multiple-species vivarium started operations in 1975, with a sizable portion of the overall focus directed toward primate breeding colonies. A five-year funded contract (1977–1982) from the National Institutes of Health will establish SP/VRD as a unique and pioneering facility for rehabilitation of chimpanzees whose use in experimental studies at various locales has ended. They will be housed in partly free-ranging breeding harems to optimize opportunities for resocialization and reproduction. The colony size could reach 130 without expansion of facilities, although it presently totals about 40 animals.

6. *Primate Center TNO, Rijswijk, The Netherlands*

The Dutch Primate Center is physically adjacent to and collaborates with the Radiobiological Institute TNO and the Institute for Experimental Gerontology TNO in Rijswijk, between The Hague and Delft, The Netherlands. This primate breeding facility, initially utilizing rhesus monkeys, was developed in the Netherlands in 1965 to provide animals for use in several immunogenetic research programs. During the late 1960s, chimpanzees were imported from Africa and were added to it. Breeding activities were begun with chimpanzees in 1970, and by 1973 the program was well under way. Early research work emphasized the study of the major histo-compatibility complex in the chimpanzee (Balner *et al.*, 1974, 1978), and more recent activities are oriented toward hepatitis B research.

Immunogenetic studies require specifically defined families of animals, and this consideration has dictated the breeding strategy employed. In 1970 four young mature males were each grouped with five mature females in a harem-style breeding environment, using indoor-outdoor cages. The probability of blood relationships among these animals was minimal, making them appropriate for such grouping for purposes of breeding siblings and half-siblings. Since that time, addition of another harem, with one male and 10 females, has increased the breeding nucleus to about 35 animals. Table 9 summarizes the reproductive productivity of the Dutch center.

Several reports concerning breeding and management of chimpanzees at Rijswijk (Boorman *et al.*, 1972, 1973; Balner, 1975; van Vreeswijk and Koning, 1978) indicate no great departure from methodology employed elsewhere.

Pregnancies are determined by use of urinary chorionic gonadotropin test (Pregnosticon) done every two weeks (Boorman *et al.*, 1974). Not only has this reliably indicated pregnancies, but it also has served to indicate possible abortions or stillbirths. The true incidence of abortion is otherwise unknown, as the abortus may be eaten, resorbed, or remain undetected. In fact, the rather high measured incidence of abortion in this colony is likely due not only to the high proportion of adolescent animals, but also to the accuracy of detection of pregnancies ending in early abortion.

B. *Pan paniscus*

The pygmy chimpanzee is very rare even in its native habitat, the tropical forest along one side of the Zaire River in Africa, and it is almost nonexistent in zoological gardens (Jantschke, 1975). The Antwerp Zoo in Belgium has maintained a healthy colony of several animals since the 1950s, but unfortunately, no breeding activity has occurred. That group is predominantly composed of females, with some males having been sent to other zoos and others dying before reaching puberty (Mortelmans, 1972).

One male was sent to the San Diego (California) Zoo in June 1960, and a female arrived there in August 1962. This pair bred successfully and produced its first infant, a female, six years later (Hill, 1968). Six additional infants have resulted from this breeding pair, Kakowet and Linda, and the infants have remained with them to create a most admirable functional display for this species.

The application of human clinical diagnostic procedures to an infertile female of *Pan paniscus* at the San Diego Zoo was reported by Lasley *et al.* (1977).

TABLE 9

Outcome of pregnancy at Dutch Primate Center.

Year	# Breeding Females	# Stillbirths	# Live Births	Total # Births	# Abortions
1971	20	1	1	2	1
1972	20	0	4	4	2
1973	20	1	4	5	4
1974	20	1	8	9	2
1975	30	0	9	9	2
1976	30	0	8	8	7
1977	30	1	14	15	2
1978	30	1	8	9	1
Totals		5	56	61	21

In March 1975, five pygmy chimpanzees on loan from Zaire were flown to the Yerkes Primate Center as part of a project under the auspices of the United States National Academy of Sciences and the Zaire government to establish baseline medical, physiologic, and behavioral information. As a result of poor health, two of the five animals died soon after arrival, leaving one young male, presently at breeding age (approximately 10 years old), and two females, one very old (25 years or more) and the other currently reaching maturity (approximately 10 years old).

A collaborative attempt has been made between the San Diego Zoo and the Yerkes Primate Center to mix the gene pools of the two groups of *Pan paniscus*. Currently, two females of breeding age from the San Diego population are in residence with the Yerkes Primate Center male, Bosondjo, and a mating program is underway. One female infant, born alive in March 1980, died shortly thereafter. Another is expected as of this writing. A male infant, born in February 1980, has both parents as part of the Yerkes colony. Until more males of this species are produced by breeding, either by natural means or by artificial insemination using frozen semen, the existence of this species in captivity will be relatively tenuous. *Pan paniscus* lives in Zaire in a very inaccessible area, thus remaining somewhat unthreatened there by invading man so far as is known. Political and legal barriers limit our ability to import the species for establishment of research colonies; therefore, successful breeding of the few available animals is essential for long-term study of the species and for acquisition of new knowledge about the species relevant to primate and especially human evolution, biology, and medicine.

C. *Pongo pygmaeus*

Except for the orangutan rehabilitation centers in Borneo and Sumatra, the largest captive group of orangutans anywhere in the world is at the Yerkes Primate Center, although individuals and small groups exist in sizable numbers in zoological gardens. In 1963, a group of 20 animals was purchased from the private collection of Mr. Kenneth Berry in Portland, Oregon, and was flown intact to Orange Park, Florida. The group subsequently came to Atlanta with the rest of the primate facility.

The majority of the animals have all been caged in pairs at the main center building. In 1967, however, a display group of three females with their infants and one male was established at the nearby Atlanta

Zoo. Both groups of animals have bred very successfully. The first infant was born in May 1966, and since then 49 births have occurred, from 14 females and 10 males. Three females and one male are of the Bornean subspecies (*Pongo pygmaeus pygmaeus*), the rest are Sumatran (*Pongo pygmaeus abelii*). Four Bornean infants and seven hybrids are among the progeny.

Three females and one male born in the colony have reached reproductive maturity and either have become pregnant or have sired offspring. One infant was stillborn after a full-term gestation; another was born alive and healthy. A prematurely born infant sired by the colony-born male and delivered by a colony-born female died within a month after delivery. Delivery of that infant was by cesarean section due to poor health of the mother.

Although 49 orangutan births implies a successful breeding program, little is known about the reproductive process in this species. Menstrual cycle physiology is discussed elsewhere in this volume (see Graham, Chapter 1, this volume). There is no genital sexual swelling, peri-anal sex-skin coloration, or labial tumescence, and only by daily testing of urine samples for menstrual blood using commercial reagent strips can menstrual periodicity be readily monitored.

At the Yerkes Primate Center, animals are paired in cages similar in size to those for chimpanzees, and when necessary, urine is collected for pregnancy diagnosis. The Subhuman Primate Pregnancy Test works equally well with orangutans as for chimpanzees in detecting pregnancy (Woodard *et al.*, 1976; Hobson, 1976; Hodgen *et al.*, 1977; Davis, 1977). However, systematic investigations involving single-day matings and subsequent urinary chorionic gonadotropin measurement are required to indicate precisely the earliest time at which the test is effective, as well as to establish an accurate gestation period for the species.

The orangutan demonstrates an external change in genital appearance during pregnancy (Schultz, 1938) that has only limited value for pregnancy diagnosis. This change manifests itself as an increase in turgidity and size of labial skinfolds, attaining its maximum during the last month of gestation. Thus, it is quite different from the genital sexual tumescence changes often seen in the chimpanzee during its first trimester of pregnancy. Experience with the Yerkes colony animals suggests a gestation period ranging from 227 to 275 days, typically about 245 days, based upon mating records and rectal palpation.

Reports have been published of orangutan breeding practices and gestation periods by various zoological facilities. Coffey (1971), at the Jersey Wildlife Preservation Trust, reports a range of from 280 to 301 days; Asano (1967) at the Tokyo Zoo in Japan indicates a time of 261 days; and de Silva (1972) at the Sepilok Game Reserve in Sabah, Malaysia, reports 275 days. Other descriptions of orangutan breeding (Perry and Horsemen, 1972; Chaffee, 1967; Fontaine, 1968) make no estimates of the length of pregnancy. Davis (1977) used the Subhuman Primate Pregnancy Test on a single pregnant orangutan and recorded the initial positive test 256 days prior to delivery.

D. *Gorilla gorilla*

Just as with the orangutan, the gorilla has been bred with limited success in captivity (chiefly in zoological gardens) for many years (Jones, 1968). Until recently, there have been no systematic studies of menstrual cyclicity, reproductive behavior, and gestation that would provide a basis for improved breeding techniques. More than 400 gorillas exist captive in zoological facilities, the vast majority being the lowland species (*Gorilla gorilla gorilla*).

During a recent 20-year span (1956–1975), no more than 136 births occurred (R. D. Martin, 1976), and infant mortality in these facilities is high. While the lowland subspecies is still a viable species in its native habitat, the mountain variety (*Gorilla gorilla beringei*) is very threatened and numbers but a few in zoological collections.

The Yerkes Primate Center has the largest captive colony of lowland gorillas in the world. Fifteen were acquired between 1961 and 1965 from animal import companies and private collections while the facility was at Orange Park, Florida. An additional four were obtained between 1965 and 1968. The animals were immature and only recently have become reproductively active. Presently, 18 animals are at the center.

Breeding activity began naturally, since these animals were caged together in male/female pairs. No particular mating practices were instituted, and in at least one instance an infant was delivered without any knowledge that the mother had been pregnant. Studies by Nadler (1975) established the length of the menstrual cycle as 31–32 days, probably a more reliable figure than the estimate of 39 days derived by Noback from one adolescent animal (1939). Both authors have clearly described the tumescence and detumescence patterns of the external perineal labia; Nadler's observations of the very short time-duration of maximal tumescence (1–5 days) make this phenomenon an appropriate indicator of the time of ovulation and thus useful for a controlled mating program.

Since 1972, seven infant gorillas have been born, involving five mothers and three fathers. Three of the infants are males and all are presently in good health. Since three of the infants were born to females in a group of five females and three males, no really accurate gestational information is available. Estimates in the literature give values of 254 days (Hopper *et al.*, 1968), 255 days (Fontaine, 1968; Tullner and Gray, 1968), 256–259 days (Seaton, 1978), and 259–270 days (Mallinson *et al.*, 1976). R. D. Martin (1976) reports 256 ± 2 days for one timed pregnancy.

Evaluation of existing pregnancy tests has not been performed using gorillas at the Yerkes Primate Center, primarily because of the very small number of births among the caged animals during the past several years. Chorionic gonadotropin, however, can be measured using bioassay (Tullner and Gray, 1968) and radioimmunoassay (Czekala and Lasley, 1978).

Several zoological gardens throughout the world have had some success in breeding lowland gorillas; most notable are those at Kansas City, Missouri; Basel, Switzerland; Dallas, Texas; and at Howlett's Wildlife Park in Canterbury, Kent, England. Several recent reports have described experience with breeding and delivery of both lowland gorillas (Fisher, 1972; Kagawa and Kagawa, 1972; Kirchshofer *et al.*, 1971) and mountain gorillas (Tijskens, 1971). At the Cincinnati (Ohio) Zoo, two breeding pairs (King Tut–Penelope and Hatari–Matari) have produced 10 live infants (four males and six females) and one stillborn female during a period from 1967 through 1978. The 10 live-born infants are still alive, and two, born in January 1970, will shortly be reaching reproductive maturity.

III. CONTEMPORARY TOPICS OF IMPORTANCE TO GREAT APE BREEDING

A. Artificial Breeding of the Chimpanzee

Laboratory-reared great apes are sometimes deficient in their reproductive behavior. Artificial breeding techniques can be used to breed such animals to increase colony breeding efficiency and to contribute to the available gene pool. Semen banking and artificial insemination would also permit increased exchange of genetic material between colonies, reducing inbreeding.

Successful artificial insemination depends upon a knowledge of semen collection and evaluation techniques, as well as a knowledge of the time of ovulation. Ovulation time is approximately known in the chimpanzee in relation to several other parameters, such as sexual-swelling detumescence (Graham *et al.*, 1973), cervical mucus ferning patterns (Gould *et al.*, 1976), and basal body temperature shifts and urinary gonadal hormone levels (Graham *et al.*, 1972, 1977). (This subject is reviewed by Graham in Chapter 1, this volume.) Although semen collection techniques have been developed and perfected for a wide variety of primate and nonprimate species, they have been difficult to achieve in the great apes.

Investigators at the Delta Regional Primate Research Center in Louisiana were the first to apply rectal probe electro-ejaculation to semen collection in the chimpanzee (Fussell *et al.*, 1967; Roussel and Austin, 1968). However, in neither of these reports nor in any other studies involving electro-ejaculation of primates was there a sufficient study of the electrical parameters involved to be certain of safety of this procedure to the animals. Knowledge of the characteristics of the electrical stimuli, the current density within the tissues, and the impedance at the electrode/tissue interface is essential. This information allows other investigators to duplicate the stimulus regimen successfully in their laboratories.

At the Yerkes Primate Center, a detailed investigation was made with the aim of developing the instrumentation and knowledge to allow routine and safe collection of semen by electro-ejaculation. Results of initial success were published several years ago (Warner *et al.*, 1974), and details of improvements in this technology have appeared more recently (Gould *et al.*, 1978).

Simultaneous with the development of electro-ejaculation came the use of an artificial vagina for semen collection. Again, the Delta Primate Center pioneered in these efforts (Fussell *et al.*, 1973). However, investigators at the Toledo (Ohio) Zoo (Hardin *et al.*, 1975) also utilized this technique, and repeated inseminations during a four-day period were successful in initiating a pregnancy, which went to term and yielded a healthy female infant.

Shortly thereafter, workers at the Yerkes Primate Center, who had trained a few of their male chimpanzees to exchange a masturbated semen specimen for an orange slice reward, were successful in initiating a pregnancy (Martin *et al.*, 1978). The male semen donor had not sired any infants due to behavioral incompetency rather than physiological problems, and the female, although multiparous, had not successfully bred for four years. The single-day insemination allowed precise determination of gestation, 244 days, which is slightly longer than the mean of 227 days for eight single-day natural matings (Martin *et al.*, 1978).

Continuing efforts are underway at the Yerkes Primate Center to increase the frequency of successful artificial inseminations. These efforts involve four areas: (1) multiple inseminations during a given ovulatory period; (2) use of females having optimally regular menstrual cyclicity; (3) performance of inseminations and semen collections without anesthesia; and (4) detailed study of the incidence of anovulation during apparently normal menstrual cycles.

A second successful pregnancy involved the insemination of sperm from two of three donors on each of two successive days into a multiparous female who had not been pregnant for three years. Delivery of a live term infant occurred 232 or 233 days later. One of the male semen donors contributed to the aforementioned single-day successful artificial insemination. The other two males subsequently left the Yerkes colony without siring any other infants.

B. Breeding Orangutan Hybrids

Wild populations of orangutan are found in the islands of Borneo and Sumatra, and two subspecies or races have been described: the Bornean (*Pongo pygmaeus pygmaeus*) and the Sumatran (*Pongo pygmaeus abelii*). Only minor phenotypic differences occur between the races (Jones, 1969; Eckhardt, 1975; MacKinnon, 1975), and hybrid animals can be produced if specimens of different races are allowed to mate in captivity.

The chromosomes of the orangutan were first described by Chiarelli (1961); he did not demonstrate a chromosomal difference between the two subspecies. Thus, there has been little reason to believe that interbreeding of the two subspecies would bring about problems with hybrid infertility. However, recent collaborative studies between the Medical Research Council in Edinburgh, Scotland, and the Yerkes Primate Center have yielded information that is somewhat contrary to this view.

A comprehensive study of the chromosomes of both subspecies identified a complex variation at chromosome #9 (two pericentric inversions) in four Bornean and eight Sumatran (12 out of 23) animals (Seuanez *et al.*, 1976). The fact that both Bornean and Sumatran orangutans have this variation indicates that the complex rearrangement that initiated this new chromosome type must have occurred before Borneo and Sumatra separated. Approximately 800 generations have passed since the land mass constituting the two islands was separated by the South

China Sea. For this complex rearrangement to have been maintained for at least 8000 years, there must have been no selective disadvantage to its heterozygote carriers.

On chromosome #2, however, an even more important finding was made. As seen in Fig. 1, this chromosome pair in Loklok, a young Yerkes male orangutan, is in fact composed of two different chromosomes. The chromosome indicated by the arrow has not been observed in 16 Sumatran animals studied, but is very common in Bornean animals, as seen in Fig. 2, illustrating the karyotype of Abang, a young Bristol (England) Zoo male orangutan. All six Bornean orangutans studied thus far are homozygote carriers for it.

It is probable (Seuanez *et al.*, 1979) that the Bornean version of chromosome #2 may be the original chromosome of the species *Pongo pygmaeus* and that a pericentric inversion has created the Sumatran form.

Since the inversion is pericentric, involving about one-fifth of the total chromosome length, the probability of producing unbalanced gametes by chiasma formation within an inversion pairing loop is rather low, with consequent production of infertile offspring. Nevertheless, in view of the fact that the orangutan is such a highly endangered species, the further production of hybrids should be avoided until the existing hybrids, identified by their karyotype or parentage, have been thoroughly tested for fertility. In addition to the seven hybrids produced at the Yerkes Primate Center, one of whom is now 10 years of age, there is at least one other hybrid reported in the literature (Turleau *et al.*, 1975). Loklok, the male Yerkes hybrid, recently sired an infant, which died after complications from a premature delivery.

IV. SUMMARY

A small breeding population of each species of the great apes in zoos is supplemented by a sizable contribution, particularly of chimpanzees, from the several research facilities. The Yerkes Regional Primate Research Center presently houses successful breeding colonies of orangutans and gorillas in addition to both species of chimpanzee. The pygmy chimpanzee and the mountain gorilla have the most tenuous hold on survival both in captivity and in the wild. The chimpanzee, with a captive United States population of approximately 1300 animals, holds the greatest promise for survival. New demographic studies of

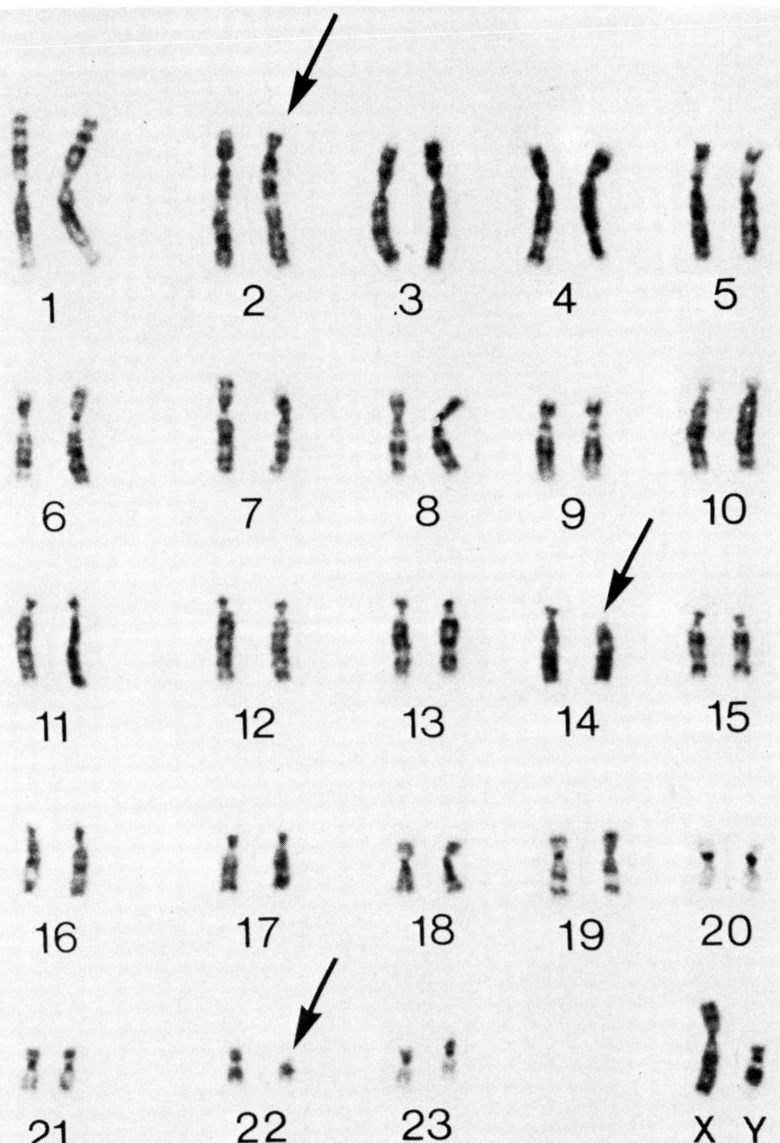

Fig. 1. Karyotype of phenotypically appearing Sumatran orangutan Loklok (Yerkes Regional Primate Center) showing three variant chromosomes: a pericentric inversion on chromosome #2, deletion of genetic material on chromosome #14, and deletion of the entire short arm on chromosome #22.

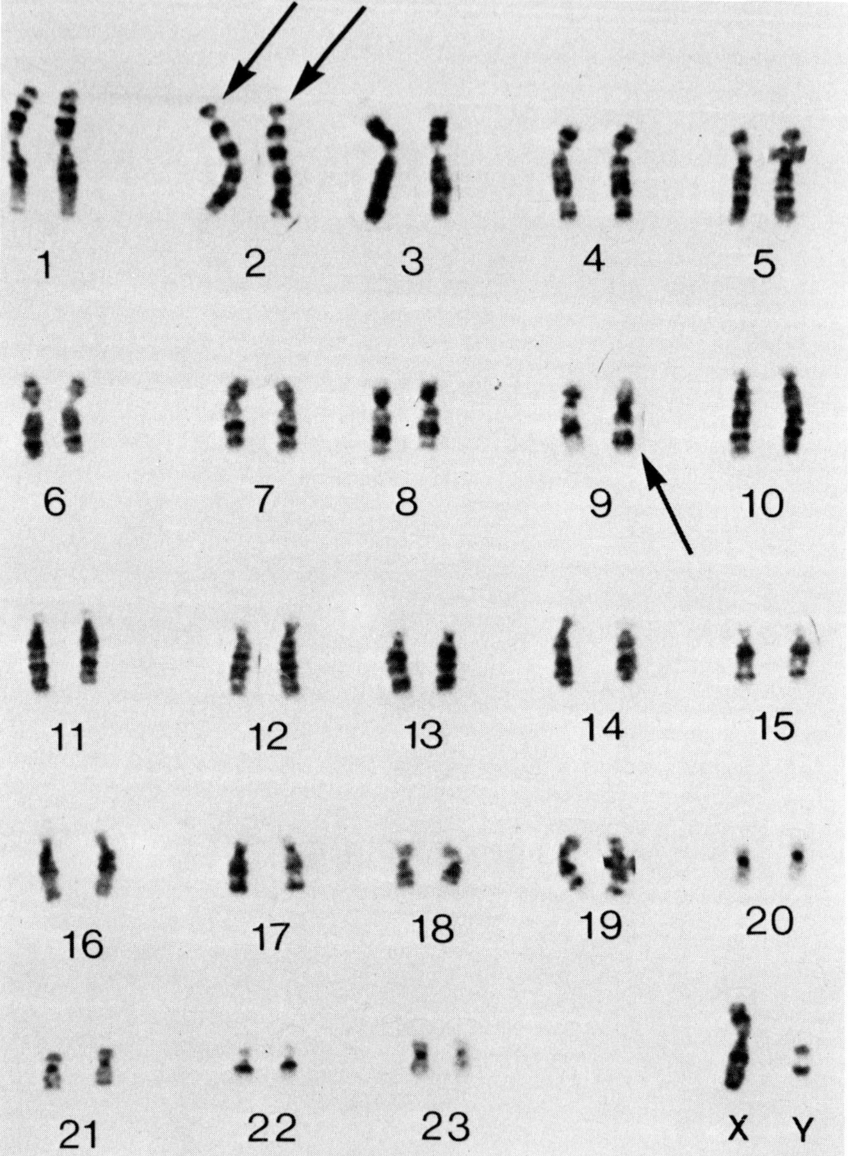

Fig. 2. Karyotype of Bornean orangutan Abang (Bristol Zoo, England) showing two variant chromosomes: #2 and #9. Chromosome #2 may be in fact the "normal" chromosome type in Bornean orangutans since it is found in all those studied thus far (Seuanez *et al.*, 1979). A pericentric inversion between two chromosome breaks explains the derivation of the Sumatran form. Thus, Loklok is probably a hybrid.

the United States chimpanzee population (Seal, personal communication) are providing an important data base for breeding management, as well as revealing new data not evident from the individual colonies, such as a trend toward a reduced summer birthrate. It is hoped that improved breeding practices, including artificial insemination, genetic studies, and practice of sound animal husbandry, will ensure the survival of all the great ape species.

ACKNOWLEDGMENTS

The author would like to thank Dr. William Hobson (ICES), Dr. Larry B. Cummins (SFRE), Dr. C. James Mahoney (LEMSIP), Dr. Hans Balner (TNO), and Dr. Hector Seuanez (MRC Edinburgh and presently University of Rio de Janeiro), who assisted substantially in the assimilation of data presented in this paper. Ms. Stephanie Linz typed the final manuscript.

REFERENCES

Asano, M. (1967). *Int. Zoo Yearb.* **7,** 95.

Balner, H. (1975). *Lab. Anim. Handb.* **6,** 31–39.

Balner, H., Gabb, B. W., D'Amaro, J., van Vreeswijk, W., and Visser, T. P. (1974). *Tissue Antigens* **4,** 313–328.

Balner, H., van Vreeswijk, W., Roger, J. H., and D'Amaro, J. (1978). *Tissue Antigens* **12,** 1–18.

Boorman, G. A., Speltie, T., Goosen, C., and Balner, H. (1972). *Proc. Conf. Exp. Surg. Prim., 3rd, Lyon* **1,** 81–87.

Boorman, G. A., Beijersbergen, E. J., Speltie, T. M., and Balner, H. (1973). *Primate Center TNO Annual Report,* pp. 223–224.

Boorman, G. A., Speltie, T. M., and Fitzgerald, G. H. (1974). *J. Med. Primatol.* **3,** 269–275.

Bourne, G. H. (1972). *In* "Breeding Primates" (W. I. B. Beveridge, ed.), pp. 24–33. Karger, Basel.

Bourne, G. H., Keeling, M. E., and Golarz de Bourne, M. N. (1975). *Lab. Anim. Handb.* **6,** 63–76.

Butler, T. M. (1972). *J. Med. Primatol.* **1,** 51–57.

Chaffee, P. S. (1967). *Int. Zoo Yearb.* **7,** 94–95.

Chiarelli, B. (1961). *Nature (London)* **192,** 285.

Clegg, M. T., and Weaver, M. (1972). *Proc. Soc. Exp. Biol. Med.* **139,** 1170–1174.

Coffey, P. F. (1971). *Ann. Rep. Jersey Wildlife Preserv. Trust* **8,** 32–36.

Czekala, N. M., and Lasley, B. L. (1978). *Abstracts Ann. Meeting Am. Soc. Primatol., 2nd, Atlanta.*

Davis, J. H., and Moor–Jankowski, J. (1971). *Proc. Conf. Exp. Surg. Prim., 2nd, New York,* pp. 947–955.

Davis, R. R. (1977). *J. Med. Primatol.* **6,** 315–318.

de Silva, G. S. (1972). *Int. Zoo Yearb.* **12,** 104–105.

Douglas, J. D., and Butler, T. M. (1970). *Lab. Anim. Care* **20,** 477–482.

Eckhardt, R. B. (1975). *Am. J. Phys. Anthropol.* **42,** 349–350.

Elder, J. H. (1938). *Yale J. Biol. Med.* **10,** 347–364.

Faiman, C., Stearns, E. L., Winter, J. S. D., Reyes, F. I., and Hobson, W. C. (1975). *Proc. Soc. Exp. Biol. Med.* **149,** 670–676.

Fisher, L. E. (1972). *Int. Zoo Yearb.* **12,** 106–108.

Fontaine, P. A. (1968). *Int. Zoo Yearb.* **8,** 115–118.

Fussell, E. N., Roussel, J. D., and Austin, C. R. (1967). *Lab. Anim. Care* **17,** 528–530.

Fussell, E. N., Franklin, L. E., and Frantz, R. C. (1973). *Lab. Anim. Sci.* **23,** 252–255.

Gould, K. G., Martin, D. E., and Graham, C. E. (1976). *Scanning Electron Microsc.* Pt. VI, 335–342.

Gould, K. G., Warner, H., and Martin, D. E. (1978). *J. Med. Primatol.* **7,** 213–222.

Graham, C. E. (1970). *In* "The Chimpanzee" (G. H. Bourne, ed.), Vol. III, pp. 183–220. Karger, Basel.

Graham, C. E. (1979). *Am. J. Phys. Anthropol.* **50,** 291–300.

Graham, C. E., Guilloud, N., and McArthur, J. W. (1969). *Recent Adv. Primatol.* **2,** 66–72.

Graham, C. E., Collins, D. C., Robinson, H., and Preedy, J. R. K. (1972). *Endocrinology* **91,** 13–24.

Graham, C. E., Keeling, M., Chapman, C., Cummins, L. B., and Haynie, J. (1973). *Am. J. Phys. Anthropol.* **38,** 211–215.

Graham, C. E., Warner, H., Misener, J., Collins, D. C., and Preedy, J. R. K. (1977). *J. Reprod. Fertil.* **50,** 23–28.

Guilloud, N. B. (1969). *Ann. N.Y. Acad. Sci.* **162,** 297–300.

Hardin, C. J., Liebherr, G., and Fairchild, O. (1975). *Int. Zoo Yearb.* **15,** 132–134.

Hill, C. A. (1968). *Int. Zoo Yearb.* **8,** 119–120.

Hobson, B. M. (1976). *Lab. Anim.* **10,** 87–91.

Hobson, W., and Fuller, G. B. (1977). *Biol. Reprod.* **17,** 294–297.

Hobson, W., Coulston, F., Faiman, C., Winter, J. S. D., and Reyes, F. (1976). *J. Toxicol. Environ. Health* **1,** 657–668.

Hodgen, G. D., and Ross, G. T. (1974). *J. Clin. Endocrinol. Metab.* **38,** 927–930.

Hodgen, G. D., Turner, C. K., Smith, E. E., and Bush, R. M. (1977). *Lab. Anim. Sci.* **27,** 99–101.

Hopper, B. R., Tullner, W. W., and Gray, C. W. (1968). *Proc. Soc. Exp. Biol. Med.* **129,** 213–214.

Hummer, R. L., May, H. C., and Knight, W. F. (1969). *Proc. Int. Congr. Primatol., 2nd, Atlanta, 1968* **2,** 1–8.

Jantschke, F. (1974). *In* "Breeding Endangered Species in Captivity" (R. D. Martin, ed.), pp. 245–251. Academic Press, New York.

Jones, M. L. (1968). *Int. Zoo Yearb.* **8,** 183–192.

Jones, M. L. (1969). *Proc. Int. Congr. Primatol., 2nd, Atlanta, 1968* **2,** 217–223.

Kagawa, M., and Kagawa, K. (1972). *Int. Zoo Yearb.* **12,** 105–106.

Keeling, M. E., and Roberts, J. R. (1972). *In* "The Chimpanzee" (G. H. Bourne, ed.), Vol. V, pp. 127–152. Karger, Basel.

Kirchshofer, R., Weisse, K., Berenz, K., Klose, H., and Klose, I. (1971). *Int. Zoo Yearb.* **11,** 121–131.

Kollar, E. J., Beckwith, W. C., and Edgerton, R. B. (1968). *J. Nerv. Ment. Dis.* **147,** 444–459.

Kraemer, D. C., Kalter, S. S., and Moore, G. T. (1975). *Lab. Anim. Handb.* **6,** 41–47.

Lasley, B. L., Kennedy, J. F., Robinson, P. T., Bogart, M. H., and Benirschke, K. (1977). *Zool. Garten N.F. Jena* **47,** 289–295.

MacKinnon, J. (1975). *Int. Zoo Yearb.* **15,** 195–197.

Mallinson, J. J. C., Coffey, P., and Usher-Smith, J. (1976). *Int. Zoo Yearb.* **16,** 189–194.

Martin, D. E. (1976). *Yerkes Newsletter* **13(2),** 20–24.

Martin, D. E., Graham, C. E., and Gould, K. G. (1978). *Symp. Zool. Soc. London* **43,** 249–260.

Martin, R. D. (1976). *New Sci.* **76,** 100–102.

Moor-Jankowski, J., and Goldsmith, E. I. (1971). *Proc. Conf. Exp. Med. Surg. Prim., 2nd, New York, 1969* 939–946.

Mortelmans, J. (1972). *Proc. Conf. Exp. Med. Surg. Prim., 3rd, Lyon* 138–142.

Nadler, R. D. (1975). *Science* **189,** 813–814.

Nissen, H. W., and Yerkes, R. M. (1943). *Anat. Rec.* **86,** 567–578.

Noback, C. R. (1939). *Anat. Rec.* **73,** 209–225.

Peacock, L. J., and Rogers, C. M. (1959). *Science* **129,** 959.

Perry, J., and Horsemen, D. L. (1972). *Zoologica* **56,** 105–108.

Roussel, J. D., and Austin, C. R. (1968). *J. Inst. Anim. Tech.* **19,** 22–32.

Schultz, A. H. (1938). *J. Mammal.* **19,** 363–366.

Seaton, B. (1978). *J. Reprod. Fertil.* **53,** 231–236.

Seuanez, H., Fletcher, J., Evans, H. J., and Martin, D. E. (1976). *Cytogenet. Cell Genet.* **17,** 327–337.

Seuanez, H. N., Evans, H. J., Martin, D. E., and Fletcher, J. (1979). *Cytogenet. Cell Genet.* **23,** 137–140.

Srb, A. M., Owen, R. D., and Edgar, R. S. (1965). "General Genetics," 2nd ed. Freeman, San Francisco, California.

Teleki, G., and Baldwin, L. (1975). *Smithsonian* **5,** 76–80.

Tijskens, J. (1971). *Int. Zoo Yearb.* **11,** 181–183.

Tullner, W. W., and Gray, C. W. (1968). *Proc. Soc. Exp. Biol. Med.* **128,** 954–956.

Turleau, C., De Grouchy, J., Chavin-Colin, F., Mortelmans, J., and van den Bergh, W. (1975). *Ann. Genet.* **18,** 227–233.

van Riper, D. C., Fineg, J., and Day, P. W. (1967). *Lab. Anim. Care* **17,** 472–478.

van Vreeswijk, W., and Koning, H. (1978). *Recent Adv. Primatol.* **2,** 279–287.

Warner, H., Martin, D. E., and Keeling, M. E. (1974). *Ann. Biomed. Eng.* **2,** 419–442.

Winter, J. S. D., Faiman, C., Hobson, W. C., Prasad, A. V., and Reyes, F. I. (1975). *J. Clin. Endocrinol. Metab.* **40,** 545–551.

Woodard, D. K., Graham, C. E., and McClure, H. M. (1976). *Lab. Anim. Sci.* **26,** 922–927.

Yerkes, R. M., and Elder, J. H. (1937). *Yale J. Biol. Med.* **10,** 41–48.

Young, W. C., and Yerkes, R. M. (1943). *Endocrinology* **33,** 121–154.

Chapter 15

VETERINARY MANAGEMENT OF GREAT APES FOR REPRODUCTIVE BIOLOGY

Kenneth G. Gould

375

Copyright © 1981 by Academic Press, Inc.
All rights of reproduction in any form reserved.
ISBN 0-12-295020-8

I. INTRODUCTION

The close phylogenetic affinity between the apes and man has been appreciated for many years. Correlations have been drawn between these species on the basis of the present and fossil anatomical record; the apparent similarities in overall size, locomotion, physiology, and cognitive behavior have led many investigators to utilize these species for research studies involved with problems related to human welfare. Although basic anatomical data has been accumulated since the turn of the century, it is in the last two decades that results of more intensive studies on great ape physiology have been published. At the same time that such data has been collected, further basic information relating to the clinical welfare of the apes in captivity has been accumulated, and such hematological and hemochemical baseline data are necessary for maintenance and utilization of apes in research. These data have been published in increasing volume during the last decade, and summary data are presented in Table 1. For complete data, the reader is referred to McClure *et al.* (1973). There is an increasing interest in the use of the great ape species involving reproductive biology and physiology. Although intuitively the value of these animals as models for research directed toward human application is evident, the practicality of their use is a matter worthy of careful consideration. The value of any one of the ape species in a given project resides in a balance between problems associated with availability, maintenance, and practicability of use on the one hand, and the significance of the physiological value of the ape as a model on the other (see Chapter 16). It is evident that the physical size of the apes permits implantation of telemetry and monitoring equipment of a practically attainable size, and techniques and equipment developed for use in the chimpanzee will be readily applicable to the human. This last situation operates in the reverse direction as well, in that equipment designed for surgical manipulations and investigation in the human are usually, with minor modification, readily applied to research projects using the chimpanzee. Throughout this discussion it is assumed that the chimpanzee is the most frequently utilized great ape model for studies applicable to the human. All the species of great apes are currently on the endangered list, and the number of orangutans and gorillas in captivity is so small that they are less practical than the chimpanzee for use as experimental models.

In light of the foregoing comments, it is appropriate to review the various requirements for maintenance, caging, and manipulation of the apes that are essential for their use in research projects. This discussion will focus on the normal maintenance, medical supervision, and an-

TABLE 1

Hematologic data for great apes[a].

	Chimpanzee			Orangutan			Gorilla		
	Mean	*Range*	*SD*	*Mean*	*Range*	*SD*	*Mean*	*Range*	*SD*
RBC, 10^6	5.02	3.05–6.23	0.48	4.36	3.17–6.64	0.63	4.21	3.34–5.02	0.34
WBC, 10^3	12.7	4.80–50.3	6.2	11.3	5.60–32.0	5.19	9.50	4.80–27.5	4.13
HCT, %	43.9	29–58	3.68	37.91	29–45	3.60	40.04	35–45	2.43
Hemaglobin, g%	13.9	7.90–18.9	1.16	11.76	8.50–15.0	1.27	12.11	10.5–13.5	0.74
Platelets, 10^3	268.1	60–975	113.12	280.53	125–508	84.82	269.84	125–455	89.34
Sedimentation rate, mm/hr	21.7	0–53	15.1	0.21	0–3	0.59	15.63	0–51	14.44
MCV, μm^3	87.8	64.9–109.1	8.0	86.83	64.8–110.4	12.28	95.37	82.0–113.8	6.93
Cholesterol, mg%	209.2	104–390	48.0	215.8	104–382	51.8	336.6	246–455	48.8
Uric acid, mg%	3.6	1.7–9.8	1.1	3.1	1.4–8.2	1.1	2.4	1.6–3.4	0.4
Total protein, g%	7.4	5.1–9.0	0.7	7.5	5.5–8.9	0.7	6.8	5.6–8.2	0.6
Albumin, g%	3.7	2.6–5.3	0.4	4.2	3.3–5.2	0.5	4.6	3.3–5.5	0.5
Globulin, g%	3.7	1.7–5.6	0.7	3.2	1.4–5.2	0.7	2.2	1.1–3.7	0.7
SGPT, S–F units	11.9	1–111	9.0	10.6	4–28	4.7	11.0	1–27	5.3
SGOT, S–F units	18.1	2–95	9.9	13.9	4–37	6.3	20.8	4–53	10.7

[a] Data derived from McClure *et al.* (1973).

Table 1 (Continued)

	Chimpanzee			Orangutan			Gorilla		
	Mean	*Range*	*SD*	*Mean*	*Range*	*SD*	*Mean*	*Range*	*SD*
Bilirubin, total mg%	0.17	0.1–0.7	0.11	0.55	0.1–3.4	0.63	0.23	0.1–0.7	0.15
Bilirubin, direct mg%	0.08	0–0.35	0.06	0.14	0–1.2	0.16	0.09	0.05–0.35	0.07
Bilirubin, indirect mg%	0.09	0.06	0.09	0.39	0–3.3	0.54	0.14	0–0.45	0.12
Calcium, mg%	9.3	6.4–12.0	0.9	9.3	5.4–13.0	0.9	9.5	7.6–10.8	0.7
Phosphorus, mg%	4.8	2.1–8.0	1.2	4.4	1.1–8.2	1.0	5.2	3.9–7.1	0.7
Sodium, Meq/liter	139.1	125–150	3.6	138.4	126–150	3.8	135.2	128–144	3.9
Potassium, Meq/liter	3.7	2.2–5.9	0.5	4.1	3.0–6.2	0.6	4.1	3.3–5.3	0.4
Chloride, Meq/liter	99.7	80–112	4.1	100.6	88–109	3.6	97.8	90–109	4.5
CO_2, Meq/liter	23.2	15.0–33.5	4.0	21.3	11.5–29.5	3.8	21.0	14.0–26.5	3.2
Glucose, mg%	77.9	36–134	16.3	78.0	44–154	17.0	79.3	54–112	12.3
Amylase, Somogyi units	68.5	14–220	39.5	113.5	20–502	66.9	39.7	12–140	24.9
Urea N_2, mg%	14.2	4–31	4.8	14.5	4–31	5.3	13.8	7–26	4.2
Creatinine, mg%	0.9	0.5–1.7	0.3	0.9	0.4–2.0	0.3	0.8	0.4–1.5	0.2

[a] Data derived from McClure *et al.* (1973).

esthetic and surgical manipulations appropriate for research into reproductive biology and physiology (Beveridge, 1969; Keeling and Guilloud, 1971).

II. MAINTENANCE OF THE GREAT APES AS EXPERIMENTAL MODELS

A. Caging

The great apes are large, and their strength is disproportionately large relative to their body weight. Individual apes, especially chimpanzees, are of mercurial temperment and often highly excitable. Their physical size and strength require caging facilities of high physical strength, and control of the animals for routine and experimental purposes requires a greater amount of medication and manpower than would be true of the lower primates. For adequate maintenance of the apes, three basic caging systems can be utilized. These are (1) compound housing, (2) large-cage housing (for two to three subjects), and (3) individual-cage housing.

Compound housing is marginally the cheapest and most conducive for natural social behavior. Individual small-cage housing is the least satisfactory in terms of the animal's welfare and well-being as a result of the inevitable restriction and isolation of the individual animal, but is virtually mandatory for successful experimental manipulation.

Although examples exist of centers in which good research is being conducted utilizing the ape species in circumstances where these three caging facilities are not all available, ideally a combination of the three alternatives should be provided to meet the various needs of the research and breeding colony, as typified by the caging choices available at the Yerkes Primate Center.

1. Compound Housing

Compound housing at the Yerkes Primate Center comprises individual compounds with a dirt floor of approximately 930 sq m area with wire-mesh fences of 250 cm height, topped with smooth sheet metal for another 250 cm beyond. Enclosed quarters are provided for each compound, which can be heated during cold weather and which permit capture and caging of individual animals when required. It is extremely difficult, however, to conduct invasive or semi-invasive research under these conditions as collection of biological fluids entails an excessive expenditure of personnel time for animal capture. It should be possible to design compound housing with improved facilities for animal manipulation and isolation, and an example of this type of housing is provided at the Center for Chimpanzee Husbandry and Environmental Breeding in west Texas.

2. Large-Cage Housing

Concrete cages at the Yerkes Primate Center, suitable for two to three adult apes, have an indoor section approximately 250 cm square, and a slightly larger outdoor section 250 cm by 300 cm with 6-gauge wire-mesh roofing. A guillotine door between the indoor and outdoor sections provides a mechanism by which one animal can be separated from its cage mates for capture and transport to restraint cages (Fig. 1). Capture of individual animals in these indoor/outdoor cages is achieved by positioning a transfer box (Figs. 2 and 3) against the sliding

Fig. 1. Indoor cage at the Yerkes Regional Primate Research Center of Emory University. Note the heavy (6 gauge wire, sliding inner door, and steel guillotine door leading to the outdoor enclosure.

Figs. 2 and 3. Transfer box. The front vertical face of the box abuts the frame of the sliding door in Fig. 1.

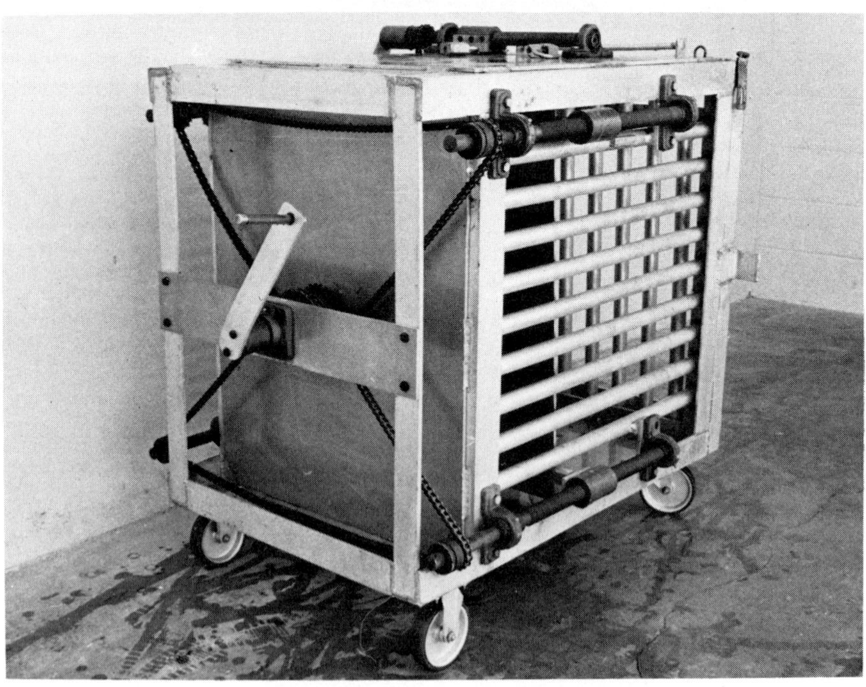

door of the cage; the animals are trained to enter this transfer box readily. The transfer box is then coupled to a mobile squeeze cage (Figs. 4 and 5), which the animal is persuaded to enter, so permitting restraint of the animal for intramuscular injection of anesthetic or other treatment. These transfer boxes and squeeze boxes are custom-built at the facility and are not available on the commercial market. Problems arise, however, in the use of such equipment when individual animals refuse to enter either the initial transfer box or the squeeze cage. Under these circumstances it is necessary to tranquilize the animals using either a Cap-chur gun (Palmer Equipment Co., Douglasville, GA) or a blow gun, to deliver a dart loaded with the drug. This requirement provides a constraint on the size of the caging provided for the animals because capture equipment cannot be used in extremely confined quarters. When large animals are involved, ketamine hydrochloride is less practical as an anesthetic for use in these methods because of the larger volume that is required to sedate the animal. Detailed description of the equipment required for capture, restraint, and transfer of chimpanzees as utilized at the Yerkes Center is available in the literature (Guilloud and McClure, 1972).

The disadvantages of this form of caging are the relatively high construction cost and inconvenience for experimental manipulation of an individual animal when compared to the accessibility provided by individual caging systems.

3. Individual-Cage Housing

For maximum animal access, the Yerkes Center utilizes specially designed and constructed metabolism cages of approximately 11.6 cu m (Fig. 6). The back of each cage can be advanced by means of a chain-driven screw drive powered by an electric motor. This feature permits the animal to be squeezed or restrained to permit examination of external lesions, administration of drugs, and so forth. Simpler designs using hand-cranked systems, or lever systems operating the back by a "scissors" mechanism, are not always satisfactory in cages of this size, as the apes can sometimes overcome the driving force of such equipment. Housing of animals in these individual cages also permits collection of 24-hour urine samples. We have found that animals caged in this manner acclimate very rapidly toward personal contact with

---

Figs. 4 and 5. Squeeze box. The front face of the transfer box (Fig. 2) joins to the front face of the squeeze box, permitting animal transfer. Note the use of screw-driven plates to provide the mechanical advantage needed to restrain the apes (Fig. 5).

Fig. 6. Metabolic cage adequate for investigation with great apes.

Fig. 7. Stainless steel cages suspended from ceiling runners and locked to the back wall. These cages are equipped with feeders and an automatic watering system. Heavy-gauge plastic beneath cages is changed daily (or more frequently, if urine is being collected).

humans and can be readily trained to cooperate in recording of menstrual cycle data and other procedures.

The cage design offered here is by no means the only practical design for experimental housing of apes. An interesting variant on this design is utilized at the Laboratory for Experimental Medicine and Surgery in Primates (LEMSIP) center, New York Medical College, in which similar metabolism cages are suspended from an overhead rail (Fig. 7), thus permitting movement of the cages for cleaning and for relocation of animals. The capital required to install such a system is considerably greater than for the free-standing, immobile cages.

It is mandatory when designing and utilizing such cages that ancillary equipment, such as electric motors, chains, gears, urine collection pans, urine coolers, and so on, be positioned sufficiently far from the surface of the cage so that the animal cannot possibly reach such equipment. If this warning is not adhered to, the animal will destroy or mutilate even the most apparently impregnable equipment in a very short space of time. With bars placed on 5 or 6.5 cm centers, the safe distance is about 40 cm.

B. Maintenance

Various maintenance systems can be used for cage care and cleaning, feeding, and handling in the experimental situation. The Yerkes Center utilizes a "traditional system" in which the cages are cleaned daily, and food and excreta are removed for disposal. The cages are then further cleaned using pressurized water or steam.

Individual experimental cages can be designed that utilize automatic cleaning systems with a controlled water flush at predetermined time intervals. This water flush removes excreta to a central drainage system.

In the so-called "dry system," cages are suspended above a catch area, which is usually covered with plastic sheeting to catch the waste for daily disposal (Fig. 7). In this situation, there is no flushing or hosing of the cages, although they have to be cleaned and sterilized periodically to remove fecal matter and food adhering to the structure of the cage. Of these three systems, the first is most expensive in terms of personnel cost, but has the major advantage in that it permits the specific identification and correction of medical problems as they occur, as it requires more careful attention to the individual animal on a daily basis. This system is also advantageous because the isolation of individual animals permits more careful routine monitoring of drug administration, food intake, and water intake. The second system is less than adequate as it is potentially associated with increased parasitic and respiratory infection of animals as a result of transfer of infectious material in the vicinity of one cage to another during the flushing procedure and is also associated with the creation of infectious aerosols as a result of the flushing water being used at too high a pressure or flow rate.

The last system is more expensive in terms of physical and material costs, but it is effective in maintaining a clean and sanitary environment and potentially requires a reduced amount of personnel for daily maintenance.

C. Diet

Within the framework of the known dietary requirements of the apes, it is possible to formulate a number of acceptable rations. The basic diet of the Yerkes Primate Center and of the Laboratory for Experimental Medicine and Surgery in Primates (LEMSIP) are provided as examples.

1. *Yerkes Regional Primate Research Center Diet*

The basic diet fed to chimpanzees, gorillas, and orangutans from age two years to adulthood is shown in Table 2. Quantities are varied according to size of animals. The Yerkes Center is now evaluating the use of Purina Monkey Chow supplemented by fruit as a sole source of nutrition. The following food preparations have been provided to animals at Yerkes Primate Center.

Milk mixture: Mix 8.2 kg whole milk powder, 0.6 kg cooked oatmeal, 0.37 kg salt, and 68 liters water. To this mixture, add 15 ml Vetzyme Vet. Powder (Phillips Yeast Products) for each animal.

Bean cake: Soak overnight 5.5–7.0 kg soybeans in water (enough to cover). Then pressure-cook 2½ hours at 7–9 kg pressure. Mash beans to paste. Add 600 ml dark molasses, 30 ml salt, and enough cracker meal (made from ground-up Chimpcrackers) to give firm consistency. Press into four pans, about 4 cm thick; chill for 24 hrs. Each full pan is cut into 20 pieces (large animals) or 30 pieces (small animals). Add 15 ml corn oil per animal per day.

Pablum cake: Five days a week a pablum cake can be routinely given to all orangutans and to those chimpanzees and gorillas that appear to be somewhat under par in health. Ingredients for 34 animals are 2.3 kg pablum mix cereal, 5 ml Moliron per animal, 15 ml corn oil per animal, 6 cc Vitamates per animal, approximately 100 mg ascorbic acid per animal, and approximately 1 liter water for mixing. The ingredients are thoroughly mixed and shaped into cakes; feed animals one cake each.

TABLE 2

Diet of apes at Yerkes Primate Center.

Food item	Wheat Bread (slices)	Milk Mixture[c]	Cabbage	Carrots	Celery	Potatoes[b]	Bean Cake[c]	Oranges	Chimp crackers
Male gorilla[a]	4	54 oz	22 oz	18 oz	25 oz	80 oz	15 oz	32 oz	9 oz
Female gorilla[a]	4	54 oz	22 oz	18 oz	50 oz	56 oz	15 oz	22 oz	9 oz
Male orangutan	4	36 oz	9 oz	6 oz	——	23 oz	10 oz	11 oz	7 oz
Female orangutan	4	27 oz	9 oz	6 oz	——	12 oz	10 oz	9 oz	7 oz
Chimpanzee	4	27 oz	18 oz	12 oz	——	28 oz	10 oz	11 oz	6 oz

[a] Gorillas receive one tablespoon wheat germ per day in addition to the regular diet.

[b] Vegetables are raw, except for sweet potatoes, which are pressure-cooked.

[c] Recipe in text.

2. *LEMSIP Diet*

    (a)   Purina 15% Protein Chow—twice daily
    (b)   Various fruits—twice weekly
    (c)   Vegetables—twice weekly
    (d)   Chewable vitamin pills—two or three times weekly

D.　Medical Supervision of Experimental Animals

1.　*Quarantine*

Routine veterinary maintenance of animals in an experimental colony begins with establishment of, and adherence to, an adequate quarantine procedure for all new arrivals. At the Yerkes Center this comprises at least six months isolation for great apes, with daily observation, and six successive negative tuberculin tests using PPD. Chest X rays are obtained every other month during the quarantine period. Fecal examinations are performed upon initial arrival of the animal into the colony and subsequently as needed to evaluate the success of treatment with appropriate medications.

While in quarantine, animals are routinely given oral polio vaccine, and the Yerkes Center is initiating a program to evaluate influenza vaccine as an aid in subsequent health management of the great ape species.

2.　*Routine Supervision and Medication*

Subsequent to successful completion of quarantine and admission to the colony, animals are observed at least twice daily for overt evidence of physical abnormality or sickness, with special regard to occurrence of diarrhea, respiratory infection, open wounds, and so on. As frequently as animal population, requirements of research protocol, and number of personnel permit, animals are scheduled for anesthesia for the purpose of undergoing a complete physical examination, chest X ray, and fecal examination. In the Yerkes colony, this is accomplished every six months for the chimpanzees and approximately every nine months for the gorillas and orangutans. Gastrointestinal parasites are controlled by routine fecal examination, administration of anthelmintics, and interruption of the parasite life cycle. Measures to achieve parasite control include an adequate cage-cleaning program using a live steam, high-pressure cleaning spray for nonportable quarters, and mechanical scrubbing. An adequate vermin control program is an essential part of disease and parasite control within an animal colony. Parasites that

TABLE 3

Common parasites and treatment (in order of decreasing frequency at the Yerkes Center).

Parasites	Treatment	Dosage
Balantidium coli[a]	Diiodohydroxyquin[b] and	30 mg/kg for 10 days
	Oxytetracycline	25 mg/kg for 10 days
Entamoeba coli	Diiodohydroxyquin[b] and	30 mg/kg for 10 days
	Oxytetracycline	25 mg/kg for 10 days
Strongyloides	Thiabendazole[d]	100 mg/kg
Trichuris	Thiabendazole[d]	100 mg/kg
	Dichlorvos[e]	5 mg/kg
Enterobius	Pyrvinium pamoate[f]	5 mg/kg
Tapeworms	Niclosamide[g]	5 mg/kg
Trichomonas	Diiodohydroxyquin[b]	30 mg/kg for 10 days

[a] Metronidazole[c] is poorly accepted and must be administered by gastric intubation. It is effective against *B. coli*.

Febantel[h] is a broad-spectrum antihelmintic that is currently being evaluated for use in great apes and appears to be effective at a dose of 30 mg/kg for two days.

[b] Yodoxin®, Glenwood Labs Inc., Tenafly, NJ.

[c] Flagyl®, Searle & Co., Chicago, IL.

[d] Omnizole-Six®, Merck & Co., Rahway, NJ.

[e] Task®, Shell Chemical Co., Houston, TX.

[f] Povan®, Parke–Davis & Co., Detroit, MI.

[g] Yomesan®, Haver–Lockhart, Shawnee, KS.

[h] Rintal Paste®, Haver–Lockhart, Shawnee, KS.

have been detected in the Yerkes colony and the currently used anthelmintics are listed in Table 3. Specific details of parasite control procedures used in this and other colonies are available in the literature (Cummins *et al.*, 1973).

Both routine maintenance and experimental procedures frequently require medication of the animal. Medications utilized are preferably suitable for oral administration, thus avoiding further problems with restraint. Utilization of oral medications requires that the drug and its vehicle be palatable to the animal. In this regard, pediatric preparations are often of value, but other drugs, even when dissolved in alcohol, can be disguised using various fruit juices or cream, or by injection into the flesh of an acceptable fruit. Examples of these vehicles and drugs include MCT* oil as a vehicle suspended in milk or pablum, or methyl cellulose for suspension of drugs that have to be administered by intubation. We have used alcohol to dissolve such agents as steroids with subsequent suspension of the alcohol in milk, grape juice, or cream to

* Medium Chain Triglycerides (Mead Johnson)

achieve acceptance on the part of the animal. The successful administration of drugs to apes on a repeated basis can sometimes become a challenging problem with the investigator/veterinarian employing his or her imagination to stay one step ahead of the animal's ability to detect and reject the drug. In this context it is pertinent to note that the intelligence of the apes is such that a change in routine is immediately suspect when they are aware that they have already been placed in an experimental situation (namely, a metabolism cage). Accordingly, it is recommended that unfamiliar vehicles be administered on a regular basis as a reward for the animal cooperating in such routine matters as examination of sex-skin tumescence and so on. If this approach is taken, and the animals are accustomed to the occasional offer of grape juice or orange juice as a part of an unpredictable routine, then the doctored vehicle will be more readily accepted.

If animals are consistently averse to accepting oral medication in whatever vehicle it can be presented, then anesthesia and delivery of medication by stomach intubation may be required. This procedure is somewhat cumbersome for routine use, but does avoid the possible drawbacks not only of animal rejection of the drug, but also the variability and uncertainty of dose administered that can result from competition between animals for a given vehicle and from regurgitation of the drug by certain animals. Where very specific dose levels are required, it is recommended that the drug be administered by intramuscular or subcutaneous injection; animals can be trained to accept subcutaneous injection of small volumes. For example, the Yerkes colony has had a diabetic chimpanzee maintained on insulin administered daily by subcutaneous injection without restraint being necessary.

Under experimental conditions, it is of value to identify drugs that are active when administered orally. The study of female reproductive physiology frequently requires administration of steroid hormones, and two orally active compounds that we have used in such studies are mestranol and chlormadenone acetate, an estrogenic and progestogenic agent, respectively. At this time, orally active gonadotropic hormones are not available. Gonadotropin-releasing factors are active when administered intramuscularly, and this is a route that could be developed for routine *in vivo* investigations.

E. Animal Records

Adequate maintenance and utilization of a great ape colony is greatly facilitated by the acquisition and maintenance of an adequate record-keeping system. Such records contain information not only on acquisition date and age of the animal together with its medical history in quarantine and subsequent residence in the colony, but also detailed records of times and reasons for experimental access, drugs administered, anesthetic used, and so on. In addition, daily records should be kept of the sexual swelling and menstrual cyclicity of the females in the colony together with dates they were exposed to, and observed to copulate with, males for the purpose of timing gestation, and for the purpose of identifying periods of amenorrhea associated with lactation, stress, or disease. It is also pertinent and of value to review periodically the records of the colony not only to identify medical problems, but also to identify original information of value that can be derived from this ever-increasing data base. Such information includes data on alteration of menstrual cyclicity with age, climate and housing environment, duration of pregnancy, identification of familial tendencies toward such reproductive aspects as twinning, amenorrhea, or recurrent obstetric problems.

F. Planning and Coordination

For productive experimental use of great ape species, mutual understanding of the medical and research requirements is required of the veterinarian and research staff involved; there should be a clearly understood set of priorities concerning the relative value of the experimental protocol and maintenance of the animals assigned to a given project. Such priorities should be established prior to the initiation of the research protocol. This groundwork is essential if later conflicts are to be avoided regarding the necessity for medical intervention at one or more stages in the ongoing experimental protocol.

Such arrangements, for example, include scheduling of routine physical examinations to coincide with experimental anesthetic incidents or ensuring that the necessary physical and routine medical work has been performed prior to the initiation of the experimental protocol to avoid interruption of experimental procedures for nonexperimental reasons.

G. Routine Handling and Anesthesia

Routine handling of the animals for transfer between cages and the smaller squeeze cages is by means of small wheeled transfer boxes that the animals have been trained to enter (Guilloud and McClure, 1972).

For certain minor procedures, including administration of oral drugs and injections, obtaining very small volumes of blood, and taking vaginal swabs, and so on, it is possible to utilize caged but nonanesthetized animals confined by means of the squeeze-cage back. Some individuals will learn to cooperate to avoid restriction by the squeeze apparatus. It is important, however, to exercise caution under such conditions because at any time the animal may become agitated and injure the operator.

Animals weighing less than 20 kg can be restrained manually without anesthetic for procedures such as blood withdrawal, anesthetic and drug administration, TB-testing, and so on. Larger animals generally require chemical restraint for even such minor procedures as venous blood withdrawal and vaginal lavage. A variety of injectable anesthetic agents have been used for immobilization of apes, including phencyclidine hydrochloride (Sernylan, Bioceutic Labs, St. Joseph, MO), peridol and fentanyl (Innovar–Vet, Pitman–Moore Inc., Fort Washington, PA), oripavine and derivatives (M–99, American Cyanamid Co., Princeton, NJ), and thiopental sodium (Abbott Labs., North Chicago, IL), but most of these have significant drawbacks (Cohen and Bree, 1978; Guilloud and McClure, 1972). Ketamine hydrochloride (Ketaset, Vetalar, Parke–Davis Co., Detroit, MI) is the drug of choice for immobilization and is administered intramuscularly at a dose rate between seven and 25 mg/kg depending on the degree and duration of restraint required. Food and water are withheld for at least four hours prior to anesthesia (Bonner *et al.*, 1972; Martin *et al.*, 1972; Kuhn, 1974).

In the last four years, animals have been restrained many hundreds of times without incident using ketamine HCl with individual periods of anesthesia being as long as 4.5 hours (Graham *et al.*, 1979). Where prolonged anesthesia is required, it can be maintained by either intravenous drip or intramuscular supplementation at appropriate intervals. Intravenous drip requires a total anesthetic dose that is approximately 50% greater than intermittent supplementation, but a more constant depth of anesthesia can be maintained. Because of the rapid availability of the drug when administered intravenously, alterations in dose rate can precipitate convulsive episodes (see the following). Ketamine HCl is not recommended as the sole anesthetic agent for major surgical procedures because it provides relatively poor muscle relaxation; however, for immobilization during repeated sequential blood withdrawal, for subcutaneous implantation of telemetry equipment, and for manipulations such as vaginal lavage, endometrial biopsy, and so on, it has proved ideal. When necessary, atropine can be given with the ketamine HCl to reduce salivation.

For prolonged or major surgical procedures, it is our routine practice to induce anesthesia with ketamine HCl, administer atropine, and intubate the animal while under ketamine. Anesthesia is maintained with a halothane (Fluothane, Ayerst Labs Inc., NY), nitrous oxide, and oxygen mixture. Induction with barbituates in general is impractical, as a prompt and predictable response requires intravenous administration. With an oxygen:nitrous oxide ratio of 40:60, the amount of halothane required for maintenance of surgical anesthesia is between 0.5 and 1.5%. Depth of anesthesia can be determined by careful observation of blood pressure, pulse and respiratory rate, and other indicators that are similar to those observed in man.

When using ketamine HCl as an anesthetic agent, assessment of anesthetic depth is relatively difficult due to maintenance of laryngeal, pupillary, and corneal reflexes. In this instance, the most useful reflex to monitor is the grip of the hand. An increasing clutching or grasping reaction upon stretching of the fingers is an indication of returning motor function.

The supplementation dose for ketamine during prolonged anesthesia is approximately 25% of the induction dose every 15–20 minutes (i.e., 2.5 mg/kg). The required dose is significantly higher in males than in females, and slightly higher in lighter animals (less than 5 mg). It is advisable to have some form of aspiration equipment readily available, although it is rarely required because the laryngeal reflexes are maintained under ketamine and provide a natural safeguard against aspiration.

During prolonged immobilization (greater than two hours), we have noted marked but transient (two to three minutes) fluctuations in blood pressure, which, in the chimpanzee, have resulted in systolic blood pressures as low as 35–30 mm of mercury. There have been no untoward sequelae from these incidents. During shorter immobilization periods, the net effect of ketamine is to depress respiration and to elevate blood pressure. Overdosage or too rapid supplementation, intravenous (i.v.) or intramuscular (i.m.), predisposes to convulsions; these can be reduced or abolished by the use of Valium 1 mg/20 kg i.m., or in severe cases, thiopental sodium i.v., at a dose of 0.1 mg/kg. It is essential, however, to distinguish between the sometimes insidious onset of minor convulsions and the spontaneous movement of an animal under lessening depth of anesthesia. Supplementation in the former case, by the intramuscular or intravenous route will very rapidly induce more severe and potentially dangerous episodes.

A secondary disadvantage of ketamine HCl for reproductive biological research is that it is actively involved in the release of certain hormones and may be contraindicated, if, for example, there is a specific requirement to measure both the steroids and prolactin. Ketamine is associated with the measurement of "baseline" prolactin levels in excess of 40 ng/ml. Blood samples drawn from conscious animals demonstrate baseline prolactin levels less than 10 ng/ml, and this is more closely in line with reported baseline values for human prolactin that vary between 2 and 6 ng/ml (Reyes *et al.*, 1975). There is still doubt, however, as to the absolute baseline value of human prolactin, as even in man, conscious venipuncture is a technique not without some associated stress: prolactin is readily released under such circumstances. Ketamine has been shown in lower primates, however, to have no effect on normal menstrual cyclicity or ovulation, even upon repeated administration (Channing *et al.*, 1977).

Ketamine is not practical for oral administration. An experimental drug, CI-744, a combination of tiletamine HCl and Zolozepan HCl (Telazol, Parke–Davis Co., Ann Arbor, MI), has been shown to be safe and effective for restraint in the apes when given orally (McNamara *et al.*, 1974; Cohen and Bree, 1978). It is to be hoped that this drug will be available in the very near future.

At the Yerkes Center we have been unable to document an increased tolerance to ketamine HCl upon repeated administration (Kuhn, 1974). There are, however, individual reactions to the drug. For instance, some animals characteristically have an abrupt recovery of motor function after ketamine anesthesia. After the initial incident that causes attendant surprise, however, appropriate precautions can be taken! We also have been unable to statistically relate apparent variations in required anesthetic dose to achieve immobilization on any given occasion with the timing, amount, or frequency of administration of ketamine HCl to that individual.

Phencyclidine hydrochloride (Sernylan) was used prior to the ready availability of ketamine, but has disadvantages in that it has been associated with severe emergent reactions including a high incidence of self-mutilation. The hallucinogenic effect of phencyclidine and ketamine have been well-documented in man, and there is some evidence that a degree of hallucination occurs in certain individual apes. In our experience, however, hallucination has only rarely complicated ketamine-induced anesthesia. Sernylan has recently had FDA approval revoked, and the drug is no longer available.

At this time, there is little general indication for the use of barbituates in the chimpanzee. There are occasional indications for the use of morphine derivatives such as Innovar and M-99 because they provide good analgesia and have the advantage that an antidote is available to provide rapid reversal of the anesthetic where required (Nalline, Merck Sharp and Dohme, West Point, PA). This provision for rapid reversal, although not frequently utilized, is of potential value where anesthesia at frequent intervals is contemplated; the necessity to withhold food and water for some hours prior to induction of anesthesia and until full consciousness is recovered interrupts the regular feeding schedule, so making adequate nutrition difficult to achieve. Rapid recovery induced by an antidote could well relieve this difficulty.

In general, tranquilizers are indicated only as preanesthetic medication in the apes. It is dangerous to utilize tranquilizers as the sole means of restraint in animals weighing above 20 kg due to possible arousal in response to external stimuli and consequent risk of injury to bystanders. The advantages and disadvantages of older anesthetics have been thoroughly discussed (Guilloud and McClure, 1972; Day *et al.*, 1965). The currently used anesthetic agents recently have been evaluated by Cohen and Bree (1978). The use of ketamine HCl as an initial immobilizing agent has made obsolete the use of a closed induction box for anesthesia with agents such as ether.

Respiratory embarrassment can occur due to restriction of the natural airways if an ape is left for any length of time on its back or is propped in such a way that the head is markedly flexed against the chest. Whenever possible an anesthetized ape should be in a lateral recumbancy with care taken to be sure that the arms are not obstructing the mouth or nares. This simple precaution not only will avoid subsequent problems associated with acute respiratory embarrassment, but when combined with maintenance of the animal in a very slight head-down position, also will reduce to a minimum the potential reflux of secretions into the air sacs, with attendant potential for air-sac infection.

When animals with nursing young are to be anesthetized, it is a matter of choice whether the infant be left with the mother or removed. If the mother is being anesthetized for a routine procedure in which personnel will be present, it is feasible to leave the infant with the mother, although suckling during the period of anesthetia should be discouraged because of potential transmission of the anesthetic agent through the milk. This is of minimal concern with ketamine HCl as the drug is of limited effectiveness when administered orally.

III. EXPERIMENTAL TECHNIQUES IN CONSCIOUS APES

A. Urine Collection

Twenty-four hour urine samples are best obtained during confinement in a metabolism cage. Cooled urine may be collected in a pan surrounded by chilled water or ethylene glycol (Fig. 8). When urine is being collected, it is important to regulate the water intake of the

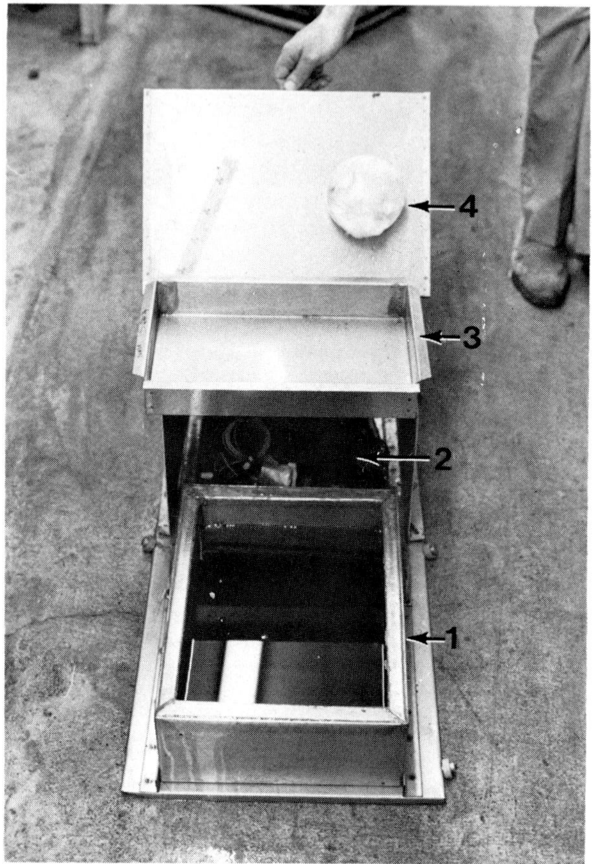

Fig. 8. Single-cage refrigerated urine collector. 1 = coolant chamber; 2 = compressor; 3 = urine pan; 4 = urine pan cover, with filter.

animal to approximately 1.5 liters/day in a divided dose so that excessive volumes of urine do not have to be handled, extracted, and utilized in the experimental procedures.

Urine can be collected in the larger concrete cages by means of a collecting device inserted into the preexisting drain in the concrete floor. This situation is not ideal as some urine is lost by concentration and evaporation in passage across the floor of the cage, and there is little facility to provide cooling of the collected urine. Also, some animals become proficient at removing the trap from the floor drain! This system is useful for collecting urine for detection of occult menstrual blood in orangutans and gorillas.

Collection of individual urines is not a practical proposition under the compound system of caging, although collection of partial urines has been successfully accomplished in smaller primates by training the animals to enter a confining cage attached to the compound in which they remain until passage of at least one urine sample (Lasley, 1978, personal communication). This approach is potentially applicable to the apes.

B. Blood Collection

It has been reported that with adequate training, adult chimpanzees will present an arm for peripheral venipuncture with subsequent collection of significant volumes of blood (Fig. 9). Such a method, however, rests very largely on the relationship between an individual chimpanzee and an individual handler, and requires regular training and positive reinforcement of the subject. Use of the method as a routine means for

Fig. 9. Venous blood withdrawal from a conscious adult chimpanzee (photo courtesy of C. E. Graham, Primate Research Institute, New Mexico).

blood collection for selected animals must be tempered with caution because of the potential risk of human injury. Nevertheless, significant studies have been accomplished using this technique (Reyes *et al.*, 1975).

When only small volumes of blood are required (50–200 µl), it is more practical to obtain them by training the animal to submit to a finger-prick with collection of blood into hematocrit tubes than by venipuncture. Such samples can be adequate for some hormonal analyses and for routine hematology. There is less risk to the technician with this procedure because it does not entail exposure of the entire arm of the animal, whereas for venipuncture the entire arm must be passed out of the cage.

C. Monitoring of Female Sexual Cycle

Chimpanzees and gorillas both show a variation in labial tumescence or perineal swelling throughout the menstrual cycle. In a chimpanzee this change in swelling is very evident (Graham, 1970), but in the gorilla it involves a more restricted alteration in the labial dimensions and in the length of the genital cleft (Nadler, 1975). All apes can be trained (by use of fruit or candy as reward) to mount the side of the cage and present the perineal region for inspection. Further training permits monitoring of menstrual flow by the insertion of hemasticks (Upjohn) into the vagina.

In an experimental colony females should be monitored on a daily basis for changes in sex-skin swelling, thus providing a continued record of reproductive status during the menstrual cycle. Such data are invaluable for subsequent design and execution of experiments related to the menstrual cycle.

D. Semen Collection

Animals have been trained both at the Yerkes Center and elsewhere to provide by masturbation and by use of an artificial vagina semen samples for a reward. The design of the artificial vagina used has been described elsewhere (Fussell *et al.*, 1973). Rewards used in such training generally comprise orange slices, apple slices, or fruit juice in a squeeze bottle.

E. Artificial Insemination

Several females in the Yerkes colony have been trained not only to present for monitoring of perineal swelling, but also to accept insertion of swabs and insemination devices into the vagina. By careful monitoring of the menstrual cycle according to the sexual swelling, tumescence and detumescence, together with rapid estimation of the LH levels in the urine (Gould and Faulkner, 1979), it is possible to inseminate the conscious female at, or close to, the time of ovulation. Artificial insemination permits use of colony males who are incompetent in their sexual behavior and who otherwise fail to contribute to the genetic pool. It is also of potential value when specifically timed matings are required or when one male is required to inseminate several females within a short period of time, since artificial insemination can avoid the logistical problems required in multiple animal transfers within the colony. Artificial insemination performed under anesthesia permits more accurate placement of the fluid portion of the ejaculate on the external os and the coagulum high in the anterior vagina (Martin *et al.*, 1978).

IV. ROUTINE EXPERIMENTAL TECHNIQUES REQUIRING ANESTHESIA

While it is not the purpose of this chapter to enumerate and describe every surgical and experimental procedure to which chimpanzees have been subjected, it is appropriate to mention certain techniques that have been used in a variety of experimental protocols applicable to study of reproductive physiology.

A. Urine Collection

It is practical to collect urine from both male and female great apes by catheterization of the bladder. The same reservations apply to this procedure in apes as apply in man, namely the techniques must be conducted under sterile conditions with great care being taken to avoid introduction of infection into the urinary tract as a result of the catheterization procedure. When required, catheterization in both sexes is accomplished using standard human equipment; Foley catheter and prepackaged units designed for human use are of value in this regard.

Urine collection also can be achieved for diagnostic purposes by direct aspiration of the bladder. In this situation, the anterior pubic area is clipped, shaved, and prepared as a sterile field, and urine is withdrawn through a 22-gauge cardiac needle inserted immediately anterior to the pubis and into the fundus of the bladder. In the orangutan especially, the relatively empty bladder can be a very small structure, and this technique should be performed only when specifically required for diagnostic purposes. It is considered, by some, a safer procedure with regard to induction of infection of the urinary tract than is catheterization via the urethra.

B. Prolonged Intravenous Fluid Infusion/Withdrawal

Routine techniques of chronic catheterization used in man for blood withdrawal can be utilized in the apes. Because there is a somewhat greater likelihood of spontaneous movement under anesthesia, it is important that flexible indwelling venous catheters be used rather than needles (Bardic Intracath, C. R. Bard Inc., Murray Hill, NJ). Such units are readily available and can be introduced into a variety of peripheral veins. When chronic catheterization is contemplated, larger veins should be catheterized to permit predictable flow rates. If intermittent withdrawal is required, repeated femoral venipuncture (using a 22-gauge needle) should be considered because maintenance of an indwelling catheter for intermittent use requires provision of a slow infusion of heparinized saline to maintain patency of the catheter.

When infusion of replacement fluids or of a vehicle for intermittent drug administration in conjunction with intermittent blood withdrawal is contemplated, a three-way tap should be used in conjunction with the catheter, thus allowing rapid reversal of flow in the catheter with minimum dislocation of the catheter. (Luer-lock fittings have a mysterious habit of irreversibly tightening themselves in critical situations.)

C. Endometrial Biopsy

Endometrial biopsy is performed under ketamine HCl anesthesia. The animal is placed in sternal recumbancy, and the vagina and perineal area cleaned and swabbed with betadine. An eye drape is placed around the vaginal area and via a Graves style speculum fabricated from rigid polyethylene tube 2.5 × 15 cm (human speculae are usually too short for use in the chimpanzee during the period of maximal swelling), a uterine sound is introduced through the cervix to determine the maximum depth of the uterine fundus. It is possible to introduce fine catheters (Foley size 5 or 7) for uterine lavage at this stage. To obtain an endometrial biopsy, we have found it most suitable to utilize a Novak endometrial curette with gentle suction applied via a 3-cc syringe; thus adequate biopsies can be obtained from any quadrant of the uterus. Care must be taken in removing the curette through the cervix in order to avoid tearing the cervical canal by the cutting edge of the curette. In our experience a variety of other curette designs have not been satisfactory.

D. Electro-ejaculation

Ketamine anesthesia is kept as light as is practical consonant with operating safety, as better response is obtained with light sedation. Atropine is not used as it inhibits the ejaculatory response. A probe of suitable dimensions (Fig. 10) is lubricated with sterile jelly (e.g., K–Y, Johnson & Johnson) and is gently introduced into the rectum. Any initial resistance by the anal sphincter is overcome by gentle, maintained pressure. It has not been found necessary to routinely

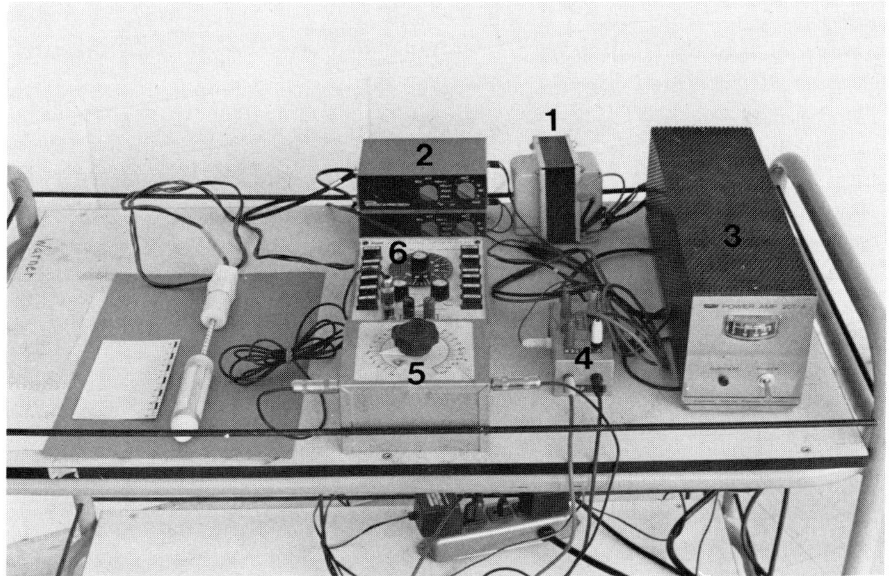

Fig. 10. Electro-ejaculation equipment currently in use. 1 = isolation transformer; 2 = digital AC meters; 3 = power amplifier; 4 = junction box; 5 = intensity control; 6 = waveform generator. Not shown is the multi-outlet terminal tip (photo reprinted with permission from Gould *et al.*, 1978).

evacuate the rectum prior to insertion of the probe, but if excessively high electrode/tissue interface resistance is measured (see the following), the probe should be withdrawn and the electrodes examined for fecal occlusion. The electrical equipment required for this procedure has been described elsewhere (Warner *et al.*, 1974; Gould *et al.*, 1978).

Electrical current is applied in a series of increments, maximum being raised in approximately 2-V steps, with six repetitions at each voltage level. The stimulus is applied by varying the potentiometer in a sinusoidal manner, with a rapid rise, a pause of 1 to 2 sec around the maximum value, and a more rapid drop. Each stimulus occupies 3 to 4 sec. When higher voltages are used, the current is not reduced to zero between stimuli.

Voltage and current readings are obtained for each series of stimuli, and careful note is taken of the values at time of initial penile erection, and seminal emission or ejaculation. It is not deleterious to pause at the maximum point of stimulus for 5 to 10 sec to obtain the necessary electrical readings.

The resistance of the tissue/electrode interface can be calculated using Ohm's law (resistance in ohms = volts/amps). This information should be obtained at the outset in order to validate the experimental conditions. It should be noted that for a constant resistance there is a proportional relationship between voltage and current. If the measured current at a given voltage is markedly lower than that expected from previous experience (>50%), an excessively high electrode/tissue resistance, usually resulting from occlusion of the electrodes with fecal material, is indicated. The probe should be removed, cleaned, and reinserted. Conversely, if any marked increase in current is recorded for a constant or increased voltage, no calculation is required to deduce that a marked drop in resistance has occurred, possibly as a result of damage to the rectal mucosa. However, stimulation parameters suggested here are well within safe limits. and this complication has not been observed.

The most effective frequency for stimulation is approximately 20 Hz. The stimuli required for seminal emission or ejaculation fall within a relatively narrow range (0.12–0.45 mamp/mm^2 of electrode area), regardless of body weight, with a mean value of 0.24 mamp/mm^2 (Gould *et al.*, 1978).

E. Testicular Biopsy

Testicular biopsy can be used as an experimental procedure very readily in the chimpanzee where the volume of testicular tissue available is relatively large. In the other species, however, the testicular volume is much reduced and, due to the endangered status of these species, this technique should be perhaps restricted to a final diagnosis of infertility in males already being evaluated for apparent absence of reproductive function. Techniques utilized for human biopsy are applicable, and biopsies can be obtained using both punch-and-wedge resection techniques.

F.   Laparotomy and Laparoscopy

For either of these procedures, water and food are withheld for several hours prior to induction of anesthesia. Anesthesia is induced with ketamine (10–20 mg/kg i.m.), and atropine (0.5 mg i.m.) is administered routinely. It is valuable to establish an intravenous drip, most usually in a cephalic vein, which can be utilized throughout the procedure. Prior to laparotomy it is valuable, and prior to laparoscopy essential, to swab the vagina and perineum with betadine and drain the bladder by means of insertion of a Foley catheter that can remain indwelling throughout the procedure. With the animal in a supine position, the abdomen is shaved and scrubbed in the routine manner for surgical preparation. The animal is transferred to an operating table equipped with shoulder rests, arm boards, and limb restraints; an endotracheal tube is inserted and inhalation anesthesia established with 2–3% halothane, 60% nitrous oxide, and 40% oxygen. The most frequently used approach to the reproductive organs is a longitudinal midline incision posterior to the umbilicus. This incision can be extended to the pubis if required. Because of the strength and potential mobility of the recovered patient, special attention must be paid to closure of the midline incision, and it may be considered necessary to use wire internally and externally. However, as a routine we use 0 gut for the internal layer closures, and wire or nylon for the skin incision. Provided adequate electrocautery equipment is available, it is practical to use a paramedian incision with the advantages that one can stagger the point of incision through the various abdominal layers, and it is also possible to use a horizontal incision such as a modified Pfannenstiel approach. Our experience with this approach in the chimpanzee has been limited, but satisfactory. For details of this surgical methodology, the reader is referred to standard human surgical texts.

Laparoscopy is a potentially valuable technique for use in reproductive investigations in the apes. At this time, our experience has been limited to the chimpanzee, and full details of the procedure are available (Graham, 1976). Again, the techniques are derived directly from those applicable to man, and the procedure is uncomplicated provided the precautions recommended in the literature with regard to use of this procedure are observed (Levinson, 1974; Williams, 1974). Routine laparoscopy in the chimpanzee involves dilation of the abdomen with carbon dioxide via a Verres cannula with subsequent insertion of the laparoscope through a 1-cm incision placed below the umbilicus. We currently use 10-mm-diameter laparoscopes and have found these to be adequate for visualization and manipulation of the reproductive organs. Laparoscopes are available in a variety of smaller sizes, and the final choice rests largely with the investigator's personal preference. The use of 10-mm laparoscopes is quite practical in animals the size of the apes, although the use of smaller two to four mm laparoscopes is feasible and has the advantage that the point of entry can be more specifically chosen with regard to the portion of the reproductive tract that it is desired to observe and manipulate.

Due to minor anatomical differences in the placement of uterus and ovaries in the apes relative to man, it is usually necessary to insert accessory instruments in the lower quadrants of the abdomen. Prior to insertion of these instruments, the interior abdominal walls are examined with the endoscope and the exterior surface observed by transillumination using the endoscope as an internal light source. Such examination permits identification of sites for penetration that will avoid major blood vessels or internal adhesions. A further accessory of value in the chimpanzee is a vaginal probe with an extension or a cup that fits over the cervix. This permits movement of the uterus from outside the animal aiding in identification and location of the organ during laparoscopy. For a detailed description of the technique of laparoscopy in the chimpanzee, the reader is referred to the paper by Graham (1976) and Graham *et al.* (1973).

Halothane and nitrous oxide can be used as anesthetic with considerable safety, but the level of nitrous oxide used must be monitored to avoid anoxia that can result from use of nitrous oxide in excess of 80% of inspired gases (Levinson, 1974). Under experimental conditions such as those involved in laparoscopy, where a steep Trendelenberg position is utilized for the animal, respiratory embarrassment, together with absorption of carbon dioxide from the abdomen, has led to acidosis and cardiac arrhythmia. These conditions can be largely avoided by adequate mechanical ventilation to counteract the increased intra-abdominal pressure resulting from insufflation and by reduction of abdominal pressure as much as possible once penetration with the trochar has been achieved (Peterson, 1971; Graham *et al.*, 1973).

V. SUMMARY

The key to successful use of the apes in reproduction research is multifaceted. Routine maintenance and husbandry of such valuable animals must be of a consistently high standard. Experimental procedures should be carefully considered with regard to the criteria of suitability of the apes for the procedure contemplated, the adequacy of facilities for handling the apes, and the potential risks to personnel. When healthy animals are carefully used in adequate facilities, the incidence of complications, both medical and logistical, is small. The treatment of medical emergencies in the apes requires a blend of veterinary and human medical expertise, and reference to human medical and surgical texts is recommended.

REFERENCES

Beveridge, W. I. B. (1969). "Using Primates in Medical Research. Part I: Husbandry and Technology," Vol. 2, Primates in Medicine. Karger, Basel.

Bonner, W. B., Keeling, M. E., Van Ormer, E. T., and Haynie, J. E. (1972). *In* "The Chimpanzee" (G. H. Bourne, ed.), Vol. V, pp. 255–258. Karger, Basel.

Channing, C. P., Fowler, S., Engel, B., and Vitek, K. (1977). *Proc. Soc. Exp. Biol. Med.* **155**, 615–619.

Cohen, B. J., and Bree, M. M. (1978). *J. Med. Primatol.* **7**, 193–201.

Cummins, L. B., Keeling, M. E., and McClure, H. M. (1973). *Lab Anim. Sci.* **23**, 819–822.

Day, P., Finig, J., and Van Riper, D. C. (1965). ARL-TR-65-18.

Fussell, E. N., Franklin, L. E., and Frantz, R. C. (1973). *Lab. Anim. Sci.* **23**, 252–255.

Gould, K. G., and Faulkner, J. R. (1979). *Proc. Congr. Int. Primatol. Soc., 7th, Bangalore* #E22.

Gould, K. G., Warner, H., and Martin, D. E. (1978). *J. Med. Primatol.* **7**, 213–222.

Graham, C. E. (1970). *In* "The Chimpanzee" (G. H. Bourne, ed.), Vol. III, pp. 183–220. Karger, Basel.

Graham, C. E. (1976). *J. Med. Primatol.* **5**, 111–123.

Graham, C. E., Keeling, M., Chapman, C., Cummins, L. B., and Haynie, J. (1973). *Am. J. Phys. Anthropol.* **38**, 211–215.

Graham, C. E., Gould, K. G., Collins, D. C., and Preedy, J. R. K. (1979). *Endocrinology* **105**, 269–275.

Guilloud, N. B., and McClure, H. M. (1972). *In* "The Chimpanzee" (G. H. Bourne, ed.), Vol. V, pp. 226–254. Karger, Basel.

Keeling, M. E., and Guilloud, N. B. (1971). Medical Primatology 1970, *Proc. Conf. Exp. Med. Surg. Primatol. 2nd, New York, 1969* 798–805.

Kuhn, U. S. G. (1974). *J. Am. Vet. Med. Assoc.* **165**, 838–839.

Levinson, C. J. (1974). *J. Reprod. Med.* **13**, 187–194.

Martin, D. E., Graham, C. E., and Gould, K. G. (1978). *Symp. Zool. Soc. London* **43**, 249–260.

Martin, D. P., Darrow, C. C., Valerio, D. A., and Leiseca, S. A. (1972). *Lab. Anim. Sci.* **22,** 837–843.

McClure, H. M., Guilloud, N. B., and Keeling, M. E. (1973). *In* "The Chimpanzee" (G. H. Bourne, ed.), Vol. VI, pp. 121–181. Karger, Basel.

McNamara, J. A., Sly, C. L., and Cohen, B. J. (1974). *Am. J. Vet. Res.* **35,** 1089–1091.

Nadler, R. D. (1975). *Anat. Rec.* **181,** 791–798.

Peterson, E. P. (1971). *Fertil. Steril.* **22,** 695–698.

Reyes, F. I., Winter, J. S. D., Faiman, C., and Hobson, W. C. (1975). *Endocrinology* **96,** 1447–1455.

Warner, H., Martin, D. E., and Keeling, M. E. (1974). *Ann. Biomed. Eng.* **2,** 419–432.

Williams, P. P. (1974). *Fertil. Steril.* **25,** 280–287.

Chapter 16

GREAT APES AS MODELS IN REPRODUCTIVE BIOLOGY

Charles E. Graham

The preceding chapters, each by leading authorities in their field, have surveyed current knowledge of great ape reproductive biology, including anatomy, physiology, sexual behavior, and husbandry. This

Copyright © 1981 by Academic Press, Inc.
All rights of reproduction in any form reserved.
ISBN 0-12-295020-8

final chapter will draw upon the preceding contributions to highlight the special values that apes possess as subjects of comparative study and as models for human reproduction. The utility of apes also will be discussed in relation to their limited availability.

I. COMPARATIVE ASPECTS OF APE REPRODUCTION

As Roy Greep intimated in his foreword to this volume, the development of research in a particular field tends to proceed from the general to the particular. Attention is usually directed first to phenomena common to a wide variety of species, which therefore are likely to reflect fundamental biological processes. Such studies can be conducted with a convenient and inexpensive model such as the rat. However, as the investigator becomes involved in even more complex levels of organization, he or she frequently encounters specializations not shared by most other species; although the basic principles of reproduction remain the same, the details are more subject to evolutionary adaptation. Variability of the luteotropic complex among species provides a good example. A primate model is often needed when data applicable to man are required. In many instances documented in this volume, even monkeys are not sufficiently similar to man, and selection of an ape species may be necessary.

The famed reproductive biologist Carl G. Hartman, modified a famous quotation of Alexander Pope (1704) to state, "The chief subject for the study of man will always be man himself" (Hartman, 1939). Recognizing the ethical and practical problems relating to limitations for experimentation on human beings as well as the limitations of lower mammals as subjects, Hartman discussed the indispensability of monkeys and chimpanzees for the study of certain aspects of human reproduction. He concluded that "Fully to understand man we must learn all about man's nearest relations, so that we can finally show what morphological, physiological, and psychological features are really peculiar to man and what others man shares with apes."

A recurrent theme of this book has been the thesis, based on phylogenetic grounds and on experimental evidence, that apes indeed resemble man to a greater extent than less closely related species, and these similarities give apes unique value as experimental animals for the study of certain human problems.

Comparative data on apes and man illuminate the unique characteristics of man, so providing insight into the significance of peculiarly human adaptations and also identifying the applicability of different primate species as models for particular human problems. When considering a potential ape model for a human problem, the species should be thoroughly investigated to document the relevant similarities to man, and also the differences, if any. The animal model does not have to be similar to man in all respects, but only in the context of the problem being investigated. One always must bear in mind that significant differences between the model and man may remain undiscovered, even though close similarities have been documented.

A temptation to avoid is to use apes for studies oriented toward man, even if documentation of general similarity to man in the particular context is lacking. A study that is conducted with no objective evidence that the ape model is suitable is of limited value and is prodigal of a scarce resource; the necessary validation studies should always be performed for the experiment to have scientific meaning.

An important aspect of comparative biology concerns differences between species. Given the background of general similarity, the differences in details of reproduction between apes and man assume special significance since they reflect details of evolutionary specialization that "fine-tune" the organism, human or ape, to its particular reproductive pattern and environment. Differences among anthropoids and man present the scientist with a natural biological experiment, in which the number of variables is limited; under these conditions, the study of the differences between closely related species can be particularly revealing with respect to the area of reproduction they affect.

Comparative study of nonhuman primates is thus basic to their use in the study of human reproduction. Not only does it yield insights into primate and especially human evolution (see Harcourt, Chapter 12; and Short, Chapter 13, both this volume), but it is also essential for the proper use of apes as experimental models.

II. CONSIDERATIONS GOVERNING THE USE OF APES AS EXPERIMENTAL MODELS

A sound rationale for the choice of a great ape species as an experimental model should consider a number of factors. First, the chosen ape species should have an essential advantage over other more available species. The advantage may be of several types: anatomical, physiological, behavioral, or a combination thereof. It may emphasize similarity with man, or dissimilarity of a particular feature against a general background of similarity; however, the advantage must be unique to apes and permit significant information to be gained that is unavailable from other sources, including man himself.

A second consideration is availability. Can a sufficient number of apes be recruited to the study to confer sufficient likelihood of obtaining an unequivocal answer to the question being asked? Does the proposed study compete for animals with other potentially more important studies?

A third factor, of the greatest importance, is the effect of the study on the breeding capability and future experimental use of the subjects. Will the study interrupt the breeding capacity of the subjects temporarily or permanently? Is the information to be gained important enough to warrant this interruption? Will the study alter the future experimental utility of the subject? If so, is this loss justifiable in terms of the potential benefit to man? Such questions usually do not have absolute answers, but they are matters to weigh carefully before committing a scarce, costly, and slowly renewed source.

Although it may be possible to evaluate proposals to use apes as models in applied research by means of criteria such as those mentioned in the foregoing discussion, basic comparative studies need to be evaluated by different criteria (except that the study must not be destructive to the subjects). Basic research has never yielded to the cost-benefit type of analysis. The scientist needs room in which to explore, to discover what can be learned from the experimental subject. Sometimes nothing may be learned of apparent or immediate value; on other occasions discoveries, by accident or by design, may be made of important new ways in which apes can help man to gain insight into humanity or to conquer human biological limitations.

III.   ADVANTAGEOUS CHARACTERISTICS OF APES AS MODELS

Apes possess a number of specific physical, physiological, and behavioral attributes that enhance their value as experimental models.

A.   Trainability

Apes are highly intelligent and can be trained to cooperate with the investigator; for instance, female gorillas and chimpanzees have been successfully trained to present genitalia for inspection. Adult chimpanzees also have been trained to present an arm for blood withdrawal without anesthesia.

B. Reproductive Patterns

The time-scale of reproductive development of apes more closely approaches that of man than does that of other primates that have been studied (Graham, 1970).

Adolescence, which occurs within the first years of life in monkeys, does not occur until six to nine years of age among the apes. Once initiated, reproductive cyclicity among female chimpanzees persists into the fifth decade, as in women. However, a postreproductive menopausal phase so far has not been demonstrated with certainty in chimpanzees, in contrast to rhesus monkeys and women (Graham, 1979); however, preliminary data suggestive of perimenopausal changes have recently been obtained (Gould, Flint, and Graham, in press). An insufficient number of gorillas and orangutans have been studied to advanced age for any comment on their status with respect to menopause. It is uncertain whether the occurrence of a postreproductive life span in laboratory rhesus monkeys and women is a product of specialized living conditions that have not yet been perfected for chimpanzees, or if there is an absolute difference between these species. Further information on this point would be of great anthropological interest.

The duration of the menstrual cycle is longer in apes than in women. Great significance was formerly attached to the identical modal menstrual-cycle length of the rhesus monkey and woman. Although this coincidence is certainly convenient, it is probably of less importance today than the question of which primates compare most closely with man in their fundamental menstrual-cycle physiology.

Chimpanzees, like women, experience menstrual cycles and births at the same frequency throughout the year in the laboratory (e.g., see Martin, Chapter 14, this volume). This is probably the case for gorillas and orangutans also, although adequate documentation is lacking. The continuing cyclicity and fertility of the chimpanzee throughout the year avoids logistical problems inherent in menstrual cycle studies with those other species that show evidence of a breeding season. Rhesus monkeys, for example, have a seasonal birth peak in the wild, and although they may revert to a year-round menstrual cycle pattern in the laboratory, there is evidence of a persistent anovulatory season.

The duration of pregnancy in apes is only a little shorter than that of man, in contrast to more distantly related primates, in which pregnancy is considerably shorter. This temporal factor may have important developmental consequences.

C. Anatomy

1. Size

The large size of apes compared to other nonhuman primates is an advantage for instrumentation and tissue availability.

Size approaching that of man is preferable for the development of surgical techniques and instrumentation for human application (see Gould, Chapter 15, this volume). Conversely, techniques developed in the human can easily be applied to apes for experimental purposes, with little or no modification. Laparoscopy as developed for humans and applied to chimpanzees provides an excellent example (Graham *et al.*, 1973). Endometrial biopsies have been applied successfully and routinely to chimpanzees using standard human procedures, and to the gorilla, adapting equipment designed for human use (Graham, Chapter 1, this volume).

The fact that specialized instrumentation does not have to be developed for such procedures in apes is a significant convenience and cost advantage. It is true that human instrumentation, especially pediatric equipment, can sometimes be applied to monkeys or baboons with success, but in such instances the size differential becomes more significant.

Apes can also accommodate rather bulky implanted instrumentation such as telemetry devices for study of basal body temperature (Graham *et al.*, 1977; Graham, Chapter 1, this volume).

Size is also a very important consideration with respect to the volume of fluid and tissue samples that can be harvested for analysis. This characteristic of apes takes on special significance when multiple and frequent serum hormone assays must be performed, especially in the perinatal period when the pool of blood available for sampling is severely limited, even in large species.

Solid tissues are also available in larger quantities from apes than from smaller species. While this aspect has many potential applications, it is worth drawing attention to the fact that several hundred milligrams of endometrium can be obtained for biochemical analysis. This fact, combined with the ease of biopsy and the renewable nature of endometrium, makes the chimpanzee of special interest for studies of uterine physiology, particularly in the context of the control of endometrial decidualization as discussed below.

2. Morphology

In many respects the close similarity between the internal and external genitalia of apes and man present obvious specific advantages. This subject has been reviewed by Gould and Martin in Chapters 5 and 6 of this volume, and only selected examples are mentioned here.

a. Endocervical canal The straight uterine canal of apes, which resembles that of man, facilitates all types of intrauterine procedure, particularly endometrial biopsy, and cannulation for collection of uterine fluid or infusion of drugs used in the study of uterine physiology. Although some macaques and the baboons also have a straight uterine canal, in contrast to the S-shaped canal of most monkeys, these species are not always the best choice because of other dissimilarities with man.

b. Uterine decidualization Extensive uterine predecidualization is spontaneous in man and chimpanzees, in contrast to all other documented species in which conception must occur before extensive decidualization can occur (Dollar and Graham, in press). Although rodents have been widely used to study implantation, only the chimpanzee is suitable for detailed study of the induction of decidualization in man; in this respect the straight cervical canal and abundant endometrial samples available from chimpanzees are of special significance.

Hartman (1939) has catalogued a variety of other similarities in placentation between chimpanzees and man, and differences between monkeys and man, which suggest that to the extent that availability permits, chimpanzees may have special value for study of implantation and early human fetal development.

The chimpanzee is the only known model of membranous dysmenorrhea, a condition of unknown etiology in which women pass an endometrial cast during painful menses. Since the condition has been observed to appear repeatedly in the same chimpanzees, this species could be used to study the etiology of the condition (Solleveld and van Zwieten, 1978).

c. Genital swelling Genital swelling of chimpanzees and gorilla, which develops as follicular estrogen secretion increases and which regresses at the time of ovulation as a result of falling estrogen levels and rising progesterone levels, is an especially valuable characteristic of these species (Graham *et al.*, 1972; Nadler *et al.*, 1979; Graham,

Chapter 1, this volume). Sexual swelling is also possessed by some monkeys and baboons, but in these species the swelling is not combined with some of the other advantages that apes offer. In the chimpanzee, knowledge of the steroid hormone control of the sexual swelling makes it a readily monitored external indicator of internal endocrine-coupled events. Hence, the swelling can be used to monitor temporal aspects of reproductive functions, including presence or absence of menstrual cyclicity, follicular development, ovulation time, onset of the luteal phase, and occurrence of abnormal cycles. Arising from these applications, the swelling pattern may be used to time mating for optimal efficiency, to monitor the efficiency of exogenous hormone administration or other experimental procedures that are incidentally expected to affect the sexual swelling, and to time experimental procedures when follicular development or regression is a useful reference point.

Similar considerations seem to apply to the gorilla based on the more limited data available. In the gorilla, the shorter period of genital swelling gives relatively late information concerning follicular development, but more precise information concerning impending ovulation, compared with the chimpanzee.

D. Antigenic Similarity of Protein Hormones

It is important to note that the relatively close immunological similarity of ape and human protein hormones permits measurement of ape pituitary gonadotropins, prolactin, and chorionic gonadotropin (Table 1) in human radioimmunoassay systems, in contrast to lower primates (Chen and Hodgen, 1976; Hobson *et al.*, 1976; Hodgen *et al.*, 1973; Nadler *et al.*, 1979; Reyes *et al.*, 1975). This is an important factor in experimental design that has cost and convenience advantages; moreover, such assays are very difficult to establish in more distantly related species due to lack of suitable antigens in sufficient quantity for antibody preparation.

Because of the close similarity between human and chimpanzee chorionic gonadotropin, the chimpanzee has special value for evaluating the contraceptive efficacity and toxicology of active or passive immunization against chorionic gonadotropin (CG), CG fragments, or derivatives (Chen and Hodgen, 1976). This contraceptive approach has the advantage that it normally would be completely specific to pregnancy. However, antisera to CG might also be effective in therapy of CG-secreting tumors.

TABLE 1

Crossreactivity of primate chorionic gonadotropins in radioimmunoassay systems using ^{125}I-h CG[a].

Antiserum	Antigen	hCG	chCG	gCG	orCG	paCG	mCG	maCG
H26	oLHβ	+	+	+	+	+	+	+
H80	hCG	+	+	+	+	±	±	−
Sb6	hCGβ	+	+	+	+	−	−	−
H93	CTP-hCGβ	+	+	+	−	−	−	−

[a] Reproduced with permission of the J.B. Lippincott Co.

Note: + = complete; ± = incomplete; − = undetectable. Chorionic gonadotropins: hCG = human, chCG = chimpanzee, gCG = gorilla, orCG = orangutan, paCG = baboon, mCG = macaque, maCG = marmoset chorionic gonadotropin.

E. Metabolism of Steroid Hormones

A number of studies have established that steroid metabolism in neonatal and adult pregnant and nonpregnant apes closely resembles that of man, exemplified particularly by the metabolism of progesterone and estradiol, which differs from monkeys. There are major differences in the metabolic capacity of the human and chimpanzee fetoplacental until compared with the monkeys and lower mammals that have been studied (Faiman *et al.* and Wright *et al.*, Chapters 2 and 7, respectively, this volume).

The difficulty of using radiolabeled steroids in human subjects and other limitations to drug experiments in humans make apes attractive for the study of toxicology and metabolism of naturally occurring and synthetic steroids, particularly potential contraceptive agents. Such studies will become increasingly important for establishment of the mode of action and safety of naturally occurring and synthetic steroids used in contraception and for therapy of conditions such as postmenopausal syndrome and breast and prostatic carcinoma. Chimpanzees are promising subjects for the investigation of the entero-hepatic metabolism of radiolabeled estrogens, since their general steroid metabolism and intestinal flora are similar to man, yet diet can be controlled and manipulated in a manner unacceptable to most human subjects. Such studies may be important in establishing the extent to which individual, diet-dependent variability in metabolic patterns determines susceptibility to diseases suspected of having estrogen as an etiologic factor, such as breast cancer.

F. Semen and Cervical Mucus

The ability to obtain semen from trained apes by automasturbation, or from anesthetized apes by electro-ejaculation (Gould *et al.*, 1978; Martin and Gould, Chapter 6, and Gould, Chapter 15, both this volume), permits use of the chimpanzee for a variety of studies difficult to perform in humans for technical or ethical reasons. In particular, the ability to test the effect of various drugs, including antispermatogenics, on semen quality is of considerable potential utility. Cervical mucus can be obtained in limited quantity by aspiration; it is a target for contraceptive methodology by manipulating its penetrability by sperm (Gould *et al.*, 1976). In addition, it can potentially serve as a medium in which to test the normality of spermatozoan motility (Gould and Martin, Chapter 5, this volume).

G. Reproductive Endocrinology

1. Postnatal Development and Adolescence

This subject has been fully reviewed by Hobson *et al.* in Chapter 4. Since the chimpanzee is the only nonhuman primate other than man known to possess a well-developed postnatal fetal adrenal zone, it is the only primate model in which the developmental significance of this structure can be investigated. This species is a model in which the existence and identity of a possible trophic hormone that controls the activity of the fetal zone can be sought (Shackleton, 1974).

The endocrinology of the newborn and infant human in relation to general and sexual development is little understood. The necessary manipulative experiments cannot be performed in the human for ethical reasons. Because of its general endocrine similarity to the human, the chimpanzee is needed as a subject in this area (Faiman *et al.*, 1972 and Chapter 2, this volume).

The chimpanzee (and possibly other apes) has an adrenarche like that of man, whereas macaques and baboons do not (Cutler *et al.*, 1978; Hobson *et al.*, Chapter 4, this volume). Thus, chimpanzees can be used to examine the endocrinology of this little-understood phase of development, hopefully leading to enhanced understanding and thus improved treatment of associated endocrinopathies such as precocious and delayed puberty and adrenarche, and congenital adrenal anomalies. The developmental role of fractionated pituitary extracts could be tested in chimpanzees to determine more exactly the pituitary trophic hormones active during adolescence and specifically to seek a possible "adrenarchal" trophic hormone. Such studies should also throw some light on the relationship between adrenarche and puberty.

2. Menstrual Cycle

Within the menstrual cycle there are many problems for which the apes, especially the chimpanzee, are preferable or indispensable, particularly in the field of contraception, because of close similarities with the human female (Graham, 1977 and Chapter 1, this volume). For example, the luteal estrogen rise common to apes and man, but absent in other primates, is of unknown significance. If elucidation of its function shows it to be essential for conception, its inhibition may represent a new approach to contraception. It has been suggested that luteal estrogen secretion is connected with luteolysis, and it seems the chimpanzee is an essential animal model for investigation of such problems as they relate to man.

3. Pregnancy

The almost identical profile of hormones throughout pregnancy in chimpanzee and man (in marked contrast to the lower primates in which steroid levels are much lower, estriol is a minor or absent metabolite, and CG is secreted for a much shorter period) shows that chimpanzees are the only feasible models for some aspects of the endocrine control of pregnancy and parturition (Hobson *et al.*, 1976; Hodgen, 1979; Faiman *et al.*, Chapter 2, this volume). Chimpanzees, therefore, have great promise for study of abortion, abortifacient agents that affect CG secretion or action, the mechanism of induction of labor, and causes and prevention of premature or delayed delivery in women.

An unexploited area of research is the use of apes to study the biology of pregnancy-associated plasma proteins. The serum concentration of such proteins increases during human pregnancy and is elevated in women taking oral contraceptives and in men treated with estrogens. Only chimpanzees and orangutans, among a variety of primate species tested, synthesized four pregnancy-specific plasma proteins immunologically identical to those produced by the human (Lin and Halbert, 1978). The significance of this observation is indicated by the fact that immunization against such proteins can reduce the pregnancy rate or induce abortion.

4. Postpartum Amenorrhea

As Nadler *et al.* showed in Chapter 3, the profile of endocrine levels of chimpanzees and orangutans in this condition is so similar to man's that the chimpanzee is a logical subject for manipulative experiments designed to prolong postpartum amenorrhea as a basis for contraception, and to evaluate the side effects of such prolonged cycle inhibition. The ability to train chimpanzees for alert blood withdrawal will permit study of the role of prolactin in postpartum amenorrhea and perhaps enable identification of the inhibitory signal and its locus of action.

5. Male Endocrinology

Research on adult male ape endocrinology has barely reached the point that applications are appearing on the horizon. Work currently in progress by Helton, Hobson, and Graham is aimed at describing chimpanzee testicular steroidogenesis more exactly, using both *in vitro* and *in vivo* methodology. Repeated use of adult males appears possible using a single pulse of steroid precursor and testicular vein cannulation to explore the effect of hormonal agents on testicular steroid metabolism. Such a system could also provide a sensitive assay for toxicological evaluation of drugs and hormones that might have undesired effects on the pituitary–gonadal axis. Only a single, brief report exists of testicular metabolism *in vitro* in the orangutan (Preslock, 1979).

The ready availability of ape semen by rectal probe electro-ejaculation provides an excellent means for evaluation of contraceptive agents that show evidence of antifertility action in primates. Thus, male chimpanzees can be used repeatedly in minimally invasive protocols (1) to analyze the mechanism of action of a wide variety of contraceptive agents that act on the pituitary or gonad, given the ability to analyze pituitary responsivity to LHRH; (2) to measure circulating gonadotropin and steroid levels; (3) to analyze testicular steroid metabolism *in vivo* (and *in vitro* when availability of material permits); (4) to examine testicular biopsies biochemically and morphologically; and (5) to analyze seminal plasma and sperm properties by biochemical, physicochemical, and morphological criteria, as discussed by Martin and Gould in Chapters 5 and 6. The large amounts of testicular venous blood that can be obtained by a simple cannulation of one of the collaterals of the pampiniform plexus (Hobson, Helton, and Graham, unpublished data), the large testis volume that permits repeated testicular biopsy without expectation of serious risk to fertility, and the adequate semen volume are all significant advantages presented by the chimpanzee as a model. Since there are many captive-raised male chimpanzees that have failed to develop copulatory behavior, there is a substantial reservoir of animals suitable for the kinds of study outlined here, without compromising the existing breeding capacity of captive populations.

Exogenous androgens, progestins, or estrogens can block sperm production, probably by feedback inhibition of LH secretion, resulting in unavailability of testosterone locally in the testis; effective steroid combinations are being sought that lack side effects. Anti-androgens (e.g., cyproterone acetate) also have promise for inhibition of spermatogenesis. Paradoxically, both gonadotropin-releasing hormone (GnRH) antagonists and agonists can inhibit spermatogenesis. Immunization against GnRH has an antigonadal effect in primates. Some of these approaches, combined with replacement steroid therapy to maintain libido and secondary sex organ function, are likely to require evaluation in a species closely related to man. There is as yet insufficient comparative data to determine which species will be most suitable; the chimpanzee may be found necessary to some extent.

Identification and synthesis of inhibin, a presumed specific natural inhibitor of FSH and hence of spermatocytogenesis, might permit contraception with affecting secondary sex organ function. This promising approach may have to be pursued in apes if inhibin is found to possess the species specificity characteristic of many protein hormones.

Ewing *et al.* (1979) have proposed an efficient experimental design, suitable for primates, which can be used to evaluate the efficacity of a large series of antispermatic drug formulations with a moderate number of animals.

6. *Hypothalamic–Pituitary–Gonadal Axis*

Unlike the rhesus monkey, the chimpanzee regularly shows a response of comparable magnitude to the human, when similar doses of gonadotropin-releasing hormone (GnRH) are injected (Graham *et al.*, 1979; Graham, Chapter 1, this volume). Moreover, the two phases of gonadotropin release that have been described in the human are more readily observed in the chimpanzee.

Differential feedback effects of ovarian steroids on LH and FSH release are demonstrable. The ways in which steroid feedback affects the two phases of gonadotropin release has yet to be fully explored. All of these problems can be investigated with relative facility in the chimpanzee. The role of maturation of the pituitary response system as a factor in human sexual maturation remains to be fully explored. Since informed consent for such studies would be difficult to obtain in prepubertal children, the chimpanzee is likely to become the essential model for such studies.

The chimpanzee, by virtue of its similar GnRH responses to man, is suitable for evaluation of GnRH analogs of potential clinical value. Antagonists and agonists of LHRH appear to have both pituitary and gonadal sites of action, and paradoxically both types of analog have contraceptive activity in both sexes. The male and female chimpanzee have recently been used for the evaluation of the potential antifertility effects of LHRH analogs (Bowers *et al.*, 1980; Gosselin *et al.*, 1979a, b; see also Graham, Chapter 1, this volume).

H. Sexual Behavior

As shown in Chapters 9 through 12, apes show a variety of sociosexual population structures and behavior patterns. Comparative study of the great apes may thus provide some insight into the role of societal factors in molding sexual behavior.

Studies of behavior in the field will continue to show how sexual, societal, and environmental factors have affected the evolution of great apes and man. Such information can yield important insights into the roots of human sexual biology and behavior, as discussed by Short in Chapter 13.

Laboratory studies of behavior (Nadler, Chapter 8, this volume) will also continue to be important, since the investigator thus can manipulate the experimental environment and control variables to an extent not possible in the field. Such controlled experiments can be used to test specific hypotheses about sexual behavior and so amplify the wealth of data that relates to the evolution of man's sexuality.

Among chimpanzees, social factors are more important than female menstrual cycle status in determining whether or not mating takes place. In this respect they trend toward the human condition in which cognitive factors are believed to be primary determinants of sexual behavior, since incidence of copulation is only weakly correlated with the phases of the human menstrual cycle. Chimpanzees are therefore favorable subjects for investigation of the endocrine basis of human sexual behavior. Since cognitive capacity is an important factor in human sexuality, the chimpanzee qualifies as an excellent model for the study of the interaction between cognitive capacity and hormonal conditioning in the expression of sexual behavior, and how imbalance in these forces may lead to abnormal sexual behavior patterns. Comparative studies of apes have potential for investigating the interaction of possible libidinal effect of androgens with cognitive factors affecting sexual behavior during the menstrual cycle. Since the duration of sexual receptivity appears to vary among the great apes and man, a correlation may be sought between the duration of receptivity and androgens levels. Nadler *et al.* (1979 and Nadler, Chapter 8, this volume) have demonstrated a correlation between the short period of sexual receptivity in the gorilla and a large, brief elevation in circulating testosterone levels. Thus, investigation of sexual receptivity as a function of androgen levels in the chimpanzee, gorilla, and orangutan is of great interest.

I. Unexplored Areas of Reproductive Biology in Apes

Doubtless many other unique similarities between apes and man remain to be discovered; some of these will certainly define presently unforeseen application of apes as models for research affecting human health and welfare. Many of these applications are likely to develop in the area of male physiology, developmental endocrinology, and pregnancy.

IV. AVAILABILITY OF APES

In order to put the question of apes as models into perspective, the availability of apes should be further discussed.

A. Distribution of Animals Available for Research

Apes are not widely available as research subjects. Gorillas and orangutans are occasionally available for study in zoos, particularly the San Diego Zoo, which has a research laboratory with a special interest in reproductive physiology. The Yerkes Primate Center maintains substantial numbers of these apes for research and breeding, and the animals are potentially available for collaborative, nondestructive research.

Chimpanzees are somewhat more widely distributed and available for research. Approximately 1000 chimpanzees are in research colonies and zoo collections in the United States. Research colonies are maintained at several locations in the United States.

Since primate supply sources were dwindling, an Interagency Primate Steering Committee was established in 1974 to assure primate supplies for scientific research in the United States. This reduction in primate availability was due to embargoes by exporting countries as well as national and international legislation restricting import of endangered and threatened species under the Endangered Species Act and the Convention on International Trade in Endangered Species of Wild Fauna and Flora. All of the apes are listed as endangered species, and they can be imported only under special permit. It is not feasible to import significant numbers of animals under these restrictions.

B. Number of Animals Needed for Reproductive Biology

The Interagency Primate Steering Committee established a Task Force on the Use of and Need for Chimpanzees. This task force met in 1978 and issued a report on the current and projected usage of chimpanzees in biomedical research. A survey of users indicated that approximately 85 chimpanzees were in use in various areas of reproductive biology, including the control of ovulation, steroid feedback mechanisms controlling gonadotropin secretion, endocrinology of adolescence, biochemistry of semen, and fertility control.

It was projected in the task force report that an additional 50 chimpanzees will be needed each year for future reproductive biology research if the full potential for use of chimpanzees in reproductive biology is realized. The extent to which this need can be met from current colonies and their progeny depends upon availability of existing animals, breeding efficiency, competing needs from other research areas, which greatly exceed the total number of animals available, and the extent to which multiple use of animals is feasible. The chimpanzee task force is currently addressing these problems.

C. Sources of Apes

It is evident that apes can no longer be imported in significant numbers, if at all. Therefore, any expansion in United States holdings can come only from increased births and reduced deaths. Hence, emphasis on improved husbandry practices would be beneficial. Collaboration of human clinicians and experienced primate veterinarians and pathologists could help to reduce the death rate in primate colonies.

D. Improvement of Breeding Efficiency in Ape Colonies

As was discussed by Martin in Chapter 14, there is considerable variability in breeding practices among the major institutions holding apes. The following are some recommendations that would contribute to improved breeding efficiency:

1. Fertility Evaluation

Unified record systems and a unified approach to evaluation of fertility among ape colonies would result in an opportunity to compare breeding practices objectively, so as to identify beneficial practices. Diagnosis and treatment of infertility in apes has received very little attention in the past. Strategies developed for treatment of human fertility should apply here. A unified approach to diagnosis and treatment of infertility between centers would permit pooling and comparison of data so that conclusions could be drawn based upon an adequate number of observations.

2. Evaluation of Social Exposure Necessary for Development of Sexual Competence

One of the most serious problems facing the future of ape breeding in captivity is that of development of normal sexual behavior patterns. Although it is known that animals raised with their mother and/or placed in social groups at an early age will develop copulatory activity to some degree, animals with severely restricted social exposure do not develop normal behavior patterns and hence fail to develop copulatory behavior (see Nadler, Chapter 8, this volume). One of the most severe forms of social deprivation is removal of the infant from the mother at birth, often done to increase conception frequency in breeder females. Although there is evidence that peer social contact helps to restore normal social competence (Chapter 8), the efficacy of such exposure in enabling the development of sexual competence is unknown.

Important progress has been made in the development of resocialization techniques that can lead to the development of adequate sexual behavior in socially inadequate animals (Fritz and Fritz, 1979).

The resocialization strategy found most effective by Fritz and Fritz incorporates a series of steps beginning with a single, gentle companion and ending with larger, more varied groups. It is an important principle that social groups of fixed composition are not optimal for resocialization since social relationships tend to become cemented and inflexible and hence fail to develop. By manipulating group composition, particularly by removal of the dominant animal, social relationships can be maintained in a more dynamic state that allows a greater degree of experimentation, adaptation, and the development of functional social relationships. It seems that rehabilitation is much more difficult for socially deprived males than for females. It appears from these considerations that:

(a)   Sexual patterns are learned in the social environment and are not innate in a complete form.

(b)   The practice of removing infants at birth from their mother and giving them limited peer contact is likely to have a severe impact on future breeding performance, especially of males.

(c)   As the existing sexually competent wild-born animals in captivity gradually die, our ability to use wild-born animals to breed and rehabilitate native laboratory-born chimpanzees will diminish.

(d)   Therefore, it is essential to establish social groups of apes whose function is to provide an environment in which at least a limited number of young females and males can have the opportunity to develop normal copulatory patterns by imitation and experimentation, as occurs in the wild. In the wild, such experimentation begins as early as two years of age. The specially reared animals would be the primary future breeding stock, as well as "trainers" for laboratory-bred chimpanzees with less social experience.

3. *Artificial Insemination*

Perfection of artificial insemination techniques (Martin, Chapter 14, this volume) would enable use of behaviorally incompetent, otherwise fertile animals to maximize production. Although artificial insemination has been proven feasible in chimpanzees, the conception rate has been extremely low, for reasons not fully understood. Artificial insemination at present should be regarded as a potential supplement to, not a substitute for, natural breeding methods.

4. *Genetic Heterozygosity*

It is doubtful if there are sufficient gorillas and orangutans in captivity to maintain genetic heterozygosity for many generations, based upon the rule of thumb that a population will survive inbreeding for about as many generations as there are effective breeding individuals in the population. The greater number of chimpanzees in captivity ensures a much better outlook for this species.

For all species it will be important to maintain heterozygosity as much as possible by outbreeding techniques. Within the colony, the breeding potential of each animal should be used as much as possible. The tendency to use preferred studs to the exclusion of others tends to increase homozygosity, and so should be avoided.

Exchange of studs between colonies is an obvious way of maintaining outbreeding. An alternative to exchanging animals is exchange of semen samples. The technology for semen collection is well-developed (Martin, Chapter 14, this volume). Unfortunately, the susceptibility of ape semen to cold shock, and the slow development of artificial insemination techniques using fresh or frozen semen has inhibited the development and evaluation of semen-banking technology and interinstitution transfer of semen for insemination. These are important problems that must be solved, particularly if gorillas and orangutans are to remain available as research models.

V. CONCLUSION

It is evident from this volume and the summary in this chapter that adequate information has been gained about the reproduction of great apes, especially of chimpanzees, to identify many areas where these animals have unique research value. It is also evident that much comparative physiological data are yet lacking, especially for the male of all ape species, and for orangutans, gorillas, and pygmy chimpanzees in general. It seems likely that usage of these latter species in research will not extend much beyond comparative studies and other work that might enhance their survival in captivity. They are too scarce to be used as biomedical research subjects, except in the most unusual circumstances of appropriateness and availability.

Chimpanzees, in contrast, have a fairly promising future in biomedical research. Sufficient animals are available for usage in some research where adequate justification for selection of the species is manifest. The demand for chimpanzees is likely to grow in step with increasing recognition of their value as research subjects in reproductive biology. Given wise management, the captive chimpanzee population will grow slowly, and although it is unlikely that there will ever be sufficient apes to meet all research needs in reproductive biology and other areas, there will be a steady increase in research and in knowledge of great ape reproduction. When one reviews the great advance in knowledge on this subject gained in the last ten years, the future must be viewed with considerable optimism.

REFERENCES

Bowers, C. Y., Humphries, J., Wasiak, J., Folkers, K., Reynolds, G. A., and Reichert, L. E., Jr. (1980). *Endocrinology* **106,** 674–683.

Chen, H-C., and Hodgen, G. D. (1976). *J. Clin. Endocrinol. Metab.* **43,** 1414–1417.

Cutler, G. B., Jr., Glenn, M., Bush, M., Hodgen, G. D., Graham, C. E., and Loriaux, L. (1978). *Endocrinology* **103,** 2112–2118.

Dollar, J., and Graham, C. E. (in press). *J. Med. Primatol.*

Ewing, L. I., Adams, R. J., and Cochran, R. C. (1979). In "Animal Models for Research on Contraception and Fertility" (N. J. Alexander, ed.), Harper & Row, New York, pp. 326–343.

Faiman, C., Winter, J. S. D., Chebib, F. S., and Butler, T. M. (1972). *J. Clin. Endocrinol. Metab.* **34,** 601–604.

Fritz, P., and Fritz, J. (1979). *J. Med. Primatol.* **8,** 202–221.

Gosselin, R. E., Hobson, W. C., and Vale, W. (1979a). Annual meeting, Society for Study of Reproduction, August 21–24, Quebec, Canada, Abstract #195.

Gosselin, R. E., Fuller, G. B., Coy, D. H., Schally, A. V., and Hobson, W. C. (1979b). *Proc. Soc. Exp. Biol. Med.* **161,** 21–24.

Gould, K. G., Flint, M., and Graham, C. E. (in press). *Maturitas.*

Gould, K. G., Martin, D. E., and Graham, C. E. (1976). In "Scanning Electron Microscopy/1976," Vol. VI, pp. 335–342. IITRI, Chicago, Illinois.

Gould, K. G., Warner, H., and Martin, D. E. (1978). *J. Med. Primatol.* **7,** 213–222.

Graham, C. E. (1970). In "Chimpanzee" (G. H. Bourne, ed.), Vol. 3, pp. 183–220. Karger, Basel.

Graham, C. E. (1977). "Progress in Ape Research." Academic Press, New York.

Graham, C. E. (1979). *Am. J. Phys. Anthrop.* **50,** 291–300.

Graham, C. E., Collins, D. C., Robinson, H., and Preedy, J. R. K. (1972). *Endocrinology* **91**, 13–24.

Graham, C. E., Keeling, M., Chapman, C., Cummins, L. B., and Haynie, J. (1973). *Am. J. Phys. Anthrop.* **38**, 211–215.

Graham, C. E., Warner, H., Collins, D. C., and Preedy, J. R. K. (1977). *J. Reprod. Fertil.* **50**, 23–28.

Graham, C. E., Gould, K. G., Collins, D. C., and Preedy, J. R. K. (1979). *Endocrinology* **105**, 269–275.

Hartman, C. G. (1939). *Am. Natur.* **37**, 139–155.

Hobson, W., Coulston, F., Faiman, C., Winter, J. S. D., and Reyes, F. (1976). *J. Tox. Environ. Health* **1**, 657–668.

Hodgen, G. D. (1979). *In* "Animal Models for Research in Contraception and Fertility" (N. J. Alexander, ed.), pp. 425–436. Harper & Row, New York.

Hodgen, G. D., Nixon, W. D., Vaitukaitis, J. L., Tullner, W. W., and Ross, G. T. (1973). *Endocrinology* **92**, 705–709.

Lin, T-M., and Halbert, S. P. (1978). *Int. Arch. Allergy Appl. Immun.* **56**, 207–223.

Nadler, R. D., Graham, C. E., Collins, D. C., and Gould, K. G. (1979). *Endocrinology* **105**, 290–296.

Nixon, W. E., Tullner, W. W., Rayford, P. L., and Ross, G. T. (1971). *Endocrinology* **88**, 702–706.

Pope, A. (1704). "An Essay on Man." Epistle II.

Preslock, J. P. (1979). Society for the Study of Reproduction, 12th Annual Meeting, —August 21–24, Quebec, Canada, Abstract #95, p. 59A.

Reyes, F. I., Winter, J. S. D., Faiman, C., and Hobson, W. (1975). *Endocrinology* **96**, 1447–1455.

Shackleton, C. H. L. (1974). *J. Steroid Biochem.* **5**, 113–118.

Solleveld, H. A., and van Zwieten, M. J. (1978). *J. Med. Primatol.* **7**, 19–25.

INDEX

Courtship display, 12, 290
  in orangutans, 290
Creatinine, 168

D

Dehydroisoandrosterone, 180
Dehydroisoandrosterone sulfate, 180
Development, 84–86, 298
  intrauterine, 84–86
    effect of exogenous steroids in, 85
    hormonal relationships in, 84–86
  sexual behavior
    in orangutans, 298
Diet
  in captivity, 386–388
Dominance, 252, 277, 289, 292, *see also*
    Competition, between males
Dominance, 252, 277, 289, 292
  in chimpanzees, 252
  in gorillas, 277
  in orangutans, 289, 292
Drug
  administration, 389–390
  intubation, 390
  orally active, 390
  solubilization, 389
  vehicle, 389–390

E

Ejaculate, *see* Semen
Ejaculatory duct, 139
Electroejaculation, 146–147, 366, 401–402
Endocervical canal
  of apes and man, 413
Endocervix, 29, 118–121
  anatomy, 109
  changes during menstrual cycle, 29
  microanatomy, 118
  mucus, 29, 119–121
    availability of, 416
    changes during menstrual cycle, 29
    composition of, 119–121
    and fertility, 121
    glycoprotein content of, 120
Endometrium, 27–29, 35, 115–118,
    121–122, 400–401
  biopsy, 400–401

cast, 121–122
cyclic changes, 27–29, 115–116
curette, 401
dating, 116
decidualization, 27, 116–118
histology, 115–116
  and ovulation, 33–35
hormonal control, 116–118
Endoscopy, *see* Laparoscopy
Epididymis, 130–135
  fluid, 130–134
  morphology of, 130–135
Estradiol, *see also* Estrogen
  during menstrual cycle, 170–173
Estriol, *see also* Estrogen
  as indicator of fetal health, 85
Estrogen, 10–22, 55–60, 74–77, 89–90,
    169–175, 184–187
  endogenous levels
    in females, 170–173
    in males, 169
    during menstrual cycle, 171–173
    during pregnancy, 185
  gonadotropin secretion, effect on,
    18–22
  during menstrual cycle, 10–17
  metabolism
    in females, 173–175
    fetoplacental unit, 186–187
    in males, 169
  negative feedback
    during infancy, 89–90
    postpartum, 74–77
  precursors
    in pregnancy, 55–57
  during pregnancy, 59–60, 185–186
  synthesis
    fetoplacental unit, 184–185
Estrone, *see* Estrogen
Estrus, 202–204, 206, 270, 284–285,
    310–311
  in chimpanzees, 202–204
  components of, 206
  duration of, 310–311
  and genital swelling, 203–204, 212
  in gorillas, 270
  in orangutans, 284–285
  suppression of
    in man, 337